Structural Aspects of
PROTEIN
SYNTHESIS

STRUCTURAL ASPECTS OF
PR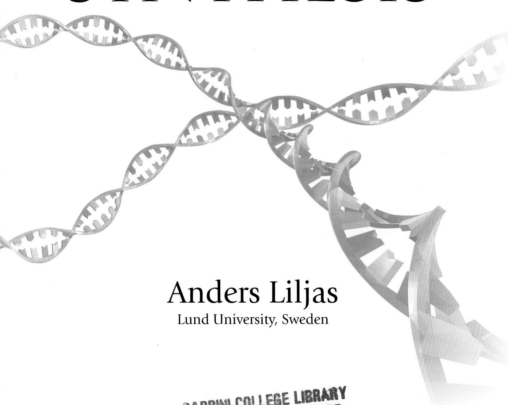TEIN
SYNTHESIS

Anders Liljas

Lund University, Sweden

 World Scientific

NEW JERSEY • LONDON • SINGAPORE • BEIJING • SHANGHAI • HONG KONG • TAIPEI • CHENNAI

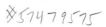

Published by

World Scientific Publishing Co. Pte. Ltd.

5 Toh Tuck Link, Singapore 596224

USA office: 27 Warren Street, Suite 401–402, Hackensack, NJ 07601

UK office: 57 Shelton Street, Covent Garden, London WC2H 9HE

QP
551
.L55
2004

British Library Cataloguing-in-Publication Data
A catalogue record for this book is available from the British Library.

ISBN   981-238-863-X
ISBN   981-238-867-2 (pbk)

Typeset by Stallion Press
E-mail: sales@stallionpress.com

Printed by Fulsland Offset Printing (S) Pte Ltd, Singapore

*To Ewa
and our family*

Preface

When I was asked whether I could write a book on the translation system, I thought it would be an interesting challenge. Now I know better; it is an almost impossible task to catch a rapidly moving target. Sometimes, it has felt like trying to get to the top of a mountain against the avalanche of new relevant literature that has moved me further and further from the goal, a finished book.

The targetted audience is anyone, undergraduate or graduate students and researchers, who privately or in doing a course, would like to get a summary of the structural knowledge of the translation system. I realize that it is a biased representation, partly due to my inability to cover the vast and rapidly growing literature. Space also necessitates a limited coverage. I have restricted the material to bacterial translation with only a few excursions into the other domains of life. Many of the black and white illustrations are cartoon-like, attempting to describe aspects of translation in the simplest way I could. The color plates are selected from the scientific literature to give correct structural representations. I am very grateful to authors and publishers for permissions to reproduce these illustrations.

I owe sincere thanks to a number of people for assistance in the production of this book. I have obtained support locally from Salam Al-Karadaghi, Michel Fodje, Martin Laurberg, Derek Logan, Suparna Sanyal, Maria Selmer and Gunilla Sökjer-Petersen. Michel Fodje has, in particular, made a number of the illustrations for me.

Through discussions and by getting manuscripts, I have learnt much from Måns Ehrenberg, Joachim Frank, Jonathan Gallant, Anatoly Gudkov, Rolf Hilgenfeld, Akira Kaji, Ole Kristensen, Charles G. Kurland, Peter B. Moore, Jens Nyborg, Venki Ramakrishnan, Tom A. Steitz, Marek Tchorzewski, Arieh Warshel and Ada Yonath.

A number of persons have given invaluable comments and assistance on the manuscript: Salam Al-Karadaghi, Anatoly Gudkov, Ewa Liljas, Derek Logan, Jens Nyborg and in particular Blanka Rutberg. I am deeply grateful for their very kind assistance.

Anders Liljas

2004

Contents

List of Abbreviations

aa	amino acid
aaRS	aminoacyl-tRNA synthetase
A-, P-, E- and T-site	Sites for tRNAs on the ribosome
ASL	Anticodon stem loop
CD	circular dichroism
CTC	General stress protein in *B. subtilis* corresponding to ribosomal protein L25 (*E. coli*) or TL5 in *T. thermophilus*
D	In front of a protein name (e.g. DL2) means it comes from *D. radiodurans*
EF2	Eukaryotic elongation factor 2 (translocase)
EF-G	Elongation factor G (translocase)
EF-Ts	Elongation factor Ts
EF-Tu	Elongation factor Tu
EM	Electron microscopy
FA	Fusidic acid
GAP	GTPase activating protein
GAR	GTPase associated region
GDP	Guanosine diphosphate
GDPNP	An uncleavable GTP analogue
GEF	G-nucleotide exchange factor
GTP	Guanosine triphosphate
GTPases	Enzymes hydrolyzing GTP
h	In front of a number means a helix in the small subunit (e.g. h44)
H	In front of a number means a helix in the large subunit (e.g. H69)
H	In front of a protein name (e.g. HL5) means it comes from *H. marismurtui*

IF1, 2, 3	Initiation factors
PCC	Protein conducting channel
PTC	Peptidyl transfer center
rRNA	Ribosomal RNA
r-proteins	Ribosomal proteins
RF1, 2, 3	Release or termination factors
RRF	Ribosome recycling factor
RRM	RNA recognition motif
RS	aminoacyl-tRNA synthetases
SD	Shine-Dalgarno interaction between mRNA and 16S RNA
SelB	EF-Tu factor for binding of Se-Cys-tRNA^{Sec}
SR	Signal recognition receptor
SRL	Sarcin-ricin loop
SRP	Signal recognition particle
T	In front of a protein name (e.g. TL1) means it comes from *T. thermophilus*
TC	ternary complex
TM helix	trans-membrane helix
tGTPases	Translation factors hydrolyzing GTP

1

The Basics of Translation

Protein synthesis is a remarkable process as the term translation indicates. The genetic information stored in the nucleic acids, RNA and DNA, expressed in a "language" of a four-letter "alphabet" and three-letter "words", is **TRANSLATED** into proteins, which have a much more diverse 20-letter alphabet. The information contained in the chromosomal DNA is transcribed into messenger RNA (mRNA). Subsequently the four nucleotides (A, C, G, U) in an mRNA are translated into the 20 amino acids that make up the proteins. For this translation, there are rules like those for any written message. There is a dictionary for the translation, the genetic code, and there are signals indicating start and stop. The fact that this dictionary or genetic code is universal is a very remarkable observation. A message from one organism can normally be correctly translated by a different organism. The evolutionary aspects of this are fascinating. Not only does all life on earth appear to have a common origin. In addition, the exchange of genetic material between different organisms over the billions of years of evolution has been extensive. Thus, it is not excluded that a cow could "infect", or transfer genetic material to, a bacterium.

The site of translation in all cells is the ribosome. Ribosomes are found in all cells and in many cellular compartments and are composed of the large and small ribosomal subunits. The transfer RNAs (tRNAs), the universal adapter molecules, are also fundamental for translation. Each tRNA is uniquely charged with a specific amino acid.

Once an mRNA has been synthesized through transcription from DNA, it is targeted to the small ribosomal subunit. Here it binds in a specific manner with the start signal, the initiator codon presented in the so-called ribosomal P-site. This start codon is recognized by the initiator-tRNA. When the mRNA and initiator tRNA are bound to the small subunit the large subunit can bind and the ribosome is ready for translation of the mRNA. It now enters the elongation phase.

In the elongation phase tRNAs charged with their respective amino acids, recognize their codons through Watson-Crick base-pairing. They are then bound to the ribosomal site for aminoacyl-tRNAs, the so-called A-site. The accurate decoding of the message is an essential feature. To obtain a satisfactory fidelity of translation, the initial recognition is combined with a step called proofreading. When the decoding has been performed, the peptide bound to the tRNA in the P-site is transferred and covalently linked to the amino acid bound to the tRNA in the A-site. The polypeptide chain is thus extended by one amino acid. Subsequently the peptidyl-tRNA, now located in the A-site, needs to be translocated to the P-site to allow for a new cycle of elongation. The elongation factors catalyze two of the basic steps in translation: the binding of an aminoacyl-tRNA to the A-site and the translocation of the peptidyl-tRNA from the A-site to the P-site. However, the central event in elongation, peptidyl transfer, is a spontaneous process where no protein factor is needed.

The translation of an mRNA is terminated when a stop codon is encountered. tRNAs do not recognize stop codons. This is rather done by proteins, which in response, hydrolyze the polypeptide from the tRNA. These proteins are called class 1 termination factors or release factors. This type of termination factor is subsequently released from the ribosome by a class 2 termination factor. After the termination of protein synthesis by the termination factors, the mRNA and the deacylated, tRNA need to be removed from the ribosome and the subunits separated from each other. This will allow the small subunit to bind a new mRNA and together with a large subunit, initiate a new round of translation. This phase of translation preceding initiation is called ribosomal recycling. The whole process of translation is illustrated in Fig. 1.1.

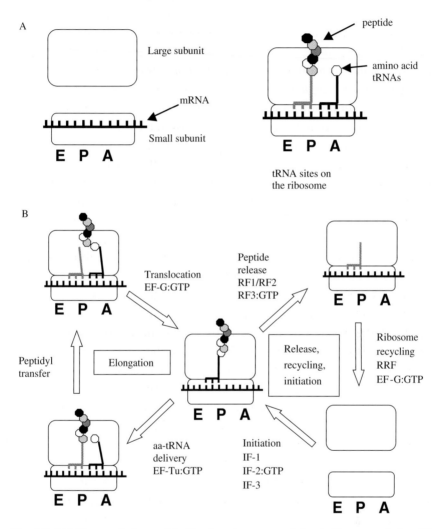

Fig. 1.1 (A) The symbols used in the drawing are explained. These symbols will be used repeatedly throughout the book. **(B)** A summary of translation on ribosomes. The start point is on the lower right, illustrating the initiation of protein synthesis from separated subunits, with the binding of the mRNA to the small subunit. In each round of the elongation cycle one amino acid is incorporated. When a stop codon is exposed in the A-site, the peptide is released. Finally, the components of the system are recycled for a new initiation.

One common view has been that the ribosome is a passive entity in translation. According to this view, it is the interplay between the mRNA, the tRNA and the translation factors on this large complex of RNAs and proteins that produces the growing polypeptide. However, it has become increasingly clear that the ribosome actively participates in many of the steps and contributes significantly to the control, rate and fidelity of translation. The ribosome is a highly active machinery moving between different states that are recognized by its many interacting partners. One can with confidence say, that the tRNA and the ribosome jointly translate the message of the mRNA.

The structure of the translation machinery has always been of central importance since, as Watson expressed in 1964, "Unfortunately, we cannot accurately describe at the chemical level how a molecule functions unless we know first its structure." This need has been evident during the whole time translation has been studied and is one of the reasons why this book has been written.

In this book, we will focus on bacterial translation. We shall only discuss archaeal or eukaryotic translation in cases where it aids insight into bacterial translation or when the process being described, is a universal one.

2

Historical Milestones

There are many shorter or longer accounts of historical developments related to protein synthesis (Perutz, 1962; Nomura, 1990; Spirin, 1999; Woese, 2001; Rich, 2001; Hoagland, 2003).

A fundamental basis for protein synthesis is the realization that DNA is the genetic material. How the information flows from DNA into proteins began to be revealed about half a century ago and is reviewed in a still very enjoyable way by Watson (1964). In eukaryotic cells, the DNA is contained in the nucleus while proteins are synthesized in the cytoplasm. Several lines of observation made it clear that RNA had to be the missing link (Chamberlin & Berg, 1962). The information contained in the DNA is transmitted to an intermediate, a messenger RNA (mRNA), which subsequently can be translated into proteins (Crick, 1958). The double-helical structure of DNA (Watson & Crick, 1953) provided an insight of fundamental importance for protein synthesis in that it explained the main interactions involved in replication as well as in transcription and translation. However, what was the machinery that could perform this remarkable process?

Casperson (1941) was first to observe the relationship between RNA and protein synthesis. A gradual improvement of techniques led to the identification of RNA-containing particles, ribosomes. Subsequent studies of *Escherichia coli* led to purified preparations and a realization that proteins are synthesized on such ribosomes (Tissièrere *et al.*, 1960; Kurland, 1960). Ribosomes are the machinery of translation. Ribosomes contained RNA (rRNA) and proteins and are composed of two subunits. In bacteria, their sedimentation constants are 30S for the small subunit and 50S for the large subunit. It was also observed that eukaryotic ribosomes are larger than bacterial ribosomes, sedimenting at 80S and 70S respectively. However, the ribosomes of mammalian mitochondria are noticeably smaller, with significantly shorter rRNAs, sometimes called mini-ribosomes.

How could the message be decoded and lead to the synthesis of a polypeptide? Crick suggested that there had to be an adaptor molecule that could read the message and incorporate the appropriate amino acid into the growing polypeptide according to the message. The adaptor would be charged with the different amino acids by specific enzymes (Crick, 1958). The adaptor was suggested to be a small RNA molecule. The message would probably be decoded through Watson-Crick base-pairing. Subsequently, the tRNA molecules (then called sRNA) were identified and characterized (Hoagland *et al.*, 1957). Crick's hypothesis was found to be correct (Fig. 2.1). These tRNA molecules could be uniquely charged with the different amino acids (see Berg, 1961 and references therein; Ibba & Söll, 2000; 2001).

One early question found a negative answer: the rRNA was not the message (Gros *et al.*, 1961; Brenner *et al.*, 1961; Jacob & Monod, 1961). The identification of the ribosomal subunits and the clarification of their separate roles were other important steps forward. It was established that the small subunit was the site of decoding of the message (Okamoto & Takanami, 1963). It was furthermore shown that the ribosome protected about 25–30 nucleotides of the mRNA (Takanami & Zubay, 1964). The large subunit was found to be the site of incorporation of new amino acids or peptidyl transfer, but no covalent link was identified between the nascent polypeptide chain and the ribosome (Gilbert, 1963). It also became clear that there must be at least two sites for tRNA on the ribosomes, one for a tRNA connected to the nascent peptide and one for the incoming aminoacyl-tRNA (Warner & Rich, 1964; Watson, 1964). These sites would then have to contain elements from both subunits since the mRNA is bound on one subunit and the nascent peptide on the other. These tRNA sites were subsequently called the A-(aminoacyl) and P-sites

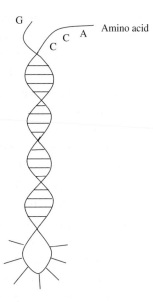

G

C

C A Amino acid

Fig. 2.1 An early so-called "hairpin" model of the tRNA molecule (after Watson, 1964). The unpaired bases at the lower end contain the anticodon. The length of the molecule was estimated at 120 Å.

(peptidyl). In the early phase of studies of translation, it was not clear why there was a need for two subunits that separated after termination and reunited upon a new initiation (Watson, 1964). The possibility of a movement between the subunits was identified (Spirin, 1969).

Purified ribosomes and cellular fractions containing tRNAs as well as the possibility to make synthetic mRNAs, allowed the genetic code to be explored (Nirenberg & Matthaei, 1961). Initially the nature of the code was not known. Code words of two of the four possible nucleotides could not code for the 20 naturally occurring amino acids ($4^2 = 16$). The code words would need to be at least three nucleotides in length ($4^3 = 64$). Furthermore, the code could be punctuated or overlapping. Neither of these possibilities turned out to be true. The genetic code is a three-letter code without overlap or punctuations (Crick *et al.*, 1961). In the 1960s, the complete genetic code was elucidated. The meaning of the 64 three-letter code words or codons was established (Nirenberg & Leder, 1964; Crick, 1966a).

In the early phase of studies of translation, it was observed that streptomycin caused extensive misreading even though the rate of protein synthesis remained the same (Davies *et al.*, 1964). A large number of antibiotics were found to target different components and sites of the translation machinery to enable organisms to eliminate their competitors (Cundliffe, 1980; 1990).

The ribosomes did not only contain RNA but also protein. Initially, comparing ribosomes with some viruses that were better characterized, one could expect that the proteins would be few but that they would occur in multiple copies. However, with the aid of gel electrophoresis and a more detailed characterization, it became clear that there were a large number of different ribosomal proteins (Kaltschmidt *et al.*, 1967; Kaltschmidt & Wittmann, 1970). The stoichiometry of almost all ribosomal proteins was subsequently found to be one (Hardy, 1975).

Since most cellular functions are carried out by proteins, one expected that the ribosomal proteins would carry the main functions of the ribosome. The rRNA was primarily considered to be a scaffold. An extensive effort was thus focused on characterizing the role of the proteins (see reviews by Kurland, 1972; Garrett & Wittmann, 1973; Wittmann, 1982). The amino acid sequences of the *E. coli* proteins were determined at an early stage (see Giri *et al.*, 1984 and references therein), different attempts to associate ribosomal functions with the proteins were made (see Liljas, 1982 for a review) and the assembly map for the small subunit was established (Held *et al.*, 1974). Early indications that the ribosomal RNA had important functions came from the observation by Noller & Chaires (1972) that kethoxal modification of ribosomes could inhibit tRNA binding. Furthermore, Shine & Dalgarno (1974) found that an mRNA would bind to the bacterial ribosome through base pairing with the 3'-end of the RNA of the small subunit. The elucidation of the nucleotide sequences of many rRNAs showed large regions of sequence conservation, hinting at important functions (Noller & Woese, 1981). Some ribosomologists then turned to explore whether the RNA components were the main sites of ribosomal functions (Noller & Woese, 1981), i.e. whether the ribosome could be a ribozyme. Many ribosomal functions have since been identified as depending on the ribosomal RNA. In addition, several of the key partners in translation, mRNA and tRNA are also RNA molecules. The possibility of an early RNA world has been discussed (Gesteland *et al.*, 1999). Biochemical experiments have come very close to proving that the ribosome is a ribozyme, even though a small uncertainty has remained until very recently (Noller *et al.*, 1992; Noller, 1993).

Protein synthesis on ribosomes is catalyzed by different soluble protein factors (see an early review by Lipmann, 1969). They are grouped into initiation, elongation and termination factors. The number of factors proteins known to interact with the ribosome may still be incomplete (see Sections 9.7 and 9.8). Two main factors that were discovered at an early stage are the elongation factors EF-Tu and EF-G, which are both involved in each cycle of elongation (see Lipmann, 1969 and Kaziro, 1978 for early reviews). Different versions of these two factors are found in all cells or cellular compartments investigated (Pandit & Srinivasan, 2003).

The fidelity of translation was an enigma from early days. How can at most three base-pairs give a fidelity that is in the order of one error per 10 000 amino acids incorporated. It was a milestone in the understanding of translation when it was suggested that the codon-anticodon recognition occurred in two steps, the initial recognition step and the proofreading step (Hopfield, 1974; Ninio, 1975). The fidelity was found to be the product of the fidelity of these two steps. This process did not assume any active role for the ribosome.

Protein synthesis is a very complex process. Nevertheless, the ribosome can be understood as a complex enzyme. For regular enzymes, structural insights are extremely valuable in the efforts to understand function. Thus, through the decades of work on protein synthesis, structural information about ribosomes has been high in demand. Numerous methods have been employed or developed to gain structural insights. Initially, information was obtained with blunt tools like genetics, chemical crosslinking, chemical labeling, and enzymatic digestion of proteins or RNA. Such methods can give invaluable insights when combined with structural information at higher resolution. The interpretation of such results alone cannot give the accurate model that is needed for the functional understanding of a complex enzyme system.

Electron microscopy techniques have been used from an early stage and have provided information of ever increasing detail (Palade, 1955; Hall & Slayter, 1959; Huxley & Zubay, 1960; Lubin, 1968). The identification of the shapes of the ribosome and its subunits was an important step in our familiarity with ribosomes (Fig. 2.2; Plate 2.1; Stöffler & Stöffler-Meilicke, 1984; Gornicki *et al.*, 1984; Lake, 1985; Spirin & Vasiliev, 1989). Furthermore, when different components could be placed within the envelope, we obtained a structural framework for the ribosome that was easier to comprehend than featureless blobs. The modern technique of

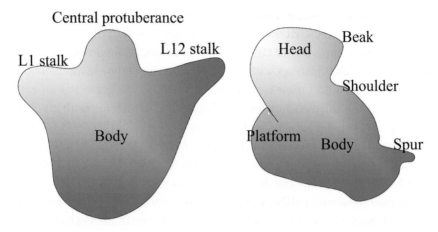

Fig. 2.2 The outline of the large (*left*) and small (*right*) subunits as observed by electron microscopy (Lake, 1976; 1985). This view of the large subunit is called the crown view. The two subunits are seen from the same side. To form the 70S ribosome, the small subunit is placed on top of the large one in this orientation.

cryo-electron microscopy (cryo-EM) single particle reconstruction has given a rich harvest of insights at constantly improving resolution (Stark *et al.*, 1997; Agrawal *et al.*, 1998). The old question of whether the two subunits move with regard to each other during translation (Spirin, 1969) has recently received a positive answer (Frank & Agrawal, 2000).

At an early phase, X-ray crystallography provided the structure of yeast phenylalanine tRNA (Robertus *et al.*, 1974; Kim *et al.*, 1974; Stout *et al.*, 1976). The secondary structure of tRNA, with the shape of a clover-leaf, was found to fold into a tertiary structure with the shape of an L. This clarified that the two functional parts of tRNA, the anticodon and the amino acid acceptor end of the molecule, are about 75 Å apart. Evidently, the sites on the ribosomal subunits interacting with each of the two ends of the tRNAs, the decoding site and the peptidyl transfer center, would also be expected to be this far apart.

Much crystallographic work has also been devoted to studying isolated ribosomal components (Ramakrishnan & White, 1998; Al-Karadaghi *et al.*, 2000a,b). Complete ribosomes or subunits initially seemed inaccessible to crystallography due to their flexibility and lack of homogeneity. Ada Yonath was the first to surmount these difficulties and has systematically explored possibilities to crystallize ribosomes and subunits (Yonath *et al.*, 1980; 1982; 1998). Improved crystallographic techniques have now allowed the determination of structures of ribosomal subunits at atomic resolution and whole ribosomes at a resolution where detailed insights can be gained (Ban *et al.*, 2000; Wimberley *et al.*, 2000; Schlünzen *et al.*, 2000; Harms *et al.*, 2001; Yusupov *et al.*, 2001). The ribosome and its subunits remain the largest asymmetric objects studied by crystallography.

The crystallographic and cryo-EM studies of ribosomes have completely changed the field of protein synthesis and brought us into a new era. Now we can with certainty discuss the conformation of tRNA when bound to the ribosome (Yusupov *et al.*, 2001; Valle *et al.*, 2003a) or specific hydrogen bonds engaged in the discrimination of cognate and non-cognate anticodons of tRNAs in the decoding center of the small subunit (Ogle *et al.*, 2002). Structural details can explain the biochemical observations of initial recognition and proofreading (Ogle *et al.*, 2003). We can also identify the atomic interactions in the peptidyl transfer center (PTC) of the large subunit (Ban *et al.*, 2000; Harms *et al.*, 2001). The observation that there is no protein component in the vicinity of the catalytic center finally demonstrates that the ribosome is a ribozyme.

It is evident that the structural work does not eliminate the need for biochemical and kinetic analyse. Such studies can now be performed from a firm structural basis. However, further detailed information about the dynamic relationships of the components of the system as well as the location of chemical groups in the functional sites of the ribosomes is needed in order to fully comprehend the intricacies of protein synthesis.

The ribosome is central for many of the cellular activities, primarily for the production of all proteins utilized in the cell or exported to other compartments. This makes it into a central locus for cellular control and inhibition (VanBogelen & Neidhardt, 1990). From many points of view it is thus of fundamental interest to clarify its functional mechanisms in detail.

3

Methods to Study Structure

In order for readers to appreciate the research presented in this book, a brief summary of some of the structural methods may be useful. The goal of structural methods is generally to obtain a molecular model at some level of detail. For this, one needs observations that relate to the molecular structure. In deriving a model from such observations, one cannot determine more parameters of the model than the number of independent observations. To obtain accurate atomic coordinates of ribosomes or ribosomal subunits, one thus needs hundreds of thousands of independent observations.

Several of the methods described below were initially developed for the work on ribosomes. During the long time that the structure of the translation system has been analyzed, most methods of structure analysis have evolved significantly. Some of the methods give only a low-resolution picture. Thus, crosslinking without the analysis of which groups are crosslinked is a low-resolution method. However, when the reacting components are identified at the residue level, or even better at the atomic level, crosslinking can provide valuable insights if combined with complementary high-resolution information. In general, low-resolution methods can give unique insights when combined with results from high-resolution methods.

The primary method in recent times has been the crystallographic analysis of ribosomal subunits and whole ribosomes (Ban *et al.*, 2000; Wimberley *et al.*, 2000; Schlüntzen *et al.*, 2000; Harms *et al.*, 2001; Yusupov *et al.*, 2001). The cryo-EM investigations have taken advantage of the high resolution models in the interpretation of complexes with factors and tRNA and have provided remarkable insights into the dynamics of the system that is revealed by several states (Klaholz *et al.*, 2003; Rawat *et al.*, 2003; Stark *et al.*, 2002; Valle *et al.*, 2002; 2003a,b). NMR has also made important contributions in the past and it is felt that certain dynamic aspects of the translation machinery will be best accessible by NMR.

3.1 LOW RESOLUTION METHODS

Assembly

One of the early ways to explore the organization of the ribosomal sub-units was the assembly of purified components into functional particles. This field was pioneered by Nomura (Mizushima & Nomura, 1970; Nomura, 1973; Held *et al.*, 1974) who determined the interdependence of the proteins of the small subunit of *E. coli* in associating with the 16S rRNA. It was found that certain proteins had to first bind to the rRNA in order for subsequent proteins to be able to associate. This suggested that the rRNA was not able to fold properly without the presence of specific proteins. A certain order of assembly was obtained for groups of proteins. It was later established by other methods, such as crosslinking and neutron scattering, that the interdependence of assembly is primarily related to the proximity between the proteins in the particles, even though there is little protein-protein contact within the ribosome (Capel *et al.*, 1987).

Surface Accessibility to Enzymes and Chemical Modifications

There are numerous methods to explore the exposed surfaces of rRNA or proteins in the ribosome. One can expose them to modifying chemicals or hydrolytic enzymes. It is also possible to analyze protection by these modifications by binding different types of ligands to the ribosome. These methods have been used primarily for the studies of the rRNAs. Thus, in addition to the secondary and tertiary structures of the rRNAs, the binding sites for ribosomal proteins, tRNAs and factors have been investigated (Moazed & Noller, 1986; 1987; 1989; Moazed *et al.*, 1988). Protection against chemical modification of certain nucleotides is usually called "footprinting" (Stern *et al.*, 1988; Culver & Noller, 2000; Joseph & Noller, 2000). Footprinting has given valuable results, but they may sometimes be difficult to interpret. It is evident that binding a ribosomal protein, a tRNA or a factor leads to loss of accessibility of a certain surface of the ribosome. If the ribosome undergoes conformational changes upon such binding, the access of enzymes or reagents to sites distant from the binding site may also be changed.

The method of mild proteolysis by using low concentrations of proteolytic enzymes under natural conditions gives a record of flexible regions not involved in firm secondary structures (Gudkov & Bubunenko, 1989; Bubunenko & Gudkov, 1990). This approach has been applied to the study of different conformations of elongation factors, but ribosomal proteins *in situ* can give interesting and variable accessibility depending on the state of the ribosome as well. For ribosomal proteins in isolation, mild proteolysis has been a method to approach the question of domain organization.

Proximity Information by Chemical and Spectroscopic Methods

Bifunctional crosslinks and affinity labels of different lengths can be used to analyze structural proximity, whether permanent or transient. The method has been used to explore the secondary structure of the rRNA and the proximity between different nucleotides or different proteins within the ribosome. The binding sites of tRNA and mRNA molecules or translation factors on the ribosome have also been examined this way. The method remains a very important tool for exploring structural and functional proximity on the ribosome.

Numerous crosslinks with reactivity to different groups and with different lengths have been developed to explore the ribosome (see Baranov *et al.*, 1998; 1999 for the ribosome crosslink database) Furthermore, the introduction of cysteines by site-directed mutagenesis into ribosomal proteins has been used to create binding sites for crosslinking or affinity labeling reagents.

Enzymatic and chemical modification and cleavage methods have also been very useful when combined with primer extension methods to identify rRNA structure and changes of it (Stern *et al.*, 1988). Particularly the Fe(II)-EDTA or hydroxyl radical method has given large amounts of valuable structural and functional insights (Culver & Noller, 2000; Joseph & Noller, 2000). This method uses the possibility to bind Fe-EDTA to selected sites of the ribosome by cysteine mutagenesis. The hydroxyl radicals produced can react with sites on the ribosome within a range of about 20 Å.

Great care has to be exercised to avoid accidental and misleading covalent reactions between the reagent and components of the protein synthesis system. Sergiev *et al.* (2001) give a critical evaluation of the correlation between chemical and crystallographic observations.

Fluorescence energy transfer is a method by which one can obtain information about the distances between chromophores. The range is quite large, but parameters concerning the orientation and flexibility of the components limit the accuracy. A number of groups on tRNAs and other components of the translation machinery have been used to explore structural relationships (Huang *et al.*, 1975).

Electron Microscopy

The identification of the shapes of the ribosome and its subunits was an important step in our familiarity with ribosomes (Plate 2.1; Fig. 2.2). Electron microscopy has been and remains an important tool to study the structure of particles as large as ribosomes (Hall & Slayter, 1959; Huxley & Zubay, 1960; Lubin, 1968). It has been applied in different forms and with increasing resolution. One early aim was to obtain the shape of ribosomes and the subunits. Here several groups made seminal contributions (Stöffler & Stöffler-Meilicke, 1984; Gornicki *et al.*, 1984; Lake, 1985; Spirin & Vasiliev, 1989). They identified not only the shape of the ribosomal particles but also the way they associated.

Subsequently the immune-EM technique was developed (Wabl, 1974; Tischendorf *et al.*, 1974; Lake *et al.*, 1974). Here antibodies were used to label selected entities on the ribosome (Stöffler & Stöffler-Meilicke, 1984; Lake, 1985). A distance relationship between epitopes could be established using ribosomes doubly labeled with antibodies (Kastner *et al.*, 1981; Lake, 1982). Much of the results concerning the shape of the ribosomal subunits and the locations of ribosomal proteins still remain valid. The preliminary information on their elongated shapes has much later been confirmed. However, immune-EM has a subjective component that has led to wrong conclusions.

For the analysis of the ribosomal shapes, improved methods have been used. Even though two-dimensional crystals are a successful way of analyzing structures, only some early attempts have focused on this option (Milligan & Unwin, 1986; Yonath *et al.*, 1987). Due to the difficulties in obtaining two- and three-dimensional crystals and analyzing them, methods focusing on single particles were developed. One of them, EM-tomography (Öfverstedt *et al.*, 1994), works by recording a series of tilted images of ribosomes *in situ* in sectioned bacteria. This method offers the possibility of studying ribosomes without any purification or handling, but the resolution remains quite low at the present time.

The most recent application of electron microscopy to the study of ribosomes is single particle reconstruction using cryo-EM (Frank *et al.*, 1981; van Heel, 1987; Dubochet *et al.*, 1988; Frank *et al.*, 1996). Here large numbers of randomly oriented ribosomes in vitreous ice give different projections of the particle. These can be combined into a three-dimensional picture of the ribosome. Resolutions of better than 10 Å are achieved (Valle *et al.*, 2003a,b). A rapidly growing number of complexes of ribosomes with tRNAs and factor proteins are becoming available and will be discussed in the later chapters. The resolution has been defined in two different ways (Harauz & van Heel, 1986; Gabashvili *et al.*, 2000). The former method gives a better value of the resolution than the other (Matadeen *et al.*, 1999). In comparison with the resolution values used in crystallography, the more conservative estimates in cryo-EM seem more appropriate.

Neutron Scattering

Neutron scattering has provided extremely valuable insight into the localization of ribosomal proteins before the time of crystallography on whole ribosomes and subunits. The difference in scattering of neutrons by hydrogen and deuterium is used. Thus, the neutron scattering of protonated or deuterated pairs of proteins has been studied in a background of the other hydrogen isotope. Distances between proteins as well as information on the shape of proteins have been obtained. From the pairwise distances of ribosomal proteins, a three-dimensional map of the 30S subunit from *E. coli* was constructed and compared with the electron microscopic shape (Capel *et al.*, 1987). A number of proteins from the 50S subunit have also been located (May *et al.*, 1992; Willumeit *et al.*, 2001). Most of these results have turned out to be highly reliable even though they only provide low-resolution information. Contrary to the immune-EM technique, the method does not suffer from subjective interpretations of images.

Tritium Bombardment

If particles are bombarded with tritium, only the outer layer of residues will exchange protons for tritium. Thus, it is possible to identify which components make up the exposed parts of a ribosome and which are buried in the interior. Yusupov & Spirin (1986) have used this method to show that tritium in this type of experiment labels essentially all ribosomal proteins (Agafonov *et al.*, 1997) This suggests that hardly any ribosomal proteins are located entirely in the interior or the ribosomal interface, a conclusion that has subsequently been verified by crystallography.

Mass Spectrometry

A new and useful method to study ribosomes and ribosomal states is electrospray mass spectrometry (MS). Not only can the full ribosomal particles and the individual components be identified, but under different solution conditions and ribosomal states different complexes can be seen and the ease by which they dissociate from each other in the gas phase gives a wealth of information (Benjamin *et al.*, 1998; Rostom *et al.*, 2000; Hanson *et al.*, 2003).

3.2 HIGH RESOLUTION METHODS

Nuclear Magnetic Resonance

Nuclear magnetic resonance (NMR) can provide structural information at atomic resolution. One limitation, which is severe when it comes to the study of the ribosomes, is the size. Molecular masses in excess of

40 000 Da are difficult to study. However, ribosomal proteins or fragments of rRNA are very suitable for studies using this method. In particular, studies of well-chosen fragments of rRNA have given very good insights into ribosomal structure and function (Fourmy *et al.*, 1996).

NMR is also a method well suited for characterizing the dynamics of molecules and to identify surfaces involved in transient interactions. It is remarkable that even complete ribosomes give an NMR spectrum despite having a molecular mass of about 2.5 MDa. The observed spectrum is primarily due to the ribosomal protein L12 that is particularly flexible even though it is bound to the ribosome (Morrison *et al.*, 1977; Tritton, 1980; Gudkov *et al.*, 1982; Cowgill *et al.*, 1984).

X-ray Diffraction

Crystallography on ribosomal components as well as ribosomal subunits or whole ribosomes has been performed for several decades. One prime limitation of this method is the absolute requirement for well-ordered crystals diffracting to a satisfactory resolution. However, the size of the object is no limitation. Macromolecular crystals are sensitive to the energetic X-ray radiation. Cryo-cooling of the crystals to around $100°$ K has become a very important solution to this problem and has turned out to be essential for ribosomal crystallography (Hope *et al.*, 1989). Despite the cooling, a number of crystals may be needed for a complete data set. This may require the inspection of numerous crystals since not all of them diffract equally well. Furthermore, the crystals used for one data set need to be isomorphous with each other. This means that they should have the same unit cell dimensions and the same molecular orientations within the cells. This is not always the case.

In all crystallographic work, resolution is a very important factor (see Section 3.2). To acheive a certain resolution, essentially all diffraction data to that Bragg spacing has to be accurately measured. The number of independent observations needed for the structure of the large ribosomal subunit at 2.4 Å resolution is more than 600 000. These are normally measured repeatedly to achieve accuracy. Another problem is the need to obtain the so-called phase angles in order to reconstruct the electron density maps. These cannot be directly measured experimentally; however, a number of methods can be applied in order to determine them. The classical method uses heavy atoms that scatter X-rays strongly, usually with compounds containing a single heavy atom. Due to the large size and lack of symmetry of the ribosome, it has been advantageous to use large clusters of heavy atoms such as tungsten to obtain phase angles at low resolution (Yonath *et al.*, 1998; Ban *et al.*, 1998; Clemons *et al.*, 1999, Cate *et al.*, 1999). If the structure of a related molecule is known, it is possible computationally to determine the orientation and position of the known molecule in the unknown crystal lattice. This makes it possible to calculate the phase angles. This method is called molecular replacement (Rossmann & Blow, 1962). In case of ribosome crystallography, cryo-EM structures were used to obtain phase angles with molecular replacement at low resolution (Ban *et al.*, 1998; Weinstein *et al.*, 1999; Cate *et al.*, 1999). This approach has been valuable to identify or verify the positions of heavy atoms or heavy atom clusters. A different method takes advantage of the tunability of the X-ray wavelength at synchrotrons to record data across absorption edges for selected heavier atoms. The method is called multiple wavelength anomalous dispersion (MAD; Hendrickson, 1991). All these methods have been applied to determine the structure of the ribosome.

Once an electron density has been obtained, its interpretation demands great care. This warning is not only valid for crystallography but also for cryo-electron microscopy. Here it is important to be aware of the resolution, the quality of the measurements and the phase angles obtained. In addition, some regions of the molecule may be more flexible than others leading to poor or no density. In contrast to virus crystallography, where the high symmetry of the particles leads to high redundancy of the crystallographic information, ribosomes have no symmetry and thus the interpretability of maps at similar resolution will be very different.

An electron density map of whole ribosomes or ribosomal subunits at around 6 Å resolution cannot be interpreted in terms of the details of individual nucleotides or amino acids, particularly if they deviate from well known secondary structure arrangements. The RNA backbone, on the other hand, can easily be followed due to the heavier electron density of the phosphates and the clear helical paths (Ban *et al.*, 1998; 1999; Cate *et al.*, 1999; Harms *et al.*, 1999; Clemons *et al.*, 1999). Protein structures cannot be determined *de novo* at such low resolution even though α-helices can be located. Structures determined from other investigations should be possible to locate and deviations in structure should be seen.

At resolutions around 3 Å, the RNA structures can be interpreted with much greater certainty, even though the hydrogen bonding may be difficult to identify. Protein structures could be interpreted without previous knowledge, but one can get out of register along the polypeptide chain. If homologous structures are known, careful alignment of the multiple sequences will assist in the homology modeling needed to prevent wrong assignments.

At resolutions below 2.5 Å, there is usually no difficulty in determining the atomic structure of proteins and nucleic acids. The main uncertainty is the accuracy of the atomic positions and as a consequence, the distances between atoms that are not covalently linked. In some instances, it may be difficult to determine the strength of hydrogen bonds.

The results from structural studies are generally accessible from databases such as the Protein Data Bank (PDB; http://www.rcsb.org/pdb). Wilson *et al.* (2002) have compiled a summary of all PDB files concerning ribosomes and ribosomal subunits before 2002.

Most scientific journals now adhere to the principle that the coordinates of structure determinations need to be deposited in PDB before a manuscript can be accepted for publication. In many cases, the original diffraction data are also available for further analysis. It is anticipated that the electron microscopists will follow the rules adopted by the crystallographers.

4

The Message — mRNA

The messenger RNA (mRNA) is a central molecule in the translation of a genetic message into protein. Genomic DNA is transcribed into mRNA that can bind to the ribosome and be translated. In RNA viruses, the genetic information is stored in a form that can immediately be translated.

4.1 THE GENETIC CODE

The genetic code (Fig. 4.1) is the universal dictionary by which the genetic information is translated into the functional machinery of living organisms, the proteins. It is remarkable that, regardless of species, the genetic code is the same. Some minor deviations are found, but it is evident that all life on earth has a common origin.

The words or the codons of the genetic message are three nucleotides long (Crick *et al.*, 1961; Crick 1962; 1963). Three nucleotides, or one codon, correspond to one amino acid. The bases are read in a sequential manner, without overlap or punctuation between fixed starting and termination points (Crick *et al.*, 1961). Since there are four different nucleotides used in the messenger RNA (mRNA), this leads to a dictionary of 64 words. There

2nd base in codon

		U	C	A	G		
		Phe F	Ser S	Tyr Y	Cys C	U	
	U	Phe F	Ser S	Tyr Y	Cys C	C	
		Leu L	Ser S	STOP	STOP	A	
		Leu L	Ser S	STOP	Trp W	G	
		Leu L	Pro P	His H	Arg R	U	
	C	Leu L	Pro P	His H	Arg R	C	
		Leu L	Pro P	Gln Q	Arg R	A	
		Leu L	Pro P	Gln Q	Arg R	G	
		Ile I	Thr T	Asn N	Ser S	U	
	A	Ile I	Thr T	Asn N	Ser S	C	
		Ile I	Thr T	Lys K	Arg R	A	
		Met M	Thr T	Lys K	Arg R	G	
		Val V	Ala A	Asp D	Gly G	U	
	G	Val V	Ala A	Asp D	Gly G	C	
		Val V	Ala A	Glu E	Gly G	A	
		Val V	Ala A	Glu E	Gly G	G	

1st base in codon (left margin)   3rd base in codon (right margin)

Fig. 4.1 The universal genetic code. The trinucleotide codons are translated into the 20 amino acids given with their three and one letter codes.

are 20 amino acids that are normally used in proteins and which are translated. In addition, the translation needs a definition of start and stop. The start codon (generally AUG) defines the start of translation as well as the reading frame or the sequence of nucleotide triplets that is to be translated. The start or initiator codon is identical to the methionine codon (AUG). Special mechanisms are used to identify the correct initiation site (see Section 11.3). In addition, there are three stop codons (UAA, UAG and UGA), which makes 61 codons available for the 20 amino acids. Thus, the genetic code is degenerate. In the case of leucine, serine and arginine there are as many as six codons, whereas methionine and tryptophane have only one codon each (Nirenberg & Matthaei, 1961; Nirenberg & Leder, 1964; Khorana *et al.*, 1966; Crick, 1966a). Different organisms use the degenerate genetic code differently. The codon usage is coupled to the availability of tRNAs that can translate them. Thus, the codon usage can differ to the extent that a gene transferred from one organism to another cannot be translated unless the new organism is supplemented with extra tRNAs.

The universal genetic code deviates slightly from the one in verte-brate mitochondria. The most prevalent variants concern methionine and tryptophane that have two codons instead of one. Knight *et al.* (2001) provide an interesting discussion of the more wide range of deviations that have been observed. A database providing information about the genetic code and its variations is provided by Elzanowski & Ostell at http://www.ncbi.nlm.nih.gov/Taxonomy/Utils/wprintgc.cgi?mode=c

4.2 TRANSCRIPTION

The genomic DNA cannot be translated but has to be copied or transcribed into RNA by different RNA polymerases. This occurs by the classical mechanism discovered by Watson and Crick (1953). One strand of the double-stranded DNA (the negative one) is copied through Watson-Crick base-pairing into a positive strand of RNA. This occurs in the 5' to 3' direction. The double-stranded DNA is opened up in a "bubble" that travels along the duplex during transcription. Only transiently is a DNA-RNA hybrid formed. The process of transcription is strongly regulated. Some genes are transcribed frequently, whereas others are transcribed only rarely. Again, some genes are transcribed during a brief period in the life of the cell whereas others are copied more or less continuously. The regulation of transcription is controlled by a great number of transcription factors. The RNA polymerase interacts with these transcription factors to perform the synthesis of an mRNA. Structures of such transcription factors and of RNA polymerases are known (Plate 4.1; Murakami & Darst, 2003).

In eukaryotes, transcription is performed in the nucleus and the transcript is transported into the cytoplasm to be translated. Transcription and translation in mitochondria and chloroplasts is performed in these cellular organelles. However, some proteins are encoded by the nuclear DNA, synthesized in the cytoplasm and finally transported to the organelle (see Chapter 12; Frydman, 2001). In the case of bacteria and archaea, the whole process takes place in the cytoplasm. Bacterial mRNAs can be polycistronic with information from more than one gene contained in one mRNA.

4.3 PROCESSING OF THE TRANSCRIBED RNA

A number of the transcribed RNAs are never translated but have their functions as RNA molecules. These are primarily the ribosomal RNA (rRNA) and transfer RNA (tRNA). The transcribed RNA, called the primary transcript, frequently has to be processed to become an mRNA. Several different processes are involved. The processes in eukarya differ from those in bacteria. The eukaryotic primary transcripts normally contain longer or shorter regions, which are not translated. They form so-called introns, while the translated regions are called exons. The splicing machinery removes the introns by cutting and ligation (Tarn & Steitz, 1997; Doudna & Cech, 2002). Eukaryotic mRNAs are also modified by the addition of a poly(A) tail at the 3′ end of the message.

In eukaryotes, the primary transcripts are also frequently edited to become mRNAs. This is sometimes done by changes of U to C or vice versa (Wedekind *et al.*, 2003). More extensive editing occurs in the mitochondria from trypanosomes, where the mRNAs are extensively modified by large enzymatic particles that use templates called guide RNAs (Estevez & Simpson, 1999).

4.4 READING FRAME AND USAGE OF THE GENETIC CODE

The control of protein synthesis occurs both on the levels of transcription and translation. Feedback regulation of the translation of ribosomal proteins has been thoroughly investigated (Nomura *et al.*, 1980; 1984). Some ribosomal proteins produced in excess over the rRNAs will bind the polycistronic mRNA and inhibit further translation. A recent review extends the picture (Romby & Springer, 2003).

The initiator AUG codon not only defines the start but also the reading frame of an mRNA. Translation proceeds from this start in steps of three nucleotides (one codon) by binding a cognate tRNA through base pairing. The frequent occurrence of termination codons out of frame prevents translation in the wrong frame for more than short stretches. However, there are mRNAs whose correct translation needs a change of reading frame (Baranov *et al.*, 2002a). This is the case for termination or release factor 2 (RF2) from most bacteria (Baranov *et al.*, 2002b). The stop codon (UGA) that is uniquely identified by RF2 occurs early in the gene and in the correct reading frame. Thus, if there is enough RF2, further synthesis will be stopped. If there is a shortage of RF2, an internal so-called Shine-Dalgarno sequence will assist in the +1 frameshifting. In other cases, a phenomenon called ribosome hopping has been identified

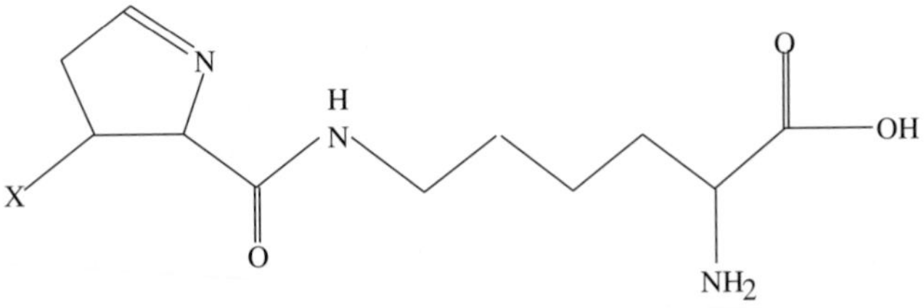

Fig. 4.2 The structure of the 22nd amino acid, pyrrolysine as identified from a crystal structure of a methyltransferase (Hao *et al.*, 2002). It is coded for by the UAG stop codon. X could be methyl, hydroxyl or ammonium.

(Herr *et al.*, 2001). In such a case the ribosome can hop or slide from a codon without a matching tRNA (a hungry codon) to a codon downstream that matches the bound peptidyl-tRNA (Gallant *et al.*, 2003).

A suppressor tRNA can read a stop (nonsense) codon and incorporate a specific amino acid instead of causing termination. A few proteins in bacteria and eukarya contain seleno-cysteine (Se-Cys). This is not achieved by a post-translational modification as in the cases of other non-standard amino acids. Se-Cys is rather incorporated during translation in response to one of the stop codons (UGA). Thus, Se-Cys is the 21st amino acid incorporated into proteins according to the genetic information (Leinfelder *et al.*, 1989). The mechanism for this involves a special tRNA (tRNA^{Sec}), which reads the stop codon. A set of enzymes has specific functions in this system. tRNA^{Sec} cannot bind to EF-Tu due to an antide-terminant sequence in its acceptor stem (Rudinger *et al.*, 1996). A special form of elongation factor Tu called SelB that uniquely binds tRNA^{Sec} is therefore needed (Forchhammer *et al.*, 1989). SelB has the property of identifying a specific secondary structure of the mRNA that precedes the stop codon that corresponds to Se-Cys. This leads to the suppression of the stop codon and the incorporation of Se-Cys (Thanbichler *et al.*, 2000).

Yet, another amino acid incorporated into proteins according to genetic information was found in a methyltransferase in the Achaean *Methanosarcina barkeri*. The amino acid is named pyrrolysine (Fig. 4.2) and is incorporated in response to the UAG stop codon (James *et al.*, 2001; Hao *et al.*, 2002; Srinivasan *et al.*, 2002). The pyrrolysine is probably functionally important. The genome contains information for a tRNA$_{CUA}$ (or tRNA^{Pyl}) that reads the UAG codon and for a special tRNA synthetase that may charge or modify the tRNA. However, it has been shown that the tRNA is charged in complex with classes I and II LysRS (see Section 5.5; Polycarpo *et al.*, 2003). tRNA^{Pyl} can bind to EF-Tu in contrast to tRNA^{Sec}. There are indications of pyrrolysine occurring in other organisms (Ibba & Söll, 2002). The mode by which the UAG is read, instead of causing termination, remains to be explored.

How the ribosome is able to maintain the reading frame and move the mRNA by three nucleotides at a time is not yet understood. Likewise, it is not understood how, in specific cases, the reading frame is altered (Atkins & Gesteland, 2001).

5

The Adaptor — tRNA

The transfer RNAs (tRNAs) are central molecules in protein synthesis. Historically, even after the structure of DNA and the basics of protein synthesis were clarified, the existence of tRNAs or its chemical nature was not known. In 1956, Crick pointed out the significant problem in assembling a polypeptide from an RNA template (Crick, 1958; Woese, 2001). A stereochemical complementarity between codons and amino acids seemed impossible. He suggested that adaptors were needed and that small RNA molecules could function as such adaptors. They could be charged with specific amino acids by enzymes and would decode the mRNA by Watson-Crick base-pairing, thereby participating in the incorporation of the amino acids into a growing polypeptide. More or less in parallel, such adaptors were identified (Hoagland *et al.*, 1957; Hoagland, 2003). They were initially called soluble RNA (sRNA), but are now known as tRNA molecules. Each tRNA is specific for one amino acid. Normally, they contain about 75 nucleotides.

5.1 STRUCTURE

When the first nucleotide sequence of a tRNA was determined, the possibility for a base-paired secondary structure was examined. Among the

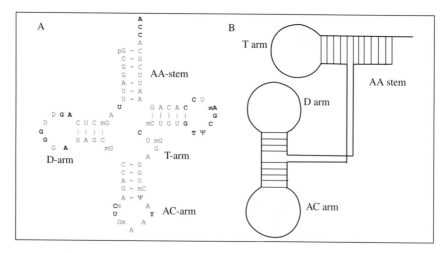

Fig. 5.1 (A) The cloverleaf secondary structure of tRNA[Phe] from yeast. Conserved nucleotides are shown in boldface. The acceptor or the aminoacyl (AA) stem with the CCA sequence is at the 3′ terminus. The D-arm contains a dehydrouridine. The anticodon (AC) arm contains the anticodon at the bottom (normally bases 34–36). The helix and loop to the right is called the TΨCG or T-arm. Finally, there is an extra loop of variable length. **(B)** A schematic representation of the way the cloverleaf structure is organized in three dimensions.

structures suggested, only one, the now classical cloverleaf (Fig. 5.1), was consistent with subsequently determined sequences (Holley *et al.*, 1965). In this structure, the stem regions contain four to seven base pairs. The cloverleaf is arranged in such a way that the 5′- and 3′-termini are base-paired to each other. The three leafs are formed by three base-paired regions, each forming a loop. The middle loop contains the anticodon of the tRNA. At an early stage, there were thoughts that there could be a structural complementarity between the anticodon and the amino acid and that they could be close together in the structure of the tRNA.

The tRNA molecules have a number of conserved features. The cloverleaf is the basis, even though the number of base pairs in the differ-ent stems can vary somewhat. Furthermore, the 3′-end has a conserved sequence, CCA, which is not base-paired. The 2′-hydroxyl or 3′-hydroxyl of the terminal ribose (A76) at the 3′-end is charged by amino acids.

In 1973, the first three-dimensional structures of tRNA were deter-mined (Robertus *et al.*, 1974; Kim *et al.*, 1974). The structures were surpris-ing in several ways. Firstly, the secondary structure of the cloverleaf was confirmed, but it was folded into the shape of an L (Fig. 5.2). The cloverleaf

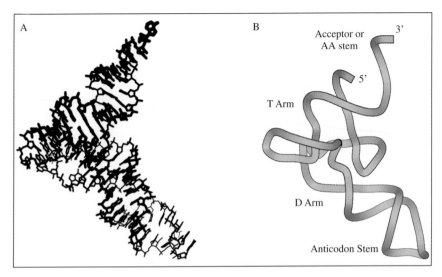

Fig. 5.2 The L-shaped three-dimensional structure of tRNA. The end of the acceptor stem, the single-stranded CCA-end, where the amino acid is attached, is at one extreme end of the molecule about 75 Å from the anticodon at the opposite end. The D- and the T-arms participate in forming the elbow of the molecule. The figures have been kindly produced by Dr. M. Fodje using the program MOLSCRIPT (Kraulis, 1991).

secondary structure was folded in such a manner that the D- and T-loops form the elbow region through tertiary structure interactions. Secondly, the anticodon was found at one end and the amino acid acceptor or the CCA sequence, at the opposite end, approximately 75 Å apart. This meant that the anticodon has no possibility of interacting with the amino acid. It also means that when the tRNA incorporates the amino acid into the growing polypeptide on the ribosome, the mRNA and the decoding site are far from the site for peptidyl transfer. This was consistent with the fact that decoding is performed on the small ribosomal subunit, whereas peptidyl transfer is performed on the large subunit (Okamoto & Takanami, 1963).

The L-shape of the tRNA molecule is a fundamental feature. It is not only observed in the crystal structure of tRNA, but also in complexes with tRNA synthetases, EF-Tu and when bound to the ribosome. Some minor variations have been seen. On the ribosome, however, the tRNA is observed to have a fair amount of conformational variability (see Sections 5.5, 8.2 and 11.5).

There is a certain amount of variation in the structures of different tRNAs. The variable loop, at the junction of the T- and anti-codon stems, does not have a single base pair in some tRNAs, but in others, it can be quite long. This is the case for tRNA^{Ser} and tRNA^{Leu}. Mitochondrial tRNAs from higher eukaryotes have tRNAs that are dramatically simplified by the removal of one of the stem-loop structures. This does not affect the three-dimensional structure in any major way, as these tRNAs maintain the L-shape.

Recently, a different conformation of tRNA was discovered (Ishitani *et al.*, 2003), called the λ form. It was observed in a complex with a tRNA-modifying enzyme. The nucleotide, which is modified G15 in the D-arm, is not accessible in the L form. The profound transformation involves a disruption of all the base pairs and tertiary interactions of the D-arm.

5.2 tRNA CONFORMATIONS WHEN BOUND TO THE RIBOSOME

The functional significance of the L-shaped three-dimensional arrangement of the tRNA and the extension of the CCA-end is gradually becoming understood with the insights of how the design of the tRNA is utilized during protein synthesis on the ribosome.

The most obvious point of flexibility in the tRNA is the elbow of the L-shaped molecule. Indeed, a range of larger angles has been observed in the elbow region in some tRNA structures (Moras *et al.*, 1980) and in a complex between EF-Tu and tRNA^{Cys} (Nissen *et al.*, 1999). In addition, the tRNA in the ribosomal E-site (see Section 8.2) has a more open structure (Yusupov *et al.*, 2001).

In the discrimination between the cognate and non-cognate tRNA, the anticodon stem-loop structure (ASL; see Fig. 5.2B) has to make a kink with regard to the D-stem. The kink occurs at nucleotide 26, which is not base-paired in any crystal structure of tRNA, but is partially intercalated between nucleotides 44 and 45 of the variable loop (Valle *et al.*, 2003a). As has already been inferred at the examination of the first crystal structure of tRNA, this point could allow some flexibility (Robertus *et al.*, 1974; Valle *et al.*, 2003a). There have not yet been any studies of how the initiator tRNA may differ from elongator tRNA in this respect, since the fMet-tRNA binds directly to the P-site of the small subunit and does not have to go through the same type of screening through kinking.

The single-stranded region of the CCA-end is obviously also a region allowing some flexibility. It is a flexibility that is needed in the interactions with the ribosomes during elongation. A comparison of the tRNA in the A-site with the one in the P-site shows that the main body of the tRNA has moved sideways, while the CCA-end has rotated by about 180° (Nissen *et al.*, 2000; Yusupov *et al.*, 2001; Schmeing *et al.*, 2002; Hansen *et al.*, 2002b; Bashan *et al.*, 2003; Agmon *et al.*, 2003). Obviously, the single-stranded character of the 3'-end of the tRNA must have been a constraint in evolution and developed in order to facilitate the attack of the next aminoacyl-tRNA on the bond between tRNA and the peptide.

From an evolutionary point of view, the design of the tRNA is now beginning to be better understood. Firstly, the distance between the bound aminoacyl moiety and the anticodon fits very well with the location of the decoding site and the peptidyl transfer center (PTC) on the ribosome (Yusupov *et al.*, 2001). Secondly, the flexible regions, the elbow, the possibility of making a kink between the anticodon stem and the D-stem (Robertus *et al.*, 1974; Valle *et al.*, 2003a) as well as the CCA-end have all been noted to have essential functional roles in translation.

5.3 SYNTHESIS

The genes for tRNA molecules are dispersed throughout the genomes. There are sometimes several genes for a certain tRNA, and tRNA genes are frequently found as clusters. Some of them are located among the genes for ribosomal RNAs. The tRNA genes do not code for the final functional molecule. The transcript contains sequences that are removed by specialized tRNA-processing nucleases (for a recent review see Ferre-D'Amare, 2003). Likewise, in most species the 3'-terminal CCA residues are added, edited or repaired by an enzyme, nucleotidyltransferase (Schürer *et al.*, 2001). The structure of one of these, the human nucleotidyltransferase, has recently been determined (Augustin *et al.*, 2003).

In addition, the mature tRNA molecules are extensively modified (see Björk, 1995; McCloskey & Crain, 1998 for reviews). Some modifications are so typical that they have given the names to the parts of the structure they belong to. Thus, the D-loop is named after 5,6-dehydrouridine. The anticodon loop is also frequently modified.

5.4 CODON-ANTICODON RELATIONSHIPS

tRNA reads the genetic code, with its 64 "words" or codons (see Section 4.1). Since the code is degenerate, the codon usage is different in different organisms. The codons used are correlated to the set of tRNAs expressed (for a recent discussion see Elf *et al.*, 2003). In some cases, the codon usage is limited to a small set of tRNAs (minimally 20), while in other species there are tRNAs corresponding to most codons.

The anticodon, normally at positions 34–36 of the tRNA reads the codons of the mRNA primarily by Watson-Crick base-pairing (Fig. 5.3). However, the first position of an anticodon of a tRNA can base pair with different nucleotides in the third position of a codon. Thus, non-canonical base pairs occur in the third, so-called "wobble" position of the codon (Crick, 1966b). This also includes modified bases of the tRNA and allows a tRNA with a certain anticodon to read several codons. Thus, a limited set of tRNAs can read a larger set of codons.

5.5 CHARGING — THE tRNA SYNTHETASES

The enzymes that charge the tRNAs with amino acids, the aminoacyl-tRNA synthetases (aaRS), are specific for one amino acid each. Only a brief description will be provided here.

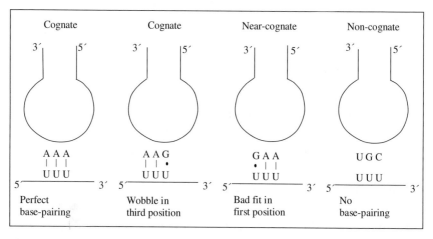

Fig. 5.3 Different types of interactions between anticodon of tRNA (*top*) and mRNA (*bottom*).

Since there are 20 amino acids, there are normally 20 tRNA synthetases in an organism, even though there are some deviations (see below). These enzymes have to be specific not only for the amino acid but also for the corresponding tRNA or a family of the so-called isoacceptor tRNAs. The charging of an amino acid onto its cognate tRNA molecule is a two-step process:

1. aa + ATP $\longrightarrow$ aa-AMP + PP
2. aa-AMP + tRNA $\longrightarrow$ aa-tRNA + AMP

The correct amino acid has to be bound by the enzyme and activated by an ATP molecule to form the reactive intermediate aa-AMP. In a second step, the amino acid is transferred to the correct tRNA that is also bound to the enzyme. The 2'- or 3'-OH group of the terminal adenosine of the conserved CCA motif directly attacks the high-energy ester bond in aa-AMP, resulting in attachment of the amino acid to the ribose. Since the distance between the amino acid and the anticodon is large, not all tRNA synthetases can contact the anticodon, but they select the correct tRNA using different structures along the length of the tRNA molecule (Giege *et al.*, 1998). The fidelity of translation depends primarily on the synthetases. Errors in the charging of the tRNAs will not be detected in subsequent steps. Thus, the synthetases must recognize their specific amino acid and the tRNA with high accuracy (Loftfield & Vanderjagt, 1972). Since some amino acids are very similar in size and structure, special editing mechanisms have evolved (Alexander & Schimmel, 2001; Schimmel & Ribas de Pouplana, 2001).

Classes and Subclasses

The molecular weight and the oligomeric states of the aaRS vary considerably, and this was originally the source of some confusion as to their evolutionary origins (Table 5.1). However, in 1990, two classes of tRNA synthetases were identified on the basis of the three-dimensional structures and sequence similarities: classes I and II. There are ten aaRSs in each class (Table 5.2, Cusack *et al.*, 1990; Eriani *et al.*, 1990).

Enzymes of the two classes have entirely different structures (Fig. 5.4). While the ATP-binding domains of class I aaRS is a Rossmann fold with parallel β-strands, the corresponding domain in the class II aaRS is built from anti-parallel β-strands (Cusack *et al.*, 1990; Eriani *et al.*, 1990).

Table 5.1 Oligomeric States of Aminoacyl-tRNA Synthetases

Class I

RS	L	I	V	C	M	R	E	Q	K	Y	W
Oligomeric state	α	α	α	α	α_2	α	α	α	α	α_2	α_2

Class II

RS	S	T	G	A	P	H	D	N	K	F
Oligomeric state	α_2	α_2	$(\alpha\beta)_2$	α_4	α_2	α_2	α_2	α_2	α_2	$(\alpha\beta)_2$

Table 5.2 Characteristics of the aaRS Enzymes

	Class I	Class II
Sequence motif	HIGH KMSKS GXGXGXER	FRXE/D R/HXXXF
Subclass a	L, I, V, C, M, R	S, T, G, A, P, H
b	E, Q, K	D, N, K
c	Y, W	F
Amino acylation	2'OH	3'OH
Fold of ATP domain	Rossmann ($//\beta$)	Antiparallel β
Amino acid binding	Surface	Deep pocket
tRNA acceptor end	Bent	Straight
tRNA recognition	Minor groove	Major groove

Different consensus sequences are characteristic for the two classes (Cusack, 1995). The aaRSs of the two classes recognize the tRNAs from opposite sides and charge the tRNA on the 2'-OH (class I) or the 3'-OH (class II) of the terminal riboses of the tRNAs, which may suggest an interesting evolutionary background (Schimmel & Ribas de Pouplana, 2001).

It is not only the ATP-binding domains that display characteristic features, but the other domains building up these enzymes do as well. Thus the two classes of aaRS can be divided into subclasses a, b and c based on sequence homology and domain architecture (Cusack, 1995). The aaRSs are built with modular domain arrangements, where the very different ATP-binding domains define the class. The enzymes within each subclass have similar domain arrangements.

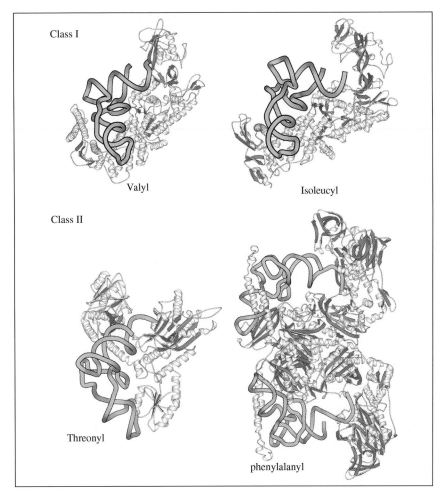

Fig. 5.4 Two structures of each of the two classes of tRNA synthetases in complexes with tRNAs. The structures and mode of tRNA binding are entirely different. The active sites for charging with amino acids are at the single stranded acceptor end. The figures have kindly been produced by Dr. M. Fodje using the program MOLSCRIPT (Kraulis, 1991).

Amino Acid Recognition: Not a Simple Matter

The identification of the different tRNAs by the aaRS is only partly done using the anticodons. Since there are up to six codons for some amino acids, this would not be a very useful way to discriminate between the different tRNAs. Instead, the identification of a cognate tRNA among the different non-cognate ones is due to a number of features of the individual tRNAs called the "identity set" (Giege *et al.*, 1998). Most elements of the identity set are localized in the anticodon and acceptor stems, the majority on the side of the tRNA that faces the aaRS (Fig. 5.4).

Deviations

Normally there are 20 aaRS in an organism. However, there are deviations that have been detected through the analysis of whole genomes. The most dramatic deviation is found in two methanogenic archaea, *Methanococcus jannashii* and *Methanothermobacter thermoautothropicus*. In these species only 16 aaRSs are found (Bult *et al.*, 1996; Smith *et al.*, 1997). The enzymes that seemed to be lacking were AsnRS, CysRS, GlnRS and LysRS (see review by Stathopopoulos *et al.*, 2001). Obviously there must be mechanisms for charging tRNAs with asparagine, cysteine, glutamine and lysine since these amino acids are used in these archaea.

Even though these archaea were lacking in normal lysyl-tRNA synthetases, the charging activity was present. When the enzyme performing the activity was purified and its amino acid sequence determined, it turned out that this LysRS did not belong to class II but that it was an aaRS of class I (Ibba *et al.*, 1997). LysRS1, as the enzyme is denoted, is found in most archaea but also in some bacteria. One organism is found to have both forms of the enzyme (Stathopopoulos *et al.*, 2001).

GlnRS or AsnRS are missing from all the archaea and many bacteria. A nondiscriminating GluRS misacylates tRNA^{Gln} with Glu and subsequently a tRNA-dependent amidotransferase (AdT) converts Glu-tRNA^{Gln} to Gln-tRNA^{Gln} (Tumbula *et al.*, 2000). A similar mechanism can also be employed for AspRS. The enzyme does not discriminate between tRNA^{Asp} and tRNA^{Asn} but charges both with Asp. Also, in this case, a transamidase converts Asp-tRNA^{Asn} to Asn-tRNA^{Asn} (Curnow *et al.*, 1996). Two different AdT enzymes have been found: GatCAB and GatDE. The former, which is heterotrimeric, can transamidate both Glu-tRNA^{Gln} and Asp-tRNA^{Asn} (Curnow *et al.*, 1998; Becker *et al.*, 2000; Tumbula *et al.*, 2000), while GatDE, which is heterodimeric, can produce only Gln-tRNA^{Gln} (Tumbula *et al.*, 2000). Both enzymes require ATP for catalysis (Wilcox, 1969). The misacylated Glu-tRNA^{Gln} and Asp-tRNA^{Asn} do not participate in protein synthesis, since they do not bind to EF-Tu in species where the possibility could occur (Stanzel *et al.*, 1994). The mechanism for this discrimination by EF-Tu is not known.

As mentioned above, CysRS was also found to be missing in two archaea. When purified, the enzyme that charged tRNA^{Cys} with cysteine was identified as ProRS. This unique aaRS is able to aminoacylate both tRNAs with the correct amino acid without cross-reactivity (Stathopouplos *et al.*, 2000, Lipman *et al.*, 2000). The Pro/Cys tRNA synthetase has also been found in some eukaryotes as well as in bacterial species (Stathopouplos *et al.*, 2001). ProRS from *T. thermophilus* has also both activities (Feng *et al.*, 2002). The structure of this enzyme has been determined and this has led to a suggestion for how the discrimination is achieved (Yaremchuk *et al.*, 2000; 2001). However, subsequent structural determinations with substrate analogues explain how the ProRS misacylates the tRNA^{Pro} with cysteine (Kamtekar *et al.*, 2003).

Editing

It is not only that the tRNAs may be difficult to differentiate, but also some amino acids are very similar in shape and nature (Jakubowski & Goldmann, 1992; Alexander & Schimmel, 2001; Schimmel & Ribas de Pouplana, 2001). Thus, valine would fit nicely into a pocket designed for isoleucine, threonine fits into a pocket defined for valine, and serine fits into pockets for threonine. The affinities for the non-cognate amino acids may be lower but the discrimination is not sufficient. These amino acid similarities are found within subclasses Ia and IIa. Misacylated tRNAs or misactivated amino acids need to be eliminated. This is done through an editing mechanism that certain aaRS possess. Thus, the class Ia enzymes LeuRS, IleRS and ValRS all have a homologous editing domain called CP1. It is conserved throughout evolution (Schimmel & Ribas de Pouplana, 2001). However, the editing domains of ThrRS and ProRS are not conserved. Structural information on the editing mechanism of IleRS (Nureki *et al.*, 1998; Silvian *et al.*, 1999) and ThrRS (Dock-Bregeon *et al.*, 2000) has been obtained (Sankaranarayanan & Moras, 2001).

6

The Workbench — Ribosomes

6.1 THE COMPOSITION OF RIBOSOMES

The ribosome is built from two subunits, the large subunit and the small subunit. The size of the ribosome and its subunits vary with the species (Table 6.1). The ribosome is composed of RNA and protein molecules. The ribosomal RNA (rRNA) is not translated. The rRNAs are large molecules that form the core of the ribosomal subunits. The majority of the ribosomal proteins have less than 20 kDa in molecular weight. In most ribosomes, the mass of the rRNA is significantly larger than that of the ribosomal proteins. It is therefore not surprising that the protein-RNA interactions are extensive and the protein-protein interactions are more limited. The ribosomes from mammalian mitochondria, that may have evolved most rapidly, have less rRNA and a larger complement of proteins (Tables 6.2 and 6.3). It is not unlikely that some proteins replace and mimic parts of the deleted rRNA structurally and perhaps also functionally. The ribosomes from trypanosomal mitochondria are the smallest found (see Tittawella *et al.*, 2003 and references therein).

Table 6.1 Sedimentation Coefficients for Ribosomes and Ribosomal
Subunits from Different Sources

Source	Ribosomes	Small Subunit	Large Subunit
Bacteria	70S	30S	50S
Chloroplasts	70S	30S	50S
Archaea	70S	30S	50S
Mitochondria			
(plant)	70S	30S	50S
(yeast)	74S	37S	54S
(mammals)	55S	28S	38S
(trypanosomes)	35S?	nk*	nk*
Eukarya	80S	40S	60S

* not known

Table 6.2 Ribosomal RNA from Different Sources

Source	Small Subunit	Large Subunit
Bacteria	16S	23S, 5S
Chloroplasts	16S	23S, 5S, 4.5S
Archaea	16S	23S, 5S
Mitochondria		
(plant)	18S	26S, 5S
(yeast)	15S	21S
(mammals)	12S	16S
(trypanosomes)	9S	12S
Eukarya	18S	5.8S, 25–28S, 5S

Table 6.3 Number of Ribosomal Proteins from Different Sources

Source	Small Subunit	Large Subunit
Bacteria	21	33
Chloroplasts	21	33
Archaea	28	40
Mitochondria		
(yeast)	31	46
(mammals)	29	48
Eukaryotes	32	46

6.2 rRNA

The ribosomal RNAs (rRNA) form the core of ribosomes. The two ribosomal subunits have a large RNA molecule each, which provides the
binding sites for ribosomal proteins. These proteins serve to stabilize

the rRNA and to organize its proper functional three-dimensional structure. Several of the ribosomal functions are completely associated with the rRNA.

The small ribosomal subunit has one rRNA molecule. In bacteria, it is called the 16S RNA from its sedimentation velocity. In other organisms, the size of the corresponding RNA molecule varies (see Table 6.2). In the large ribosomal subunit, there is one small RNA molecule called the 5S RNA composed of about 120 nucleotides. The large RNA molecule in bacterial large subunits is the 23S RNA. Its size also varies, depending on the species and sometimes it occurs as several pieces (Table 6.2).

The expression of the genetic material through protein synthesis is fundamental for all forms of life. Since protein synthesis is performed on ribosomes and all organisms have ribosomes, the nucleotide sequences of rRNA have been used for studies of evolution. It was revealed that rRNA is a tremendous source for evolutionary studies (Woese & Fox, 1977). The first complete rRNA sequences were obtained from *E. coli* (Brosius *et al.*, 1978; 1980) and could be used to align oligonucleotide sequences available from hundreds of species (Noller & Woese, 1981). From these studies, it could be established that archaea are not a subgroup of bacteria but form their own kingdom (Woese & Fox, 1977; Woese *et al.*, 1990).

The possibility of obtaining one unique scheme of base pairing for a large RNA molecule is close to impossible. Even for the tRNA molecule, this was initially difficult (see Chapter 5). However, since structures are normally conserved, the sequences of a number of molecules could give a good consensus model of the secondary structure of the rRNAs (Glotz & Brimacombe, 1980; Glotz *et al.*, 1981; Noller & Woese, 1981; Noller *et al.*, 1981). This gave rise to consensus secondary structures that showed the arrangement of the rRNAs into helices and domains (see Chapter 7). The conservation of the sequences of the base-paired regions was frequently less than for the single-stranded regions. This suggested that the rRNA could contain functional regions.

The studies of the secondary structure of rRNA have led to the conclusion that the 5S RNA has a secondary structure with the shape of a Y, where the 5'- and 3'-termini form one of the five short helices of the molecule (Plate 6.1B; Ban *et al.*, 2000; Yusupov *et al.*, 2001). The large RNA molecule of the large subunit forms six domains (I–VI) all originating from one central region (Plate 6.1B). In addition, some of the domains have a structure, which has excursions from a central region. The 23S RNA has

101 helices denoted H1–H101. The rRNA of the small subunit, the 16S RNA, has 45 helices (h1–h45). They are organized into four different domains, the 5' domain, the central domain, the 3'-major, and 3'-minor domains (Plate 6.1A). In the 16S RNA, the domains also originate from a central part of the structure. It is interesting that in essentially all of the RNA molecules, the 5' and the 3' ends are close in the secondary structure. The 16S RNA deviates in that the 3'-minor domain extends from the connection of the two ends.

The organization of rRNAs from all species follows this general pattern of domain arrangements, but the number of helices varies greatly. Despite the large variation in size of the rRNAs, the core of the secondary structure remains. The differences are found in the large variation in the size of the loops. Thus, the smaller rRNAs from mammalian mitochondria lack various loops (Cavdar Koc *et al.*, 2001a,b; Plate 6.2).

The organization of the rRNA in three dimensions has been explored by chemical methods (Sergiev *et al.*, 2001; Green & Noller, 1997). RNA can be cleaved with enzymes or reacted by a number of chemicals and the position of the cleavages or modifications can be established. Bifunctional crosslinking can also be used for studies of RNA and the residues involved can be identified. This gives information on proximity. Furthermore, base-paired regions, permanent or temporary can be analyzed as well as protection against labeling or cleavage by ribosomal proteins or other components of the system. Labeling of the rRNA from other components of the protein synthesis system has also been highly informative (Culver & Noller, 2000; Joseph & Noller, 2000).

The rRNA folds into a fully functional particle with the aid of ribosomal proteins (see Section 6.4). A certain order of assembly has been identified (Held *et al.*, 1974; Röhl & Nierhaus, 1982). The binding sites for the ribosomal proteins have been established by chemical and enzymatic methods and have given further insights into the organization of the rRNAs (Stern *et al.*, 1989; Powers & Noller, 1995; Mueller & Brimacombe, 1997; Mueller *et al.*, 1997). The pattern of protein interaction with the rRNA is sometimes very complex, with several domains of rRNA involved.

Detailed information on the structure of fragments of the rRNAs has been obtained using NMR spectroscopy and X-ray crystallography (Fourmy *et al.*, 1996; Yoshizawa *et al.*, 1999; Wimberley *et al.*, 1999; Agalarov *et al.*, 2000). More recently, the crystal structures of complete ribosomal subunits from different species have provided extensive information about the organization of the rRNAs (see below in Chapter 7). These studies show that the structures of the rRNAs are highly complex and deviations from standard A-type RNA are numerous. The hydrogen bonding between bases as well as between bases and ribose hydroxyls gives extensive possibilities for variation beyond the classical Watson-Crick interaction. On the other hand, bulges and loops in the secondary structures are frequently accommodated into the helical structures.

6.3 RIBOSOMAL PROTEINS

The Identification and Number of Ribosomal Proteins

A large number of usually small proteins are bound to the ribosomal RNA. The exact enumeration of the ribosomal proteins has met with some difficulties. When ribosomes are purified, different washing procedures lead to a variable number of proteins attached. Thus, several of the proteins are found in less than stochiometric amounts (Hardy, 1975). On the other hand, proteins that cannot be classified as ribosomal proteins may stick to the ribosome artificially during the purification. In addition, some of the ribosomal proteins were not initially identified as such due to their limited size and high positive net charge that made them run out of the classical two-dimensional gel (Kaltschmidt & Wittmann, 1970).

The small subunit from *E. coli*, which has become the reference organism, has proteins S1–S21 and the large subunit contains proteins L1–L36 (Table 6.3). Some numbers of the large subunit have been deleted. These are L7, which is a modified form of L12 not found in all species; L8 is a complex of L7/L12 and L10 (Pettersson *et al.*, 1976); L26 is identical to S20. Since L7 is an oddity of a few species, we will not use the name L7/L12 for this protein but rather L12.

The realization that many of the ribosomal proteins in bacteria occur in operons mainly containing ribosomal proteins has helped to verify the ribosomal nature of several proteins. In total, there are around 54 proteins in bacteria and chloroplasts while there are between 70 and 80 in eukarya and mitochondria (Table 6.3). Archaea have an intermediate number of proteins. It has been observed that a large number of the ribosomal proteins can be deleted from the *E. coli* ribosome without apparent effects on cell viability (Dabbs *et al.*, 1983; Dabbs, 1986). On the other hand, a comparison of the completely sequenced genomes shows that the few proteins that are universally conserved are primarily ribosomal proteins

Table 6.4 Index of Ribosomal Proteins Classified by Families on the Basis of Sequence Similarities (Large subunit L; Small subunit S)

Family Name	Alternative Name[1]	Taxonomic Range[1]	Universal Constant[2]	Deletion in Mutant[3]	Ribosomal Function
L1b	L10Ae	A B E C c M	X	X	L11 operon[6]. tRNA exit domain.
L2b	L8e	A B E C M	X		
L3b	L3e	A B E C	X		Assembly[4].
L4b	L4e	A B E C	X		Assembly[4]. S10 operon[6]. Exit tunnel.
L5b	L11e	A B E C M	X		
L6b	L9e	A B E C M	X		
L9b		B C c			
L10b	P0	A B E	X		β operon[6]. At GAR.
L11b	L12e	A B E C c	X	X	At GAR.
L12b	P1/P2	A B E C c			At GAR. GAP?
L13b	L13Ae	A B E C c	X		Assembly[4].
L14b	L23e	A B E C M	X		
L15b	L27Ae	A B E c	X		
L16b	L10e	A B E C M	X		
L17b		B	m		
L18b	L5e	A B E C M	X		
L19b		B E C M	X		
L20b		B C			Assembly[4].
L21b		B C c			
L22b	L17e	A B E C c	X		Assembly[4]. Exit tunnel.
L23b	L23Ae	A B E C	X		Binds SRP.
L24b	L26e	A B E C c	X	X	Assembly[4].
L25b		B			
L27b		B E C		X	
L28b		B C c	m X		
L29b	L35e	A B E C		X	
L30b	L7e	A B E		X	
L31b		B C M			
L32b		B C			
L33b		B C		X	
L34b		B C c	m		
L35b		B C c			
L36b		B C	m		
L6e		E			
L7Ae		A B E			

Table 6.4 (Continued)

Family Name	Alternative Name[1]	Taxonomic Range[1]	Universal Constant[2]	Deletion in Mutant[3]	Ribosomal Function
L13e		A E			
L14e		A E			
L15e		A E			
L18e		A E			
L18Ae		E			
L19e		A E			
L21e		A E			
L22e		E			
L24e		A E			
L27e		E			
L28e		E			
L29e		E			
L30e		A B E			
L31e		A E			
L32e		A E			
L34e		A E			
L35Ae		A E			
L36e		E			
L37e		A E			
L37Ae		A E			
L38e		A E			
L39e		A E			
L40e		A E		Ubiquitin C-terminal CEP52	
L41e		A E			
L44e	L36Ae	A E			
LX		A			
S1b	S1e	B E C c M		X	Interacts with mRNA.
S2b	SAe	A B E C M	X		
S3b	S3e	A B E C M			
S4b	S9e	A B E C M	X		Assembly[5]. α operon[6].
S5b	S2e	A B E C	X		
S6b		B C		X	
S7b	S5e	A B E C M	X		Assembly[5]. Str operon[6].
S8b	S15Ae	A B E C M	X		Assembly[5]. Spc operon[6].
S9b	S16e	A B E C	X	X	
S10b	S20e	A B E C M	X		
S11b	S14e	A B E C M	X		

Table 6.4 (Continued)

Family Name	Alternative Name[1]	Taxonomic Range[1]	Universal Constant[2]	Deletion in Mutant[3]	Ribosomal Function
S12b	S23e	A B E C M	X		In decoding site.
S13b	S18e	A B E C c M	X	X	
S14b	S29e	A B E C M	X		
S15b	S13e	A B E C M	X		Assembly[5].
S16b		B C m			
S17b	S11e	A B E C c m	X	X	Assembly[5].
S18b		B C			
S19b	S15e	A B E C M	X		
S20b		B C		X	S20 operon[6].
S21b		B			
S22b		B			Stationary phase.
THX		B			
S3Ae		A E			
S4e		A E			
S6e		A E			
S7e		E			
S8e		A E			
S10e		E			
S12e		E			
S17e		A E			
S19e		A E			
S21e		E			
S24e		A E			
S25e		E			
S26e		E			
S27e		A E			
S27Ae		A E		Ubiquitin C-terminal	
S28e		A E			
S30e		A E			
S30Ae		B c			
S31e		B c			
var1		M			

"A" = archaea; "B" = bacteria; "E" = eukaryotes; "C" = chloroplast encoded; "c" = chloroplast, but nuclear encoded; "M" = mitochondrion encoded; "m" = mitochondrion, but nuclear encoded
L7b = L12b; L8b = L10b:$(L12b)_4$; L26b does not exist = S20b; S22e = L32e; L10e = L16b;
L16e = L12e; L25e = L30e; L33e = S24e; L20e does not exist

(1) Bairoch; http://www.expasy.org/cgi-bin/lists?ribosomp.txt
(2) According to Harris *et al.* (2003)
(3) Dabbs *et al.* (1983), Dabbs (1986)
(4) Röhl & Nierhaus (1982)
(5) Held & Nomura (1974)
(6) Translational control. See Nomura *et al.* (1984)
(7) Wada (1998)

(see Table 6.4; Lecompte *et al.*, 2002) and translation factors (Pandit & Srinivasan, 2003). Evidently, a subgroup of these proteins is completely essential. One can note (see Table 6.4) that few of the totally conserved proteins have been deleted in mutants (Dabbs *et al.*, 1983). Furthermore, there seems to be a disagreement between the notion of universally conserved proteins and their occurrence in different types of organisms.

Copy Number of Ribosomal Proteins

Ribosomal proteins are normally present at one copy per ribosome (Hardy, 1975). However, there is one exception, the acidic protein L12 in *E. coli,* which is found at four copies per ribosome. In archaea or eukaryotes, this protein corresponds to an acidic protein of the same size but with quite different sequence. This protein is also found at four copies per ribosome. In eukaryotes, there are several forms of this protein encoded by different genes. The total number of these proteins, however, remains at four per ribosome.

6.4 THE ASSEMBLY OF RIBOSOMES

It has been established that, on the one hand, the ribosomal RNAs need some of the ribosomal proteins to be correctly folded. On the other hand, several of the ribosomal proteins cannot bind to their sites on the ribosome unless the RNA is correctly folded. Thus, there is a succession in the binding of ribosomal proteins to the rRNA. This has been explored for the ribosomal subunits (Table 6.4; Held *et al.*, 1974; Röhl & Nierhaus, 1982). Only two proteins, S17 and L24, of the early assembly proteins have been deleted in mutants (Dabbs *et al.*, 1983; Dabbs, 1986). No additional ribosomal protein binds to the ribosome once the subunits have joined.

7

The Structure of the Ribosome

Elucidation of the structure of ribosomes has long been a significant challenge due to its central role in the cell and also due to its size and lack of symmetry. A primary interest is focused on the binding sites for mRNA and tRNA molecules, the decoding site and the site for peptidyl transfer as well as the sites for interactions with the translation factor proteins. A large number of approaches have been developed in order to obtain a detailed structure of the subunits or even better the whole ribosome. Structural studies of individual components or complexes of ribosomal components have also been performed.

7.1 EARLY STUDIES OF THE STRUCTURE OF RIBOSOMAL SUBUNITS AND RIBOSOMES

Electron microscopy studies have provided outlines of the bacterial ribosome and its subunits (Lake *et al.*, 1974; Lake, 1985; Tischendorf *et al.*, 1974; Stöffler & Stöffler-Meilicke, 1984; Vasiliev *et al.*, 1974; Spirin & Vasiliev, 1989). The small subunit has the shape of a right-hand mitten (Fig. 2.2; Plate 2.1). It has features called the body the thumb or the platform and the head, which corresponds to the finger parts of the mitten. Irrespective of the species, this general shape seems to hold (Lake, 1985). Subsequent cryo-electron microscopy studies using 3D reconstitution techniques have provided many further details. Some of these features have been given names. Thus, the end of the head is called the beak or the nose (Plate. 2.1, Gabashvili *et al.*, 2000). The upper part of the body on the side opposite to the platform is called the shoulder. The bottom part of the body has a minor protuberance called the toe or the spur. The interface side of the small subunit, the inner side of the mitten, interacts with the large subunit, whereas the opposite side is called the back or the solvent side.

The large subunit was early on observed to have a structure like a crown when seen from the side of the interface between the subunits (Plate. 2.1). The three protuberances on one side of the particle are called the central protuberance, the L12 stalk on the right-hand side (Strycharz *et al.*, 1978) and the L1 stalk on the left-hand side (Lake & Strycharz, 1981; Dabbs *et al.*, 1981) when seen from the interface side. When seen from the side, the particle has a hemispherical shape. The flatter surface is the interface side.

The positions of the individual proteins were identified in the early EM studies using polyclonal antibodies specific to the different proteins (see Section 3.1; Lake, 1985; Stöffler & Stöffler-Meilicke, 1984). The technique is naturally of a low-resolution nature and the interpretation of the images has a subjective component. Nevertheless, a significant fraction of the protein locations has subsequently been confirmed by the recent high-resolution crystal structures (Ban *et al.*, 2000; Wimberley *et al.*, 2000; Schlünzen *et al.*, 2000; Harms *et al.*, 2001). At an early stage, many of the ribosomal proteins were interpreted to have a highly elongated shape. Even though this could not be safely established at such a low resolution, it has in general terms been verified by the crystallographic investigations.

In the early phase of identifying the locations of ribosomal proteins in the subunits, a more objective technique was used that was based on the difference in neutron scattering between hydrogen and deuterium (see Section 3.1). Thus, the distances between pairs of proteins, differently labeled than the rest of the small ribosomal subunit, were measured and a map of the subunits could be created (Plate. 7.1; Capel *et al.*, 1987). Similar studies have also been carried out on the large subunit (Willumeit *et al.*, 2001).

An early observation of specific interest by electron microscopy is the identification of a tunnel through the large subunit going from the interface side to the external surface (Milligan & Unwin, 1986; Yonath *et al.*, 1987). Recent cryo-EM studies have verified this and reported observations of several tunnels or cavities in the large subunit (Gabashvili *et al.*, 2001). This is the exit tunnel for the nascent polypeptide (see Section 8.4), as has been confirmed by crystallographic investigations (Ban *et al.*, 2000; Schlünzen *et al.*, 2001).

At the resolutions now achieved, individual rRNA helices and ribosomal proteins can be observed. Furthermore, the locations of tRNAs bound to the A-, P- and E-sites and translation factors have provided significant and important insights into the function and dynamics of the ribosome (Stark *et al.*, 2002; Valle *et al.*, 2003a,b).

7.2 CRYSTAL STRUCTURES OF RIBOSOMES

Some Steps in the History of Ribosome Crystallography

Until recently, much of the structural work on ribosomes has been performed with blunt tools. The information gathered could be confirmed by other methods but to the extent one tried to generate atomic models, they were uncertain. To crystallize whole ribosomes or even subunits seemed like a hopeless endeavor due to the heterogeneity of the material. Thus, factors, different tRNAs, or proteins with uncertain role in translation may adhere to the ribosomes. Furthermore, it is difficult to obtain subunits or ribosomes with a full complement of the ribosomal proteins (Hardy, 1975). In addition, ribosomes go through a number of conformational states, which may be trapped in the purification. However, spontaneously ordered arrays of ribosomes had been observed at an early state (Morgan & Uzman, 1966; Kingsbury & Voelz, 1969; Unwin, 1977). Milligan & Unwin (1986) analyzed two-dimensional ribosomal crystals of hibernating lizard oocytes.

Despite bad odds, Yonath started to explore the possibility of obtaining three-dimensional crystals of ribosomes or ribosomal subunits (Yonath *et al.*, 1980). The analysis of a broad range of species is generally a useful strategy (Glotz *et al.*, 1987). Improvements in ribosome treatment, crystal growth conditions and handling (Ban *et al.*, 1998; Clemons *et al.*, 2001; Gluehmann *et al.*, 2001), crystal freezing (Hope *et al.*, 1989), synchrotron equipment and methods (Hendrickson, 1991; Helliwell, 1998) have led to the successes we enjoy, and benefit from, today.

The early successful crystallization experiments by Yonath's group included primarily 50S subunits (Yonath *et al.*, 1980; 1982) from a number of thermophilic bacterial species like *Thermus thermophilus* and *Bacillus stearothermophilus* and the halophilic Achaean *Haloarchula marismortui*. The crystals of *H. marismortui* 50S subunits obtained with the initial recipe (Shevack *et al.*, 1985) led to crystals diffracting to resolutions of about 3 Å (von Böhlen *et al.*, 1991). However, the crystals had different defects and problems. They were thin and fragile, not isomorphous, and sensitive to radiation damage (Yonath *et al.*, 1998; Harms *et al.*, 1999). In addition, the crystals were twinned (Ban *et al.*, 1999). Such problems can make the crystallographic work very difficult if not impossible. Bigger crystals could be grown and with a different method of handling (Ban *et al.*, 2000), the crystals could be improved, phase angles could be determined and electron densities at resolutions of 9 Å (Ban *et al.*, 1998), 5.5 Å (Ban *et al.*, 1999) and finally 2.4 Å could be obtained (Ban *et al.*, 2000). Using electron density maps at 9 Å resolution, the well-known shape of the large ribosomal subunit was identified with recognizable features such as double-helical RNA (Ban *et al.*, 1998). Crystallographic analysis of the large subunit from *Deinococcus radiodurans* has given a structure of the 50S subnit from a bacterial species (Harms *et al.*, 2001).

Successful crystallization of 30S subunits and 70S ribosomes from *T. thermophilus* began in a Russian laboratory (Trakhanov *et al.*, 1987) and was continued by the Yonath group (Glotz *et al.*, 1987). The initial resolution was limited. Different approaches were used and the outcome was improvement of the resolution to around 3 Å. One of the approaches was the inclusion of a heavy atom cluster, the W-18 cluster (Tocilj *et al.*, 1999). Ramakrishnan's group found that the crystals did not contain protein S1, whereas the purified ribosomes did. The purification of ribosomes that lacked S1 as well as the introduction of the heavy atom compound osmium hexamine helped in improving the resolution in this case (Clemons *et al.*, 2001).

The work on whole bacterial ribosomes from *T. aquaticus* has not yet reached atomic resolution (Cate *et al.*, 1999; Yusupov *et al.*, 2001). However, since atomic structures of the subunits are available these structures could be fitted into the density of the whole ribosome and a credible atomic interpretation could be obtained.

Elucidation of the crystal structures of ribosomes and ribosomal subunits has indeed been highly rewarding and has opened a new era for all future work on translation (Ban *et al.*, 2000; Wimberley *et al.*, 2000; Schlünzen *et al.*, 2000; Harms *et al.*, 2001; Yusupov *et al.*, 2001). The structures have first of all confirmed a number of previous expectations and results but have also eliminated others. All studies of translation now have a firm and growing base of structural information for the formulation of hypotheses; the planning of new experiments; and the interpretation of results. For the analysis of function and dynamics of the translation system, the structural insights are mandatory like for any enzyme, whether simple or complicated.

The Small Subunit

The structural determination of the small subunit from *T. thermophilus* has proceeded in steps of resolution. Clemons *et al.* (1999) and Tocilj *et al.* (1999) presented the structure of the subunit at 5.5 Å and 4.5 Å, respectively. The structures are now available at around 3 Å resolution (Wimberley *et al.*, 2000; Schlünzen *et al.*, 2000). The subunit structure confirms previous observations with EM and cryo-EM. All proteins from the small subunit, except S1 and S21 that are not found in all species, have been located in the crystal structures. The location of S1 has been identified by cryo-EM (Sengupta *et al.*, 2001). THX, a protein present in *T. thermophilus* but not in most other bacteria (Choli *et al.*, 1993), has also been observed in the crystal structure (Brodersen *et al.*, 2002).

The 5′ domain of the 16S RNA is located in the body of the subunit; the central domain is located in the platform; and the 3′-major domain is located in the head of the subunit (Plate 6.1C). The 3′-minor domain runs as a long helix (h44) from the region between the head and the body down to the bottom of the subunit. The domains of the 16S RNA are separate and are more or less globular. In addition, a single helix (h28) connects the head to the body of the subunit. This leads to a significant flexibility between the domains as has also been seen by other methods (Gabashvili *et al.*, 1999). This flexibility is essential for the function of the small subunit (Ogle *et al.*, 2003).

The ribosomal proteins are easily distinguished in the electron density and their structures can be interpreted. The interface side of the 30S subunit is distinctly poor in proteins as has previously been identified from tritium bombardment (Yusupov & Spirin, 1986). However, most proteins are located on the exterior side (Wimberley *et al.*, 2000;

Schlünzen *et al.*, 2000). The structures of many ribosomal proteins were known in isolation and their structures *in situ* in the ribosome are not much different (Schlünzen *et al.*, 2000; Brodersen *et al.*, 2002).

The primary role of the small subunit is to bind the mRNA and to engage in the decoding of the message. Based on a combination of structural and chemical knowledge, it was possible to identify the binding sites for A- and P-site tRNAs and thus the site of decoding. In fact, a part of one subunit involved in the crystal packing (the spur or h6) imitated part of the P-site tRNA and bound to a mimic of the mRNA (Carter *et al.*, 2000). The tRNA sites are closely associated with the top of the 3'-minor domain or the penultimate helix (h44) of the 16S RNA (Wimberley *et al.*, 2000; Carter *et al.*, 2000; Schlünzen *et al.*, 2000).

To study the mechanism of decoding, fragments of the anticodon stem-loop (ASL) of tRNA have been investigated when bound to the small subunit (Ogle *et al.*, 2001; 2002). The ribosomal RNA was seen to interact with the base pairing of tRNA to mRNA, to distinguish cognate from near-cognate codons. These studies will be further described in Chapters 9 and 11. Furthermore, the flexibility of the small subunit became evident and is related to function. Thus, cognate ASLs induce a conformational change in the small subunit that can be described as a closure of the subunit (Ogle *et al.*, 2002). The changes involve a rotation of the head of the small subunit towards the shoulder and subunit interface. The shoulder (S4, the G530 loop and S12) moves towards the subunit interface and h44. With near-cognate codons, the conformational changes are not induced unless the antibiotic paramomycin is present (see Chapters 9 and 11).

The Large Subunit

Most of the structure of the large subunit has been clarified from crystallographic studies, but some parts of the subunit are nevertheless not well characterized (Ban *et al.*, 2000; Harms *et al.*, 2001; Klein *et al.*, 2004). At low or moderate resolution all that had been seen earlier from EM was evident (Ban *et al.*, 1998; 1999). However, at higher resolution, the two side protuberances, L1 and the L12 stalk, became too low in electron density to be safely interpreted. As will be discussed below, these regions of the subunit are functionally important. As long as they are not properly engaged in function, they remain flexible. Evidently, the two ribosomal subunits have different types of flexibility. The small subunit has domain flexibility, while in the large subunit the protuberances are mobile.

Contrary to the situation in the small subunit, the six domains of the 23S RNA, identified from the analysis of its secondary structures, are thoroughly interwoven (Plate 6.1D; Ban *et al.*, 2000; Harms *et al.*, 2001). This leads to a stable structure of the core of the large subunit. The 5S RNA is located in the central protuberance and could be said to form the seventh domain of the RNA in the large subunit. On average, the large subunit proteins contact twice as many RNA domains as the proteins from the small subunit (Klein *et al.*, 2004). As for the small subunit, most of the ribosomal proteins are located on the external surface, leaving the side of the interface relatively poor in proteins (Ban *et al.*, 2000, Harms *et al.*, 2001). As in the case of the small subunit, several proteins are small globular entities with a stable structure that is the same whether in isolation or when bound to the ribosome. However, a significant number of proteins have multiple domains or long extensions at their N- or C-termini or sometimes long internal loops (see Section 7.4; Ban *et al.*, 2000; Harms *et al.*, 2001; Klein *et al.*, 2004).

The central function of the large subunit is to perform peptidyl transfer. This is done in the peptidyl transfer center (PTC). Here the acceptor ends of the tRNAs are stably bound close to each other, with the nascent peptide in the P-site and the incoming amino acid in the A-site. This site was easily identified with the aid of previous biochemical observations (Ban *et al.*, 2000; Nissen *et al.*, 2000; Harms *et al.*, 2001; Schlünzen *et al.*, 2001). Nucleotides of the 23S RNA known to be important for binding of the A- and P-site tRNAs, were identified in a groove across the interface side of the subunit. Crystallographic investigations of bound antibiotics or different types of tRNA fragments or inhibitors have also led to the identification of the PTC (Nissen *et al.*, 2000). In addition, the tunnel from the PTC at the interface side of the subunit to the external side observed earlier, became very evident. The tunnel is wide enough along its length to allow a peptide to migrate through. Some macrolide antibiotics like erythromycin prevent the synthesis of polypeptides longer than a few residues due to the fact that they are bound in this tunnel (see Section 10.5).

Two regions of the large subunit are particularly rich in proteins: the region binding the translational GTPase (tGTPase) factors (L3, L6, L11, L10, L12, L13, L14) and the external side of the polypeptide exit tunnel (L19e, L22, L23, L24, L29, L31e; Klein *et al.*, 2004).

The 70S Ribosome

Elucidation of the structure of the complete bacterial ribosome (Cate *et al.*, 1999; Yusupov *et al.*, 2001) is an extraordinary step forward. Since the resolution remains at 5.5 Å, the detailed interpretation depends to a large extent on the available coordinates of the small and large subunits. The relative orientation of the two subunits is clarified (Plate 7.3) and the locations of the mRNA (Agrawal *et al.*, 1996; Stark *et al.*, 1997; Yusupova *et al.*, 2001) and the tRNA molecules in the A-, P- and the E-sites are identified (Yusupov *et al.*, 2001). The structure of the complete 70S ribosomes illustrates why the functional regions of the tRNA molecules are 75 Å apart. The shape and length of the tRNA molecule correlate well with the location of the decoding site on the small subunit and the PTC on the large subunit. Indeed, all the main activities of the ribosome occur at the subunit interface.

The Inter-subunit Bridges

The subunit interface of the ribosome is of crucial importance. Inter-subunit bridges hold the subunits together. Since the functional sites are all at the interface, the dynamic properties of the ribosome depend to a large extent on these bridges. The bridges were analyzed in the first cryo-EM maps (Frank *et al.*, 1995; Gabashvili *et al.*, 2000) and in the low-resolution crystallographic structure of the 70S ribosome (Cate *et al.*, 1999). The components that are engaged in these bridges could not be identified at that stage, but were given names. The current set of bridges totals 12 (Fig. 7.1; Plate 7.4). They are named B1–B8 with B1, B2 and B7 being composed of several neighboring contacts (Table 7.1; Yusupov *et al.*, 2001). A very similar set of bridges was also identified in *E. coli* and yeast (Gao *et al.*, 2003; Spahn *et al.*, 2001). The high degree of conservation emphasizes that these bridges are of functional importance (Gao *et al.*, 2003).

One main component at the subunit interface is helix h44, the penultimate stem of the 16S RNA. It runs along the length of the body of the small subunit and makes several contacts with the large subunit (B2a, B3, B5 and B6) at subsequent turns of the double helix (Cate *et al.*, 1999; Yusupov *et al.*, 2001).

The B1a and B1b bridges connect the head of the small subunit with the top part of the 50S subunit. The B1a bridge is also called the A-site finger (Frank *et al.*, 1995). It is the result of a contact between helix H38

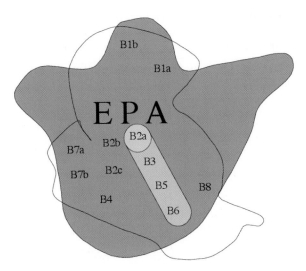

Fig. 7.1 The approximate locations of the intersubunit bridges (see Plate 7.4; Yusupov *et al.*, 2001). The approximate location of h44 is indicated.

(domain II) of the 23S RNA and protein S13 and is situated above the A- and P-sites (Yusupov *et al.*, 2001). Bridge B1b is composed entirely of proteins, S13 in the head of the small subunit and protein L5 close to the central protuberance of the large subunit (Yusupov *et al.*, 2001). Bridge B2a connects helix H69 of the large subunit with helix h44 of the small subunit at the decoding area and has contacts with the A- and P-site tRNAs (Plate 7.4; Gabashvili *et al.*, 2000; Yusupov *et al.*, 2001).

The orientation of H69 that participates in bridges B2a and B2b, differs between the isolated *D. radiodurans* 50S subunits and the whole 70S ribosome (Yusupov *et al.*, 2001; Harms *et al.*, 2001) and is disordered in the *H. marismortui* 50S subunit. This bridge is close to both the A- and P-sites and forms an important contact between the peptidyl transfer region and the decoding region (Bashan *et al.*, 2003). H69 may be important for signaling between these two well-separated functional sites.

The B4 bridge is composed of protein S15 that contacts helix H34 of the 23S RNA (Culver *et al.*, 1999). The proteins L14 and L19, which are in close contact, both participate in interactions with the small subunit (bridges B8 and B6, respectively; Yusupov *et al.*, 2001; Harms *et al.*, 2001). The bridges B1a and B7b break, and B1b changes its pattern of, contacts in the ratchet-like movement when EF-G:GTP binds to the ribosome (see Section 8.7; Woese, 1970; Frank & Agrawal, 2000; Gao *et al.*, 2003).

Table 7.1 Inter-subunit Bridges as Seen in *T. thermophilus* 70S Ribosomes (from Yusupov *et al.*, 2001). RNA helices from the small subunit are denoted "h" and those from the large subunit are denoted "H"

Bridge	30S Subunit	Residues	50S Subunit	Residues	Dynamic Character
B1a	S13	92-94	H38	886–888	Breaks at locking-unlocking
B1b	S13	N-term	L5	134–153	Changes at locking-unlocking
B2a	h44	1408–1410	H69	1913–1914, 1918	
B2b	h24	784–785	H67	1836–1837	
		794	H69	1922	
	h45	1516–1519	H69	1919–1920	
			H71	1932	
B2c	h24	770–771	H67	1832–1833	
	h27	900–901	H67	1832–1833	
B3	h44	1484–1486	H71	1947–1948 1960–1961	
B4	h20	763–764	H34	717–718	
	S15	40–44, C-term	H34	713, 717	
B5	h44	1418–1419	H64	1768–1769	
		1420–1422	L14	44–49	
		1474–1476	H62	1689–1690	
		1474–1476	H64	1989	
B6	h44	1429–1430	H62	1689–1690 1702–1705	
B7a	h23	698, 702	H68	1848–1849 1896	
B7b	h23	712–713	L2	162–164 172–174	Breaks at locking-unlocking
				177–178	
	h24	773–776	L2	177–178 198–202	
B8	h14	345–347	L14	116–119	

7.3 THE STRUCTURES OF THE RIBOSOMAL RNA MOLECULES

The crystallographic structures of the ribosomal subunits (Wimberley *et al.*, 2000; Schlünzen *et al.*, 2000; Harms *et al.*, 2000; Ban *et al.*, 2000; Harms *et al.*, 2001) have confirmed and extended the understanding of the structures of the rRNAs. The rRNAs of both the subunits form the core of the particles. The large variation in size of the rRNAs has led to a determination of a minimal structure that is shared by all (Gerbi, 1996; see also Plate 6.2). The extensions from this minimal structure as seen in the *H. marismortui* large subunit, occur primarily on the external side, suggesting that the interface region is more highly conserved (Ban *et al.*, 2000).

The crystallographic structures of the ribosomal subunits have given information on the rRNA on the secondary structure, the tertiary structure and at quaternary structure levels. The secondary structure is first of all based on double-stranded helices, but also on turns, bulges and kinks (Moore, 1999; Westhof & Fritsch, 2000). All known motifs have been observed in the rRNAs (Moore & Steitz, 2003b). Among the previously known types of secondary structure motifs, there is also a new internal asymmetric loop called the kink-turn or K-turn (Klein *et al.*, 2001). This is often associated with binding of proteins (see below).

The 45 helices of the 16S RNA are arranged in such a way that 13 of them are coaxially stacked and 23 are unstacked helices. This makes a total of 36 helical elements in the small subunit (Wimberley *et al.*, 2000). Helix h44 contains around 100 residues and runs along the length of the body domain of the small subunit. The majority of the helices in this domain are more or less parallel with h44, and due to the coaxial stacking, become quite long.

The 23S RNA of *H. marismortui* contains 2923 nucleotides; 1157 of them are in van der Waals contact with proteins, and there are only 10 sequences longer than 20 nucleotides that have no contact with protein molecules. The longest stretch without protein contact is 47 nucleotides and this is the part of domain IV that forms the ridge of the peptidyl transfer cleft (Ban *et al.*, 2000).

In the structure of the *H. marismortui* large subunit, two types of conserved RNA sequences could be identified. The first type is related to the active site regions for interactions with tRNAs or factors. The other type was involved in tertiary or quaternary structure interactions within or between the domains of RNA molecules. It turned out that adenosines are overrepresented in this class (Ban *et al.*, 2000). This is due to the frequent occurrence of the so-called "A-minor motif", where two or more stacked adenosines dock into the minor groove of adjacent RNA-helices (Nissen *et al.*, 2001; Moore & Steitz, 2003b). Distortions from regular A-form RNA-helix make the minor groove wider. The adenines that participate in these interactions form cross-strand base stacking and sugar and base hydrogen bonds stabilize these interactions (Wimberley *et al.*, 2000; Nissen *et al.*, 2001).

7.4 THE STRUCTURES OF RIBOSOMAL PROTEINS

In the early attempts to obtain structural information about the ribosome, the ribosomal proteins were found to be most interesting and also most accessible. It was the general belief at that stage that the ribosomal proteins carried out the main functions of the ribosome. Thus, the structures of individual ribosomal proteins have been studied for a long time. The primary methods used have been NMR and X-ray diffraction. The focus has gradually changed towards complexes of fragments of rRNA with bound proteins.

One problem in the early studies of ribosomal proteins was the difficulty in purifying them not using denaturing methods. Even when this difficulty was solved, (Liljas & Kurland, 1976; Liljas *et al.*, 1978; Dijk & Littlechild *et al.*, 1979) many proteins were unstable in the presence of proteolytic enzymes or at elevated temperatures or had little or no secondary structure as identified from CD-spectra (Tumanova *et al.*, 1983). The exploration of the proteins from thermophilic ribosomes was very helpful, and so was the possibility of expressing the proteins heterologously. Despite some success, ribosomal proteins have remained more difficult to handle and crystallize than normal globular proteins.

The structures of less than half of the ribosomal proteins (r-proteins) from bacteria have been determined by crystallography or by NMR (see Appendix II). In most cases, these proteins were found to be globular. Some distinctive deviations are L9 (Hoffmann *et al.*, 1996) and L12 (Österberg *et al.*, 1976; Bocharov *et al.*, 1996), which are composed of domains separated by more or less flexible regions.

The elucidation of the structures of the ribosomal subunits has revealed the structures of essentially all bacterial ribosomal proteins (Brodersen *et al.*, 2002; Schlünzen *et al.*, 2000; Harms *et al.*, 2001). In addition, we know the crystal structure of an archaeal large subunit including most of its proteins (Ban *et al.*, 2000; Klein *et al.*, 2004). When proteins from *T. thermophilus* are discussed, the letter T is put in front of the protein name (e.g. TS4). Likewise, the letters D or H represents proteins from *D. radiodurans* or *H. marismortui*.

Most of the protein structures that were elucidated in isolation could be fitted into the subunit structures without problems. These proteins were primarily globular. However, many of the protein structures that were characterized solely from the subunit structures could hardly have been determined in isolation. Many have globular domains with long extensions or tails that could hardly have a stable structure in isolation (Appendix I; Plate 7.5, Ban *et al.*, 2000; Brodersen *et al.*, 2002; Klein *et al.*, 2004). Some proteins lack any type of internal fold. This organization is reminiscent of the protein-nucleic acid interactions of histones or viral proteins (Liljas, 1986; Rossmann & Johnson, 1989; Luger & Richmond, 1998), and from amino acid sequence and mild proteolysis, this type of proteins was predicted to be found also in the ribosome (Liljas, 1991).

The subunit structures have also provided the organization of the proteins in the ribosome. One general observation is that the globular parts of the proteins are situated on the ribosomal surface, primarily the external one. The tails or long loops extend into the subunit structures (Ramakrishnan & Moore, 2001). In addition, there is a dramatic charge separation of the ribosomal proteins *in situ*. The surface parts are more acidic, while the inner parts are highly basic, neutralizing the negative charges of the RNA (Klein *et al.*, 2004). The extensions are unusually rich in arginine, glycine and lysine and they are invariably in extensive contact with the RNA.

Relationships between Ribosomal Proteins

The relationship between ribosomal proteins from the three domains (Bacteria, Archaea and Eukarya) has been analyzed (see Table 6.4). It is also continuously updated in a database (Bairoch; http://www.expasy.org/cgi-bin/lists?ribosomp.txt) and analyzed in the literature (Lecompte *et al.*, 2002; Harris *et al.*, 2003). With the elucidation of the structures of the *H. marismortui* and *D. radiodurans* 50S subunits, the relationship of L16 with L10e could also be identified from their fold and identity of position in the subunit (Harms *et al.*, 2001; 2002). Some additional proteins were positioned in the same location and evidently in some sense replacing each other without having any relationship in fold (Table 7.2).

A number of r-proteins and translation factors were found to have related structures (Table 7.3). This may suggest evolutionary connections

Table 7.2 Fold and Positional Relationship of Bacterial and Archaeal Proteins from the Large Subunit (Ban *et al.*, 2000; Harms *et al.*, 2001; Klein *et al.*, 2004)

Bacteria	Archaea	Type of Relationship
Fold and Position		
L16	L10e	Similar structure and position at A-site tRNA
L33	L44e	Similar fold and position of globular domain. Circularly permuted amino acid sequences? L31 occupies part of the space of the extended part of L44e.
Position only		
L17	L31e	Interact with the same parts of 23S RNA
L19	L24e	Both form inter-protein β-sheets with L14
L21	L32e	Loops of L21 and L32e occupy the same space. Both interact with H46.
L23	L39e	L39e in *H. marismortui* replaces the tail of bacterial L23
L27	L21e	Interact with the same parts of 23S RNA
L31	L15e	Interact with the same parts of 23S RNA
L34	L37e	Interact with the same parts of 23S RNA

Table 7.3 Classification of Ribosomal Proteins (Modified from Klein *et al.*, 2004)

α-helical	S2, S4, S13, S14, S15, S18, S20, L12, L29, L19e, L39e
Antiparallel α+β	
split β–α–β	S6, S8, S10, L1, L5, L6, L9, L12, L22, L23, L30, L10e, L15e, L31e
Mixed α+β group	S3, S4, S5, S7, S9, S16, S19, L1, L3, L4, L5, L13, L18, L7Ae, L32e
β-meander	S3, S5, S11
β-barrel	L3, L14, L25
OB-fold	S1, S12, S17, L2
SH3-like	L2, L24, L21e
Zinc-containing	S4, L32, L36, L24e
Rubredoxin-like	L37Ae, L37e, L44e
L15 group	L15, L18e
Homeodomain	L11

(Ramakrishnan & White, 1998; Draper & Reynaldo, 1999; Liljas *et al.*, 2000). Klein *et al.* (2004) classify the large subunit proteins into six groups: the antiparallel α+β group; the β-barrel group; the α-helical group; the mixed α+β group; the zinc-containing group; and the L15 group. This is a general and useful way to relate proteins. Even though members of each group can be widely dissimilar, the closer relationships should be found within these groups. In comparing the structures of different proteins, it is almost always observed that the smaller or larger motifs can be closely related, while the protein on the whole are quite different. One such motif is the RNA recognition motif (RRM). This is composed of alternating β-strands and α-helices in a pattern called split β–α–β. Sometimes, the motif is repeated and becomes a double split β–α–β (Leijonmarck *et al.*, 1988; Orengo and Thornton, 1993; Brodersen *et al.*, 2002). This motif is found in the antiparallel α+β group and quite a number of the ribosomal proteins have a single or double split β–α–β (see Table 7.3 and Appendix I). L2, L21e and L24 are partly very similar with folds of the SH3-like β-barrel type (Klein *et al.*, 2004). One domain of L3 is a β-barrel similar to the conserved domain II in the tGTPase factors. L15 and L18e not only have closely related structures but also homologous sequences (Klein *et al.*, 2004).

Another repeatedly occurring motif or fold is the oligonucleotide binding or OB-fold. It is composed of a five-stranded antiparallel β-sheet. This type of structure is a very common protein fold (Murzin, 1993; Agrawal & Kishan, 2003). Protein S1 has six consecutive OB-fold domains (see below).

One specific type of protein structure is called zinc finger. It is usually associated with DNA interactions (Klug & Schwabe, 1995). The ribosomal structures provide several examples of such zinc fingers. Thus, somewhat unexpectedly the ribosome is the binding site for a number of zinc ions in addition to magnesium and monovalent metals.

Protein-RNA Interactions

With few exceptions, the ribosomal proteins interact with the ribosomal RNA. Except for the two protuberances of the large subunit, the r-proteins do not extend beyond the envelope defined by the rRNA (Ban *et al.*, 2000; Moore & Steitz, 2003b). Thus, ribosomes are very different from viruses where proteins form a shell around the nucleic acid.

One feature of RNA that frequently has been seen associated with binding of proteins is the kink-turn or K-turn (Klein *et al.*, 2001). Here a so-called canonical stem, composed of two Watson-Crick base pairs, is followed by an internal loop and subsequently a non-canonical stem which typically starts with two nonWatson-Crick base pairs. The angle between the stems is around 120°. The K-turn occurs six times in *H. marismortui* 23S RNA and twice in *T. thermophilus* 16S RNA. These kinks bind nine proteins from the large subunit and two from the small subunit. There is no systematic way that these proteins interact with the RNA and the folds of these proteins also differ significantly (Klein *et al.*, 2001).

Few general principles for the protein-RNA interaction have been identified except for the charge neutralization that is an essential component. There are distinct differences in the way the globular domains and the tails or loops interact with the RNA (Moore & Steitz, 2003b; Klein *et al.*, 2004). The globular parts are normally located on the surface of the subunits, while the tails or loops fill the spaces and bridge between different parts of the rRNA. The ribosomal proteins, which generally are highly basic proteins, frequently display more acidic surfaces to the exterior on the globular domains while the positive charges are located on the parts extending towards the interior of the ribosomal structure to neutralize the negative charges of the rRNA. In fact, despite being a limited part of the protein component, the tails contribute in a major way to the interactions with the RNA.

Klein *et al.* (2004) make a thorough analysis in the ways ribosomal proteins from the large subunit interact with the rRNA. The proteins that interact with the 5S RNA are L5, L18, L21e, L10e and L30. In bacteria, the protein L25 with its three different versions (see below) should be added. There are at least four different main ways proteins interact with RNA: with the edges of the bases exposed in the minor groove; in the widened major grooves of the RNA helices; with the flipped out bases or bulged nucleotides; and insertion of amino acid side chains into hydrophobic crevices between the exposed nucleotide bases. The protein domains with similar structures can interact with the RNA in very different ways. Most of the interactions in the minor groove are with the 2-amino group of guanines in Watson-Crick base pairs. The interactions in widened regions of the normally narrow major groove are as frequently with the Watson-Crick base pairs as with the wobble or noncanonical base pairs. The extended tails are narrow enough to interact with such major groves. Generally, the role of the extended portions of the proteins is to interact with the RNA.

In the assembly of the ribosomal subunits, some basic principles can be assumed. First of all, the assembly must be a sequential process. Some proteins may find their binding sites while the RNA is still synthesized. The folding of the 5'-end is therefore likely to precede that of the 3'-end. The proteins that bind early in the assembly of the ribosome bind to and organize different pieces of the rRNA, perhaps from different domains. Evidently, it is unlikely that an unfolded protein can interact specifically with an unfolded RNA. Likewise, it must be highly unlikely that a protein tail can thread itself into a prefolded RNA (Moore & Steitz, 2003b; Klein *et al.*, 2004). Rather, a folded or globular part of a protein can identify features of a contiguous piece of the ribosomal RNA, and subsequently its extensions can associate with the nearby or remote parts of the rRNA. The tailed proteins or proteins without any secondary structure would generally need to associate at an early stage in the assembly, or alternatively, bring separated domains together in the later steps of the assembly for the extended part to find its final location.

The primary binders of the small subunit are all globular proteins (Brodersen *et al.*, 2002). This corresponds with the observation that a large fraction of the small subunit proteins associate with single domains of the 16S RNA. Many of the large subunit proteins interact with several domains of the 23S RNA (Moore & Steitz, 2003b). One extreme case is L22 that has a long, extended loop that interacts with all the six domains of the 23S RNA (Ban *et al.*, 2000; Schlünzen *et al.*, 2001). Several of the primary binders in the large subunit like L2, L3, L4, and L15 have several long tails or loops (see Appendix I; Klein *et al.*, 2004). However, the protein that may be most crucial for large subunit assembly is L24 that only interacts with domain I. The binding of L24 may prepare domain I for the binding of, in the first place, proteins L4, L22 and L29 that interact with other domains (Klein *et al.*, 2004). L3 is another protein important for the early assembly of the large subunit. Its globular domain only interacts with domain VI, while its extensions interact with all the other domains except domain I. A number of additional proteins depend on the previous binding of L3.

Some Specific Proteins

S1

S1 is a protein that is not found in all the bacterial species (see Table 6.4). It is the largest protein in bacterial ribosomes with a molecular weight of 61 kDa and has a high affinity for mRNA (Draper & von Hippel, 1978). The protein is composed of two domains. The N-terminal domain (residues 1–195) is involved in the association with the small subunit, while the C-terminal region is involved in mRNA-binding. The C-terminal domain has a six-fold repeated structure of about 70 amino acid residues each

(Subramanian, 1983). The structure of the repeated domain is the OB-fold and has been found as single or repeated domains in several RNA-binding proteins (Bycroft *et al.*, 1997). One of the proteins with a single S1 domain is bacterial initiation factor IF1 (Sette *et al.*, 1997; Battiste *et al.*, 2000).

For a long time S1 has been thought of as a very elongated protein (Giri & Subramanian, 1977; Laughrea & Moore, 1977). The organization of the complete protein has been observed by cryo-EM (Plate 7.6; Sengupta *et al.*, 2001). At 11.5 Å resolution, S1 is seen with a central globular unit, with two extensions and two holes. Its location on the 30S subunit is on the interface side and in the region between the head and the platform. This location coincides with the upstream part of the mRNA from the SD interaction. The characterization of S1 from *T. thermophilus* has shown that in isolation the protein behaves as a compact and globular entity as well (Selivanova *et al.*, 2003). S1 is also one of the subunits of the Qβ replicase (Berestowskaya *et al.*, 1988). Here S1 also appears to have a more or less compact structure.

S12

The S12 protein has not lent itself to structural studies in isolation. The long N-terminal tail can explain this. The binding site for S12 in the small subunit is at the decoding site (Wimberley *et al.*, 2000; Carter *et al.*, 2000; Schlünzen *et al.*, 2000; Brodersen *et al.*, 2002). It is located on the inner side of the elbow of the tRNA in the A-site where amino acids 78–80 interact with the acceptor arm near nucleotide 69 (Valle *et al.*, 2003a). This was anticipated, since different studies have shown that the protein is associated with the fidelity of translation. Resistance to streptomycin, known to affect the accuracy of translation, is frequently found to be due to mutations of S12 (see Section 10.2; Davies *et al.*, 1964; Kurland, 1992). Streptomycin and protein S12 can both affect the balance between the two conformational states of the small subunit and thereby also the fidelity of translation (see Sections 10.2 and 11.5).

S12 also interacts with EF-G when bound to the ribosome (Girshovich *et al.*, 1981; Valle *et al.*, 2003b). This leads to a rotation of S12 by 19° towards the inter-subunit space that enhances the interaction with EF-G (Gao *et al.*, 2003). Early studies indicated that S12 from *E. coli* is unusually rich in cysteinyl residues, and when modified by chloromercuribenzoate stimulate factor free translocation (Gavrilova & Spirin, 1971; Southworth *et al.*, 2002).

L1

The structure of the L1 protein has been studied in isolation (Nikonov *et al.*, 1996, Nevskaya *et al.*, 2000) and in a complex with a truncated fragment of the 23S RNA (Nikulin *et al.*, 2003). The protein is composed of two domains that are separated by a hinge region, which allows a fair amount of flexibility (Unge *et al.*, 1997).

L1 has given its name to the left protrusion of the large subunit seen from the interface side. This protrusion is composed of L1 and helices H76–H78 of domain V of the 23S RNA (Zimmermann, 1980). As mentioned above, the crystallographic investigations have met with significant difficulties in characterizing this part of the ribosome due to its flexibility (Ban *et al.*, 2000; Harms *et al.*, 2001; Klein *et al.*, 2004). L1 and its binding site on the rRNA are more visible in the structure of the 70S ribosome, albeit at lower resolution (Yusupov *et al.*, 2001). There is a difference of 30° between the orientation of the L1 stalk between the *D. radiodurans* structure of the large subunit and the structure of *T. thermophilus* ribosomes (Yusupov *et al.*, 2001).

The flexibility of the L1 stalk is related to the E-site tRNA. In cryo-EM (Valle *et al.*, 2003b) and crystal structures (Yusupov *et al.*, 2001) of the 70S, L1 interacts with the elbow of the E-site tRNA. In this location, L1 and protein S7 from the small subunit prevent the deacylated tRNA from leaving the E-site (Yusupov *et al.*, 2001). When the E-site is empty, the L1 stalk has moved into an outer position. Thus, the L1 stalk acts as a gate for the exiting tRNA, but through its interaction with the deacylated tRNA it may participate actively in translocation (Valle *et al.*, 2003b). The two conformations can be described as an open and a closed conformation, where the exit of deacylated tRNA from the E-site is possible or not possible (see Section 11.5).

L10

L10 forms a strong complex with the two dimers of L12 (see below). The molecular interactions in the L12 dimers (see below), and also in the isolated pentameric complex of L10 with the two dimers of L12, are strong. This complex does not dissociate but produces a unique spot, called L8, in the classical 2-dimensional urea gels (Kaltschmidt & Wittmann, 1970). This spot was identified as a complex of four copies of L12 with L10

(Pettersson *et al.*, 1976; Österberg *et al.*, 1977). L12 binds to the C-terminal domain of L10 (van Agthoven *et al.*, 1975; Koteliansky *et al.*, 1978; Gudkov *et al.*, 1980). The dimers of L12 bind to two different sites on the C-terminal part of L10 (Gudkov *et al.*, 1980). After deletion of the last ten residues of L10, only one of L12 dimers can bind (Griaznova & Traut, 2000). L10 binds to the 23S RNA via its N-terminal region (Gudkov *et al.*, 1980). L10 as well as L11 are bound to domain II (H43 and H44) of the 23S RNA (Edebjerg *et al.*, 1990). L10 is a protein that is unstable and difficult to handle on its own. In the complex with L12, it is significantly stabilized (Newcomer & Liljas, 1980).

L12

L7/L12 is called L12 in this book. The name L7/L12 originates from the situation in *E. coli*, where L12 can be acetylated on its N-terminus to give rise to the L7 spot in 2D gels (Terhorst *et al.*, 1972). This modification is rarely seen in other bacteria. Numerous reviews have been written about L12 (Liljas, 1982; 1991; Gudkov, 1997; Möller & Wahl, 2002; Gonzalo & Reboud, 2003). Much of the details has been discussed in these reviews and here we can only draw attention to some basic facts and give some key references. Despite decades of work and the structures of both the ribosomes and ribosomal subunits, neither the structure nor the functional role of L12 is well understood.

L12 and its corresponding proteins in other species form the so-called L12 stalk on the right-hand side of the large subunit viewed from the interface side (Fig. 2.2; Plate 2.1; Boublik *et al.*, 1976; Strycharz *et al.*, 1978). L12 is essential for the function of translational GTPases and may be involved in the induction of GTP hydrolysis (see Section 9.1). L12 is the only repeated protein and occurs as two strongly coupled dimers in the ribosome (Österberg *et al.*, 1976; 1977). Both dimers are bound to protein L10 (see above). The N-terminal domains of L12 are involved in the strong dimer interaction (Gudkov & Behlke, 1978). The N-terminal domain was the first part of the ribosome to be crystallized (Liljas & Kurland, 1976) and the C-terminal domain was the first part for which the structure was determined (Leijonmarck *et al.*, 1980; 1987). Knowledge of the crystal structures of the subunits or of whole ribosomes has not revealed the organization of proteins L12 and L10 in the ribosome. The high flexibility of the proteins has precluded this.

From different observations, it is clear that the two domains of L12 are connected by a very flexible link or hinge primarily involving residues 36–51 (Table 7.4; Bushuev *et al.*, 1989). The hinge makes the protein highly flexible as has been shown by different methods (see Table 7.4). The L12 hinge in different organisms varies considerably in length and composition (Liljas *et al.*, 1986; Bushuev *et al.*, 1989). It is particularly rich in alanyl and glycyl residues. Spontaneous mutations in the hinge region (Kirsebom *et al.*, 1986) lead to a decreased proofreading (Ruusala *et al.*, 1982). If the length is significantly shortened, the ribosomes behave as if there was no CTD, but essentially doubling the length of the hinge has no effect (Gudkov *et al.*, 1991; Bubunenko *et al.*, 1992).

Table 7.4 Domain Flexibility of L12

Method	Observation	Reference
Sequence	L12 proteins are difficult to align between *E. coli* residues 36 and 50 due to different lengths and composition of hinge. The region is rich in glycyl and alanyl residues.	
NMR	Spectra of ribosomes or 50S subunits show that L12 is highly flexible. The region 33–51 was identified as the source of flexibility.	Tritton, 1980 Gudkov *et al.*, 1982 Cowgill *et al.*, 1984 Bushuev *et al.*, 1989, 2004; Mulder *et al.*, 2004
Proteolysis	Spontaneous proteolysis of L12 in the region 36–50 in crystallization attempts.	Liljas *et al.*, 1978 Wahl *et al.*, 2000
EM	The L12 stalk appears very variable in different states of the ribosome if at all seen. Nanogold labeled C-termini were observed in four different locations in the subunit interface.	Agrawal *et al.*, 1999 Valle *et al.*, 2003a,b Montesano-Roditis *et al.*, 2001
Crystallography	The subunit or ribosome structures show no or little evidence of the protein due to its mobility.	Ban *et al.*, 2000 Harms *et al.*, 2001 Yusupov *et al.*, 2001
Crosslinking	L12 can be crosslinked to proteins far away from the L12 stalk	Dey *et al.*, 1998
Mutations	Certain decreases in the length of the hinge can lead to poor function. Changes of residues or increasing length has little effect.	Gudkov *et al.*, 1991 Bubunenko *et al.*, 1992

The N-terminal domain was found by NMR methods to be composed of two α-helices (Bocharov *et al.*, 1996; 1998). From investigations of the dimer (Bocharov *et al.*, 1996; Mulder *et al.*, 2004) or the pentameric L8 complex (Bocharov *et al.*, 1998), it is entirely clear that the C-termini are completely free from each other or the N-termini due to the highly flexible linker region. The definition of the hinge is done from the NMR observations and is localized to residues 33–51 (Bocharov *et al.*, 2004; Mulder *et al.*, 2004).

The crystal structure of L12 from *Thermotoga maritima* (Wahl *et al.*, 2000) provided important insights. The crystallographic asymmetric unit contains two full molecules and two N-terminal fragments (Fig. 7.2). The N-terminal domains are composed of two α-helices (α1 and α2) as in the NMR analysis (Bocharov *et al.*, 1996). However, the flexible hinge region adopts a helical conformation called α3. Wahl *et al.* (2000) suggested that the two full molecules might display the proper dimer interaction. The main interaction between the monomers in such a dimer is due to α3, the hinge. Since this region is observed to be structurally and functionally

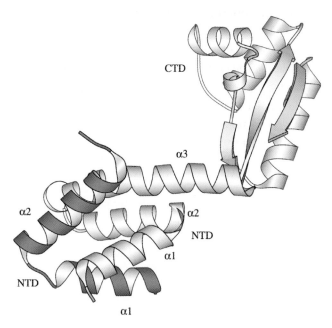

Fig. 7.2 The crystal structure of L12 (Wahl *et al.*, 2000). The picture shows one full molecule (*white*) and one N-terminal fragment (*gray*). This is found to be the proper dimer interaction. The figure has been kindly provided by Dr. M. Fodje using the program MOLSCRIPT (Kraulis, 1991).

very flexible (see Table 7.4), it is highly unlikely that this interaction could form the stable and functional dimer or relate the dimers to each other. The second dimer alternative derived from this crystal structure is one full monomer that interacts with an N-terminal fragment (Wahl *et al.*, 2000). The structural determination of the L12 dimer in solution using NMR shows this model to be the correct one (Bocharov *et al.*, 2004). Such a dimer is anti-parallel, where the two N-terminal helices form hooks that are coupled (Fig. 7.3). The NMR analysis also identifies the possibility for the temporary existence of the helix α3 in the N-terminal part of the hinge.

The observation that the hinge not only can be extended but also form helix α3 gives further insight into the dynamic possibilities of this molecule (Sanyal & Liljas, 2000). One additional interesting insight from this dimer interaction is that only one of the hinges can bind to the N-terminal domain as an α-helix in the way observed in the crystals (Fig. 7.3; Wahl *et al.*, 2000). The binding site is over the two-fold axis in the dimer.

NMR studies of L12 in the 50S subunits or 70S ribosomes indicate that only two of the four hinge regions of L12 have high mobility. The mobile residues start from residue 40, in the middle of the hinge (Mulder *et al.*, 2004). Whether the flexible L12 molecules belong to the same or separate dimers and whether the four molecules can alternate in the flexible form remain to be clarified (Sanyal & Liljas, 2000).

The range that L12 can cover on the ribosome due to the flexible hinge is very large. Different methods such as crosslinking (Dey *et al.*, 1998) and fluorescence spectroscopy (Zantema *et al.*, 1982a,b) have identified a substantial part of the subunit interface.

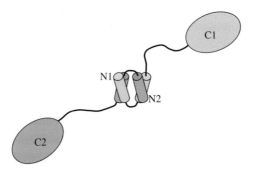

Fig. 7.3 A simplified illustration of structure of the L12 dimer. The antiparallel arrangement of the N-terminal domains seen in the crystal structure (Wahl *et al.*, 2000) is also seen in the NMR structure (Bocharov *et al.*, 2004). The hinge regions give the C-termini a large flexibility.

The removal or modification of L12 has detrimental effects on GTP hydrolysis (Kischa *et al.*, 1971; Hamel *et al.*, 1972; Koteliansky *et al.*, 1978; Pettersson & Kurland, 1980; Mohr *et al.*, 2002). The removal of the C-terminal domains has the same effect as removing the whole protein (Agthoven *et al.*, 1975; Koteliansky *et al.*, 1978). Thus, the C-terminal domains of L12 participate somehow in the GTPase activity of the translational GTPases. A C-terminal deletion of L10 eliminated the binding of one dimer from the L8 complex. Ribosomes with only one L12 dimer are still active (Griaznova & Traut, 2000). Ribosomes where one C-terminus of each L12 dimer was removed are also active (Oleinikov *et al.*, 1998). Evidently, there may be a functional overcapacity of L12, or it could be that the *in vitro* assays used have not fully illuminated the physiological needs.

The removal of L12 is shown to delay the release of inorganic phosphate after GTP hydrolysis by EF-G (Mohr *et al.*, 2002). Two surprising observations concerning L12 have been made using electrospray MS (Hanson *et al.*, 2003). Among the molecules that dissociate from the ribosome are L12 molecules that seem to have the added molecular weight of a phosphate moiety. Such modifications have not been observed before. In addition, a large fraction of the L8 complex is released from the ribosome in a complex with a tRNA. These are entirely new observations that need to be controlled and analyzed by different methods before an understanding can be obtained.

There are numerous questions concerning the structure and role of L12 that remain to be clarified. One of these is: why are there always four copies of the protein when ribosomes seem to manage with less? When and how does L12 interact with other partners of the translation system? For a continued discussion of the functional role of L12, see Section 9.1.

The P proteins

The proteins that correspond to L12 and L10 in eukarya and archaea are called the P-proteins since they can be phosphorylated (Zinker & Warner, 1976). The most easily observed similarity is that they are all acidic and that there are two dimers of the L12 analogue bound to the L10 homologue at the right-hand stalk of the large subunit (Uchiumi *et al.*, 1987; for reviews see Liljas, 1991; Ballesta *et al.*, 2000). The stalk has been observed in ribosomes with EF2 bound at low resolution by cryo-EM (Gomez-Lorenzo *et al.*, 2000). However, in the crystallographic structure of the archaeal large subunit at high resolution, the P-proteins are not seen due to their high flexibility (Ban *et al.*, 2000).

There are two different proteins, P1 and P2, which correspond to L12, and P0 which corresponds to L10 (Wool *et al.*, 1991). As in bacteria, there is one copy of P0 and four of the L12 related proteins, two each of P1 and P2. In yeast, there are four different versions of the P1 and P2 proteins, called P1A, P1B, P2A and P2B (Planta & Mager, 1998). In plants, there is one additional group, P3 (Bailey-Serres *et al.*, 1997).

It has not been possible to align the sequences of L12 to those of the P1 and P2 proteins. There may be no real sequence homology or structural similarity. P1 and P2 are difficult to align with each other for most of their lengths, but they both have N-terminal domains of about 70 residues followed by a flexible region or hinge of about 30 amino acid residues (Wool *et al.*, 1991; Bailey-Serres *et al.*, 1997; Tchorzewski, 2002). As in L12, this hinge is highly variable in length and amino acid sequence. It is rich in alanyl and glycyl residues, and the C-terminal part of the hinge is highly charged but dominated by acidic residues. The C-terminus of the protein is a highly conserved stretch of 10-13 residues. Several protein kinases can phosphorylate a conserved seryl residue of this region (Ballesta *et al.*, 1999). Similar to bacterial L12, it is the C-terminal region of the P-proteins that is associated with interactions with translational GTPases (Bargis-Surgey *et al.*, 1999). The N-terminal domains of P1 and P2 are involved in the dimerization as well as in the binding to P0 (Ballesta & Remacha, 1996). The dimers are heterodimers of P1 and P2 (Tchorzewski *et al.*, 2000a; Ballesta *et al.*, 2000). In yeast, where there are four different versions, P1A forms a dimer with P2B and P2A forms a dimer with P1B (Tchorzewski *et al.*, 2000b; Guarinos *et al.*, 2001). The P1 but not the P2 proteins are associated with the binding to P0 (Zurdo *et al.*, 2000; Gonzalo *et al.*, 2001). The heterodimers are highly helical (Tchorzewski *et al.*, 2003).

The sequence of P0 is 100 amino acid residues longer than that of L10. The N-terminal part is homologous to the whole of L10 (Shimmin *et al.*, 1989). In addition, the C-terminal region of P0 contains a homologue of P1 or P2 (Santos & Ballesta, 1995). Thus in eukarya and archaea there are five proteins related to L12. If all five proteins are deleted, the ribosomes cannot function. If only the P1 and P2 proteins are missing, the ribosome can still manage with P0 alone (Santos & Ballesta, 1994; Remacha *et al.*, 1995). A different subset of mRNAs is translated.

However, if the C-terminal conserved residues are also removed from P0, the ribosome is no more functional. Sordarin is an antibiotic that locks EF2 on the ribosome. Mutants in P0 can induce resistance to sordarin (Gomez-Lorenzo & Garcia-Bustos, 1998; Justice *et al.*, 1999).

The yeast system offers a unique opportunity to study the role of the four copies of the stalk proteins. The four different yeast proteins have unique binding sites on P0 and could be mutated individually. The question is whether they have unique roles or whether they with the high flexibility can substitute for each other. The fact that the C-terminus of P0 alone can induce GTP hydrolysis suggests that the proteins may substitute for each other.

The only ribosomal proteins for which there is a pool free in the cytoplasm are the acidic stalk proteins (Ramagopal, 1976; van Agthoven *et al.*, 1978; Zinker, 1980). A large number of patients with systemic lupus erythematosus develop autoantibodies against ribosomal P-proteins (Elkon *et al.*, 1985). The range of locations and functional aspects of the P-proteins remain to be fully clarified.

L22

It was possible to determine the structure of L22 in isolation despite its very extended shape (Unge *et al.*, 1997). The protein has a globular domain and an extended β-ribbon about 30 Å long. L22 is one of the first proteins to bind to the 23S RNA in the assembly of the large subunit (Röhl & Nierhaus, 1982). Its role in the assembly is obvious from the fact that it is in contact with all six domains of the 23S RNA (Ban *et al.*, 2000; Harms *et al.*, 2001). In the large subunit, it is bound with the globular domain to the exterior surface and the long β-ribbon essentially along the exit tunnel (Ban *et al.*, 2000). The β-ribbon of the protein has been observed to adopt several different conformations in the ribosome. One of these conformations blocks the exit tunnel (Berisio *et al.*, 2003b). The structure of an erythromycin resistant mutant has also been determined (Davydova *et al.*, 2002).

L25–CTC

L25 is a protein that binds to the 5S RNA (Horne & Erdmann, 1972). It is highly variable in size and nature. In *E. coli*, it is a relatively small protein containing 94 amino acid residues and arranged into one domain (Stoldt *et al.*, 1998; 1999). This version is rarely observed in other species. In *T. thermophilus*, it is called TL5 and is twice as large with 206 residues (Gryaznova *et al.*, 1996). It is composed of two domains (Gongadze *et al.*, 1993; 1999; Fedorov *et al.*, 2001). In *D. radiodurans*, the protein is even larger and composed of three domains (Harms *et al.*, 2001). In *B. subtilis*, the protein was identified as a general stress factor called CTC, which is expressed at high levels as an effect of different stress conditions (Völker *et al.*, 1994). It is not known whether the protein acts as a stress factor in

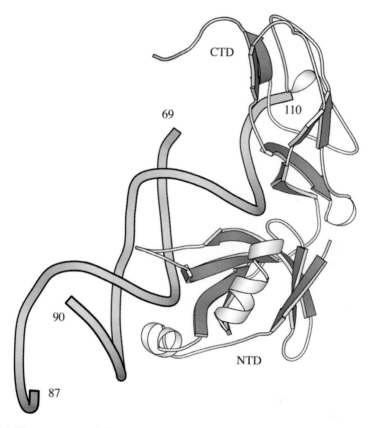

Fig. 7.4 The structure of the TL5 protein from *T. thermophilus* bound to a fragment of *E. coli* 5S RNA (nucleotides 69–87/90–110; Fedorov *et al.*, 2001). The protein is L-shaped and is called CTC in *B. subtilis* where it is known to be a stress factor (Völker *et al.*, 1994). The interactions with the 5S RNA are with the NTD that corresponds to the single domain protein L25 in *E. coli* (Stoldt *et al.*, 1998; 1999; Lu & Steitz, 2000). In *D. radiodurans*, the corresponding protein has three domains (Harms *et al.*, 2001). The figure has kindly been provided by Dr. M. Fodje using the program MOLSCRIPT (Kraulis, 1991).

other species. The protein has a binding site with high affinity, on the ribosome, where it binds under normal conditions. It may have a second binding site with low affinity where it binds at the high concentrations reached at stress conditions. The structure of all the three versions of the protein is known (Fig. 7.4). They can be characterized as having one, two and three domains, respectively. The structure of the CTC version has two domains and mimics a tRNA. Whether CTC binds to a tRNA-binding site to exert its function at stress is not known. In the ribosome, it binds to the 5S RNA. The C-terminal domains have an interesting location at the subunit interface close to the A-site tRNA.

8

Ribosomal Sites and Ribosomal States

The functional sites of the ribosome are located between the ribosomal subunits. This is true for the decoding site on the small subunit and the peptidyl transfer center (PTC) on the large subunit. In addition, the binding sites for the translation factors are also at the interface between the subunits (Plate 8.1). The ribosomal functions are to a large extent closely related to the rRNA. In fact, the central function of the ribosome, peptidyl transfer, is associated only with the ribosomal RNA (Noller *et al.*, 1992; Ban *et al.*, 2000). The binding of tRNAs and factors to the ribosome involves extensive parts of the rRNAs. It has long been believed that the ribosome is a ribozyme and this has now been firmly established with the crystal structures (Ban *et al.*, 2000). Here, we will focus on some of the main sites for ribosomal interactions with mRNA and tRNA as well as with the translation factors and the nascent polypeptide.

The ribosome goes through a number of different states during protein synthesis. This is an area that is only partly explored. Some states have been identified. The best known examples are the pre- and post-translocation ribosomes. The main difference between them is the position of the peptidyl tRNA in the A- and P-sites, respectively, as well as the empty A-site in the post-translocation ribosome. A number of gross conformational differences have been observed and one movement has been called a ratchet-like movement (Woese, 1970; Frank & Agrawal, 2000). Only some details of these differences are known. It is evident from numerous observations that the ribosome is not a passive partner in the translation process, but it actively responds to the ligands and participates in the process.

Some of the ribosomal states are known at close to atomic resolution due to the crystallography of the ribosomal subunits. The highly resolved subunit structures form the basis for the interpretation of the different states studied by lower resolution crystallography or the cryo-EM analysis of the 70S ribosome.

8.1 THE BINDING OF mRNA

The Recognition of mRNA

In bacteria, a nucleotide sequence rich in purines is usually found upstream of the initiator codon of the mRNA's. These sequences are complementary to varying extents to the extension from h45 at the 3′-end of the 16S ribosomal RNA (Fig. 8.1A; Plate 8.2). The binding of this region of the mRNA to the 3′-end of the 16S rRNA is called the Shine-Dalgarno (SD) interaction after the discoverers (Shine & Dalgarno, 1974).

In eukaryotic systems, the binding site on the mRNA for the ribosome is recognized quite differently. The eukaryotic mRNAs have an N^7-methylated GTP linked by a 5′-5′-pyrophosphate bond to the terminal nucleotide. The cap binding proteins that mediate the binding of the mRNA to the small subunit recognizes this so-called cap.

The Binding Site for mRNA — The Decoding Site

The binding site for the mRNA has long been known to be located centrally on the small subunit between the platform and the head (Shatsky *et al.*, 1991). The ribosome protects about 30 nucleotides of the mRNA by the binding (Steitz, 1969). Crystallographic investigations of the 30S subunit have given insights into the binding site for the mRNA and studies by Yusupova *et al.* (2001) give the most extensive picture. Pieces of mRNA of different lengths were bound to the 70S crystals and could be

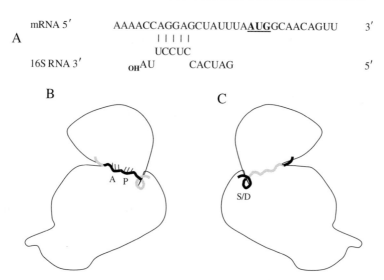

A

mRNA 5′   AAAACCAGGAGCUAUUUA**AUG**GCAACAGUU   3′
                 | | | | |
                 UCCUC
16S RNA 3′   _{OH}AU   CACUAG   5′

Fig. 8.1 The binding of the mRNA to the small subunit (Yusupova *et al.*, 2001). **(A)** One example of the base pairing between the 5′-end of the mRNA and the 3′-end of the 16S RNA. The initiator AUG codon is underlined. **(B)** The codons of the A- and P-sites are exposed on the interface side of the subunit. Only a few nucleotides in addition to the two codons of the mRNA are exposed due to the narrow groove. **(C)** The SD base pairing between the 5′-end of the mRNA and the 3′-end of the 16S RNA is located in a groove between the head and the platform. The mRNA crosses over from the interface side to the external side of the subunit in a narrow groove on the side of the shoulder.

characterized. In simple terms, the 5′-end of the mRNA contacts the back of the platform and wraps around the neck of the small subunit to place the codons to be read in the A- and P-sites, respectively (Figs. 8.1B and C; Plate 8.2). From the way the mRNA is bound, it is evident that initiation has to start with small subunits that bind the mRNA. When the small subunits are bound to large subunits, there is no simple way to introduce an mRNA into the proper place through the tunnel between the subunits. Likewise, to dissociate an mRNA from the ribosome, the subunits may need to be separated from each other.

The mRNA is threaded through two tunnels of the 30S subunit and its ends protrude up- and downstream of the decoding region (Plate 8.2; Yusupova *et al.*, 2001). The only region that is exposed, are nucleotides −1 to +7 of the mRNA counting from the first nucleotide of the P-site codon. The E-site codon (nucleotides-1–3 is only partly accessible due to its location in a tunnel. The region of the mRNA that is in contact with the ribosome, are nucleotides −15 to +16, in good agreement with previous estimates (Steitz, 1969).

The SD interaction occurs in the cleft between the platform and the head of the subunit (Yusupova *et al.*, 2001). The regions of the ribosome in contact with the helix formed by the SD interaction are h20, h28 (the neck helix), h37 and proteins S11 and S18.

The decoding site has been defined by the crystallographic investigations of the small subunit and whole ribosomes (Carter *et al.*, 2000; Schlünzen *et al.*, 2000; Ogle *et al.*, 2001; 2002; Yusupov *et al.*, 2001). It is situated at the interface side of the small subunit close to the top of the penultimate helix or h44 of the 16S RNA. The mRNA makes a kink of about 45° between the codons in the A- and P-sites (Yusupov *et al.*, 2001). Furthermore, the two sites have been defined by a number of interactions with ribosomal RNA and r-proteins (Stahl *et al.*, 2002). The details will be further discussed below in the sections on the tRNA, A- and P-sites.

8.2 THE tRNA BINDING SITES

The aminoacyl tRNAs are the substrates in the process of translation and the ribosome is a large enzyme, a polymerase. Enzymes need to distinguish correct from incorrect substrates and also bind the substrates accurately so that a chemical reaction can proceed at a physiological rate. The tRNAs and the ribosome have evolved together in order for translation to be performed with sufficient accuracy and speed.

The tRNA molecules are bound at the interface between the ribosomal subunits and form bridges between the decoding site on the small subunit and the PTC on the large subunit. The subunit arrangement fits the unperturbed L-shaped structure of the tRNAs with the anticodon at one end and the amino acid at the opposite end. Classically, two sites for tRNA molecules on the ribosome have been discussed (Warner & Rich, 1964; Watson, 1964). These are the A-site (the site for the acceptor or aminoacyl tRNA) and the P-site (the site for the donor or peptidyl tRNA). A third site has also been suggested and subsequently generally accepted. This is the E-site where the deacylated tRNA binds before it dissociates from the ribosome (Wettstein & Noll, 1965; Rheinberger *et al.*, 1981; Grajevskaja *et al.*, 1982; Kirillov *et al.*, 1983). A fourth site is also required, the entry or recognition site where the tRNA is located when bound to the ribosome with EF-Tu (Hardesty *et al.*, 1969; Lake, 1977; Robin & Hardesty, 1983, Stark *et al.*, 1997). This site (called the T-site) is evident from the observation that the aminoacyl residue cannot be engaged in peptide bond formation until it is released from the ternary complex when bound to the ribosome (Skogerson & Moldave, 1968).

These sites have been delineated with regard to the rRNA using protection from chemical modifications (Table 8.1; Moazed & Noller, 1989). Crystallographic and cryo-EM investigations of ribosomes have improved our understanding of the tRNA sites and the dynamics involved in decoding, peptidyl transfer and translocation. Generally, there is a good agreement between the chemical experiments and the structural approaches.

The tRNAs move through a tunnel between the subunits from the T- and A-sites on the side of the L12 stalk to the E-site on the L1 side of the ribosome (Fig. 7.1; Plate 8.1). The tRNAs in the four binding sites interact with the ribosome in similar ways; the anticodon stem interacts with the 30S subunit and the D-stem, elbow and acceptor arm interact with the 50S subunit (Yusupov *et al.*, 2001). The planes through the tRNAs in the A- and E-sites form angles of 26° and 46°, respectively with the plane of the tRNA in the P-site. The closest approach of the anticodons in the A- and P-sites is about 10 Å due to the 45° angle between the A- and P-site codons in the mRNA. Their acceptor ends are within 5 Å of each other. The closest approach between the anticodons of the P- and E-site tRNAs is about 6 Å. However, their elbows and 3′-ends are nearly 50 Å apart (Plate 8.3).

Table 8.1 Protection of Specific Bases in *E. coli* 16S and 23S rRNA by tRNA (after Moazed & Noller, 1989)

		16S rRNA	23S rRNA
A-site	Strong	G530, A1492, A1493	C2254, A2439, A2451, G2553, ψ2555, A2602
	Weak	A1408, G1494	G1068, G1071, U2609
P-site	Strong	G693, A794, C795 G926, G1401(N7)	G2252, G2253, A2439 A2451, U2506, U2584, U2585
	Weak	A532, G1338, A1339 G966	A1916, A1918, U1926, G2505
E-site	Strong	—	C2394
	Weak	—	G2112, G2116

Not surprisingly, the binding sites for tRNA on the ribosome involve conserved parts of the ribosomal structure. The ribosomal RNA constitutes the major interaction with the tRNAs (Yusupov *et al.*, 2001). Likewise, the parts of the tRNA that interact with the ribosome are primarily the universally conserved parts.

The T-site

The binding of aminoacyl-tRNA to the ribosome involves EF-Tu:GTP and more than one functional state. For a full understanding we must know:

(1) How does the tRNA bind to the ribosome in a complex with EF-Tu?
(2) How does the tRNA recognize the codon?
(3) How is the fidelity of decoding obtained?
(4) How does the codon-anticodon recognition trigger GTP-hydrolysis in EF-Tu?
(5) When EF-Tu GDP has dissociated from the ribosome, how does the tRNA move into the A-site to allow peptidyl transfer?

The tRNA enters the T-site of ribosome in a complex with EF-Tu. As long as the tRNA remains bound to the EF-Tu, at most its anticodon can be located in the A-site, or more precisely the decoding part of it, whereas the aminoacyl end remains bound to EF-Tu far from the PTC (Stark *et al.*, 1997).

The structure of the ternary complex is incompatible with the structure of the inter-subunit cavity of the ribosome. If a normal L-shaped tRNA were bound to the codon, EF-Tu in a ternary complex would collide severely with the large subunit and in particular with the PTC. Rather, the binding of EF-Tu to the ribosome dominates the initial binding of the ternary complex. EF-Tu binds to the GTPase associated region (GAR) at the base of the L12 stalk of the large subunit. The aminoacyl-tRNA remains bound to EF-Tu with the anticodon close to the decoding site, but with the acceptor end and its aminoacyl moiety far from the PTC (Stark *et al.*, 1997).

Cryo-EM studies at 9 Å resolution show that in the initial binding of the ternary complex, the tRNA undergoes a conformational change. Firstly, the elbow region of the tRNA moves by about 7 Å compared with its interaction with EF-Tu in the free ternary complex (Nissen *et al.*, 1995;

1999; Valle *et al.*, 2003a). Secondly, the anticodon stem makes a bend with regard to the D-arm (Valle *et al.*, 2003a). This allows the anticodon to match the codon in the decoding site (Fig. 8.2). This mode of binding of the tRNA to the ribosome in the ternary complex corresponds to the A/T-state of Moazed & Noller (1989).

The bent conformation has not previously been seen in tRNAs and may thus be strained and unstable. The kink in the tRNA is most certainly due to its multiple interactions with EF-Tu, the mRNA and the ribosome. One important observation is also that the ASL of the cognate tRNA in this A/T-state binds in the same manner as a tRNA in the A-site (Yusupov *et al.*, 2001; Ogle *et al.*, 2003; Valle *et al.*, 2003a). Interestingly, the temperature factors of the anticodon arm are the highest in a crystallographic study of a complex between EF-Tu and tRNA (Valle *et al.*, 2003a).

In addition to the interactions with EF-Tu and the codon, the aminoacyl-tRNA in the A/T-state has three interactions (Valle *et al.*, 2003a). Ribosomal protein S12 interacts with the tRNA in this state of translation and whether it undergoes a conformational change in the process has been discussed (Valle *et al.*, 2002; Stark *et al.*, 2002). In the cryo-EM studies at 9 Å (Valle *et al.*, 2003a), only one visible contact between S12 and the tRNA was observed, between the acceptor arm, near nucleotide 69, and amino acid residues 78–80. Furthermore, nucleotide C56 in the conserved TΨCG of the T-arm is close to A1067 of helix H43 in the GAR.

Helix H69 of the 23S RNA extends into the A-site, where nucleotide 1914 interacts with the minor groove of the D-arm (residues 24–25) of the

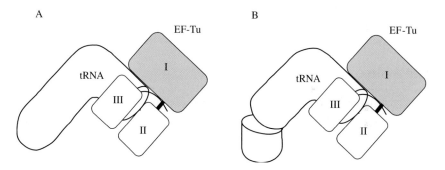

Fig. 8.2 A schematic representation of the ternary complex of EF-Tu:GTP (domains I, II and III) and an aminoacyl-tRNA. **(A)** In the normal configuration. **(B)** In the kinked conformation it adopts when bound to the ribosome in the A/T state (B; Valle *et al.*, 2003a).

EF-Tu bound aa-tRNA (Stark *et al.*, 2002; Valle *et al.*, 2002; 2003a). Helix h44 of the 16S RNA, on the other hand, interacts with the codon and the anticodon in the A-site, both in the case of the ternary complex and in the case of the A-site bound tRNA (Plate 8.4; Valle *et al.*, 2002; Stark *et al.*, 2002). The decoding site is designed to distinguish correct (cognate) from incorrect (non-cognate or near-cognate) tRNAs. The difference in affinity between the isolated cognate and near-cognate tRNAs for the codon is not enough to explain the error frequency. Nevertheless, on the ribosome the cognate interaction is favored (see Section 11.5). The ribosome plays an active role in the decoding as crystal structures of the small subunit have shown (Carter *et al.*, 2000; Schlünzen *et al.*, 2000; Ogle *et al.*, 2001; 2002; 2003).

Incorrect matches between codon and anticodon lead to the dissociation of the ternary complex EF-Tu:GTP:tRNA from the ribosome. However, with cognate codons the affinity for the tRNA is increased through induced interactions with the ribosomal RNA. Particularly G530, A1492 and A1493 alter their conformations and bind to the codon-anticodon complex provided that it is a cognate complex. EF-Tu is simultaneously induced to hydrolyze its GTP to GDP, which changes the conformation of the protein drastically. EF-Tu then looses its affinity for the tRNA as well as for the ribosome and dissociates in its complex with GDP. As a consequence, the cognate aminoacyl-tRNA remains on the ribosome and can move from the A/T state into the A-site. This would mean a major swing of the acceptor end from the interaction with EF-Tu to the PTC, while the codon-anticodon interaction is retained. The induction of the GTP hydrolysis and the dissociation of EF-Tu will be discussed in Sections 8.5, 9.3 and 11.4.

The A-site

The fact that the decoding of the mRNA by the tRNAs is done on the small subunit was established early (Okamoto & Takanami, 1963; Davies *et al.*, 1964). The kinked anticodon of the tRNA (Valle *et al.*, 2003a) forms base pairs with its corresponding codon of the mRNA while the tRNA is bound to EF-Tu as part of the ternary complex. In addition, a cognate complex is stabilized by specific interactions with the rRNA (see above). These interactions are retained when the tRNA is liberated from EF-Tu to swing the acceptor end into the PTC.

When the tRNA has moved into the A-site, the shape of the tRNA does not deviate from the classical structure of tRNA^{Phe} (Robertus *et al.*, 1974; Kim *et al.*, 1974; Yusupov *et al.*, 2001). The kinked tRNA in the A/T-state is liberated from its binding to EF-Tu and the ASL and D-stem are fused again and their bases are stacked. The orientation of the anticodon stem loop in the kinked state defines the orientation of the liberated tRNA when it moves into the A-site. In this manner, the acceptor end is able to reach the PTC (Valle *et al.*, 2003a).

A tRNA Bound to the A-site

In the A-site, the universal sequence PNSA of protein S12 (residues 44–47 in *E. coli*) is located between the 530 loop and the 1492–1493 region, and is involved in hydrogen bonding to the second and the wobble base-pair of the decoding part of the A-site (Plate 8.4A and B; Yusupov *et al.*, 2001; Ogle *et al.*, 2001; 2002). The elbow, or the D- and T-loops, of the A-tRNA interacts with protein L16 and the inter-subunit bridge B1a (the A-site finger or H38; Yusupov *et al.*, 2001). H69, the B2a bridge, interacts with the inner side of the elbow of the L-shaped tRNA (Plate 8.4C; Bashan *et al.*, 2003). Protein L25 in *E. coli* has longer forms in *T. thermophilus* and *D. radiodurans* (see Section 7.4). In these species, the protein reaches the A-site tRNA between the A-site finger and the L11 arm (Harms *et al.*, 2001). The CCA or acceptor part of the tRNA with its bound aminoacyl residue is in the PTC, in close proximity to the nascent peptide. Here, C75 of the CCA-end of the tRNA base-pairs with G2553 of the so-called A-loop (Green *et al.*, 1998; Yusupov *et al.*, 2001). A76 makes A-minor interactions with the 23S RNA (Hansen *et al.*, 2002b).

The P-site

The P-site tRNA, just like the A-site tRNA, stretches across the tunnel between the subunits. The tRNA in the P-site is somewhat kinked compared to the classical crystal structure of tRNA^{Phe} (Robertus *et al.*, 1974; Kim *et al.*, 1974). The kink occurs at the junction of the anticodon and the D-stems (Yusupov *et al.*, 2001). The P-site tRNA is related to the A-site primarily by a 26° rotation (Yusupov *et al.*, 2001). The P-site tRNA not only remains attached to the PTC, but also to the codon in the decoding site (Plates 8.3 and 8.5). The P-site is centrally located at the subunit interface. A number of specific contacts between the tRNA in the P-site and the ribosome have been identified by different methods.

A detailed structural insight into how the anticodon stem loop (ASL) is bound to the P-site comes from studies of *T. thermophilus* 30S where, in the crystal packing helix h6 or the spur of one 30S subunit was bound as an ASL into the P-site of another 30S subunit (Carter *et al.*, 2000). Other observations come from the crystallographic and cryo-EM studies of the 70S ribosomes from *T. thermophilus* and *E. coli*, respectively (Yusupov *et al.*, 2001; Stark *et al.*, 2002; Valle *et al.*, 2002; 2003a,b).

The tRNA in the P-site has numerous contacts (Plate 8.5). Ofengand and coworkers (Prince *et al.*, 1982) made the earliest observation of the proximity between the decoding part of the P-site and part of the tRNA bound to it. They found that the wobble base of the P-site tRNA and C1400 of the 16S RNA could be crosslinked. The C-terminal tails of ribosomal proteins S9 and S13 also interact with the ASL (Yusupov *et al.*, 2001). Arg128 of S9 that interacts with phosphate 36 of the tRNA in the P-site is universally conserved (Yusupov *et al.*, 2001). S13 is situated between the anticodon arms of the A- and P-site tRNAs and is in close proximity to the anticodon parts of both tRNAs (Stark *et al.*, 2002). Furthermore, protein L5 contacts the minor groove of H69 (subunit bridge B2a) that interacts with the T-loop and the minor groove of the D-stem of P-site tRNA (Yusupov *et al.*, 2001; Stark *et al.*, 2002). C74 and C75 of the CCA-end of the P-site tRNA base-pair with G2253 and G2252, respectively of the so-called P-loop, and A76 also interacts with the P-loop (Samaha *et al.*, 1995; Green *et al.*, 1998; Nissen *et al.*, 2000; Yusupov *et al.*, 2001; Hansen *et al.*, 2002b). The acceptor end makes backbone-backbone contacts with the stem of the P-loop. Furthermore, the CCA-end is close to H93 (Yusupov *et al.*, 2001; Stark *et al.*, 2002).

The proximity of subunit bridge B2a (H69) to both the A- and P-sites as well as to both the decoding and the peptidyl transfer centers make it a likely candidate for signal transmission between the decoding and GTPase associated region (Bashan *et al.*, 2003). Its inherent flexibility supports this possibility (Ban *et al.*, 2000).

The E-site

The E-site binds the deacylated tRNA molecule before it dissociates from the ribosome (Plate 8.6). The binding of tRNAs to the 50S part of the E-site requires a free CCA-end (Lill *et al.*, 1986). The E-site is at the end of the mRNA and tRNA tunnel of the subunit interface toward the L1 side.

The E-site tRNA is related to the P-site tRNA by a rotation of 46° (Plates 8.3 and 8.6; Yusupov *et al.*, 2001). In contrast to earlier findings (Moazed & Noller, 1989), it is clear that the E-site tRNA interacts with both subunits as well (Yusupov *et al.*, 2001). The tRNA that is bound to the E-site is substantially distorted (Yusupov *et al.*, 2001). The anticodon of the E-site tRNA does not retain its interaction with its codon and is situated between the platform and the head of the small subunit (Yusupov *et al.*, 2001). Protein S7 interacts closely with the ASL. The tRNA acceptor end is far from the PTC. Proteins S7 and L1 can jointly block the exit of a deacylated tRNA from the E-site. The L1 protuberance has been observed in an open as well as a closed form (Valle *et al.*, 2003b). The open form would permit the dissociation of the deacylated tRNA from the E-site. *In vitro* the E-site seems usually to be occupied by a deacylated tRNA (Yusupov *et al.*, 2001; Valle *et al.*, 2003b).

Hybrid tRNA States

The chemical footprinting by tRNAs bound to the different sites shows a characteristic pattern (Table 8.1). In the process of tRNAs going from the A/T state to the A-site, through the P-site to the E-site, they are found to go through intermediate or hybrid states (Moazed & Noller, 1989). The A/T-state is such a hybrid state, since it agrees with the A-site for the ASL but not for the rest of the tRNA.

A peptidyl-tRNA in the P-site gives the characteristic pattern in Table 8.1. When an aminoacyl-tRNA is added, peptidyl transfer occurs with interesting changes in the protection pattern. The nucleotide protections, characteristic for binding to the E-site, appear (Moazed & Noller, 1989; Sharma *et al.*, 2004). The P-site protections remain while the A-site protections are observed only for the 30S subunit. Addition of EF-G removes the A-site footprint. This suggests that a peptidyl-tRNA can be bound in a hybrid state, the A/P state, before translocation (Bretscher, 1968). While its ASL remains bound to the decoding part of the A-site, the acceptor end has moved to the P-site part of the PTC (Fig. 8.3). This may seem natural since the peptide in the exit tunnel is unlikely to shift its position depending on which tRNA it is bound to. However, it is more remarkable that the CCA-end, upon peptidyl transfer, dissociates from the A-loop and associates with the P-loop. Likewise, after peptidyl transfer, the deacylated tRNA in the P-site would interfere with the tRNA in the A/P state.

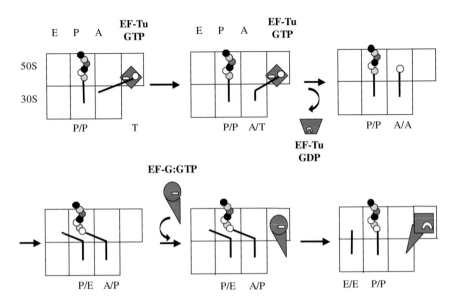

Fig. 8.3 Schematic representation of the elongation cycle with hybrid sites for tRNAs in the style introduced by Moazed & Noller (1989).

It has thus been observed to footprint in what is called the P/E state. The P/E hybrid state has also been observed by cryo-EM (Valle *et al.*, 2003b). The hybrid state model will be further discussed in Sections 8.6 and 11.4.

8.3 THE PEPTIDYL TRANSFER CENTER (PTC)

The central function of the ribosome is to transfer a nascent peptide from the P-site tRNA to the amino acid residue of the A-site tRNA. These tRNAs are bound to subsequent codons of the mRNA in the decoding site. The nature of the PTC has been the object of studies for a long time. It has long been clear that the large subunit is responsible for peptidyl transfer (Okamoto & Takanami, 1963; Monro *et al.*, 1968). A number of large subunit proteins have been suspected, but rejected on various grounds (Hampl *et al.*, 1981; Liljas, 1982). The identification of ribozymes with different catalytic activities (Guerrier-Takada *et al.*, 1983; Cech *et al.*, 1981) suggested that the ribosome could belong to the family of ribozymes. Thus, assays representative for peptidyl transfer such as the puromycin reaction have been performed with rRNA essentially devoid of ribosomal proteins, indicating that the ribosome is a ribozyme (Noller *et al.*, 1992; Noller, 1993).

A number of approaches have been used to identify the regions of the 23S RNA that are involved in peptidyl transfer (Sonenberg *et al.*, 1975). The central loop of domain V is of great importance (Barta *et al.*, 1984). Many nucleotides in this area are completely conserved; resistance against antibiotics that inhibit peptidyl transfer is found in this region; chemicals reacting with nucleotides are protected by tRNAs from reactions in this region; crosslinking from the acceptor ends of tRNAs or from the amino acid or peptide that is attached to the peptidyl-tRNA are found on this surface; mutations in this region can severely affect peptidyl transfer (Eckerman & Symons, 1978; Hummel & Böck, 1987; Vester & Garrett, 1988; Mankin & Garrett, 1991; Moazed & Noller, 1991; Green & Noller, 1997).

The peptidyl transfer center (PTC) has been identified as a cavity on the interface side of the large subunit (Nissen *et al.*, 2000). Crystallographic analyses of large subunits show that no protein is in the vicinity of the PTC (Ban *et al.*, 2000; Nissen *et al.*, 2000; Harms *et al.*, 2001). Thus, in the structure of the halophilic large subunit, there is no protein within a radius of 18 Å from the PTC (Ban *et al.*, 2000; Nissen *et al.*, 2000). In the structure of the bacterial large subunit, the protein that comes closest to the PTC is L27 (Harms *et al.*, 2001). The N-terminal tail of the protein is flexible and not seen in the electron density. It could possibly interact with the CCA-ends of the A- and P-site tRNAs (Wower *et al.*, 1998; Harms *et al.*, 2001). However, the protein L27 is not found in archaea (Ban *et al.*, 2000). The archaeal protein that is bound in a similar location on the interface side of the large subunit just below the central protuberance is L21e. Its tail, however, folds backward towards the interior of the subunit (Ban *et al.*, 2000; Harms *et al.*, 2001). Thus, L27 or its archaeal replacement can hardly be essential for the catalytic activity of peptidyl transfer. In conclusion, since there are no proteins in the PTC, it is clear that the rRNA is the catalyst of peptidyl transfer and that the ribosome is a ribozyme. Many of the nucleotides that may have functional properties in the PTC are highly conserved (Nissen *et al.*, 2000; Bashan *et al.*, 2003).

The crystallographic analysis of inhibitors and substrate analogues bound to the PTC is the most direct approach to the study of the details of this enzymatically active site. Unfortunately, the structures of the whole ribosome are at a resolution that is insufficient for detailed analysis of hydrogen bonds and thus lack the accuracy that is needed for an understanding of the enzymatic function. Thus, conclusions about the PTC must be derived from the crystallographic studies of the large subunits.

In the structure of the halophilic large subunit, the site for PTC was easily located with the aid of an inhibitor that aims at mimicking the transition state (Plate 8.7; Nissen *et al.*, 2000). The cavity at the entrance of the exit tunnel on the interface side not only has residues that are well known from chemical studies of the PTC, but a number of inhibitors affecting peptidyl transfer can be bound here. Experiments with different fragments of tRNA bound to the A- and P-sites as well as a transition state analogue have given further insight into the nature of this active center. Thus, the CCA-ends of the A- and P-site tRNAs are bound to the A- and P-loops, respectively. Whereas a sideways movement relates the tRNAs in the A- and P-sites, their CCA-ends are related by an approximate 180° rotation (Nissen *et al.*, 2000; Yusupov *et al.*, 2001; Schmeing *et al.*, 2002; Hansen *et al.*, 2002b; Bashan *et al.*, 2003; Agmon *et al.*, 2003). The difference in structure between the A- and P-site tRNAs primarily occurs between nucleotides 72 and 74 (Hansen *et al.*, 2002b). Bashan *et al.* (2003) regard nucleotides 73–76 as the rotating moiety. In fact, much of the PTC has a two-fold rotational symmetry with a small translational component in agreement with the relationship between the acceptor ends of the A- and P-site tRNAs (Nissen *et al.*, 2000; Bashan *et al.*, 2003). The two-fold symmetry includes the A- and P-loops. About 90 nucleotides around the A-site part of PTC have symmetry related nucleotides in the P-site part (Bashan *et al.*, 2003; Agmon *et al.*, 2003). The actual positions of the substrate analogues differ somewhat in the investigations of the two species. This is partly due to an inherent flexibility of the PTC. This flexibility is likely to be of functional importance (Yonath, 2003).

The details of the PTC are extensively discussed in Section 11.4. The basis of this discussion comes from the interpretation of a number of crystal structures of complexes between tRNA substrate fragments bound to the A- and P-sites. In addition, an intermediate in peptidyl transfer and a product have been investigated (Nissen *et al.*, 2000; Schmeing *et al.*, 2002; Hansen *et al.*, 2002b; Bashan *et al.*, 2003).

8.4 THE POLYPEPTIDE EXIT TUNNEL

It has long been observed that ribosomes protect a number of amino acids of the nascent polypeptide from digestion by proteolytic enzymes (Malkin & Rich, 1967; Blobel & Sabatini, 1970). It was suggested that there could be a tunnel in the large subunit through which the polypeptide exits. From an early stage, it was clear that ribosomes interacted with microsomal membranes through the large subunit (Sabatini *et al.*, 1966). A tunnel was indirectly inferred from early EM-studies in which ribosomes associated through the back of the large subunit when a dimeric protein was synthesized. Evidently, the nascent peptide appeared from the backside of the large ribosomal subunit (Bernabeau & Lake, 1982).

A suitably positioned tunnel through the large subunit was subsequently directly observed by electron microscopy at very low resolution (Milligan & Unwin, 1986; Yonath *et al.*, 1987). In more recent cryo-EM studies, a branched tunnel system was identified (Frank *et al.*, 1995; Gabashvili *et al.*, 2001). The crystallographic investigations (Ban *et al.*, 2000; Nissen *et al.*, 2000; Harms *et al.*, 2001) have made it clear that this tunnel originates from the PTC and leads to the outer surface of the large subunit (Plate 8.8). The tunnel length is about 100 Å and has a variable width of between 10 Å and 20 Å. The tunnel walls are mainly nonpolar but have a mosaic of small hydrophobic and hydrophilic patches (Moore & Steitz, 2003b). In yeast, the tunnel is continuous with the translocon tunnel (Beckmann *et al.*, 1997; Menetret *et al.*, 2000). This is likely to be the tunnel through which the nascent peptide exits from the ribosome and can be further transported through the cytoplasmic membrane. A tunnel of this type is a necessity for the transport of proteins from the cytoplasm to other compartments. This requires a stable structure that could contact a receptor structure. Such a stable structure is only found in the body of the large subunit, since the subunit interface is highly dynamic. Thus, the stable organization of the large subunit corresponds to the need for a tunnel that remains open during the different stages of translation.

A strong support for this thinking is the fact that macrolide antibiotics that block protein synthesis of peptides longer than a few residues were observed to bind in the tunnel (see Chapter 10; Schlünzen *et al.*, 2001; Hansen *et al.*, 2002a).

Gabashvili *et al.* (2001) have made a cryo-EM analysis of the tunnel system of the large subunit and summarized previous observations of variations in the tunnel. In addition to the main tunnel, they observed three more exits diverging from the main tunnel (Plate 8.8; Gabashvili *et al.*, 2001). These side tunnels are not observed in the crystallographic structures of the large subunits and may be artifacts at the low resolution of the EM studies (Moore & Steitz, 2003b).

The tunnel widths, particularly that of the narrowest part, vary significantly depending on the state of the ribosome. Thus, in the ribosomes without ligands or in the H50S subunit, the narrowest part of the tunnel is about 10 Å wide. Ribosomes with tRNA and EF-G bound have a larger opening of 16–20 Å (Gabashvili *et al.*, 2001).

The narrowest part of the peptide exit tunnel is composed of parts of the 23S RNA as well as two ribosomal proteins L4 and L22, that are intimately associated with the tunnel (Nissen *et al.*, 2000; Berisio *et al.*, 2003b). Both L4 and L22 are built of globular bodies with a long protruding loop or β-ribbon (Worbs *et al.*, 2000; Unge *et al.*, 1999). The globular parts of the proteins are located on the outer surface of the large subunit stretching their tips to the most constricted part of the tunnel (Nissen *et al.*, 2000; Harms *et al.*, 2001; Berisio *et al.*, 2003b).

Macrolide antibiotics (see Section 10.5), such as erythromycin, bind at the narrowest part of the tunnel not far from the PTC (Schlünzen *et al.*, 2001; Hansen *et al.*, 2002a). The binding is due to hydrophobic interactions with the lactone ring and to hydrogen bonds between the sugar moieties and the sides of the tunnel. In addition, the 16-membered ring macrolides were found to make a reversible covalent bond with A2062 (Hansen *et al.*, 2002a).

Among the antibiotics that bind to the tunnel are oleandomycin and troleandomycin (Berisio *et al.*, 2003b). They bind further into the tunnel and at a different angle from erythromycin (Berisio *et al.*, 2003b). Troleandomycin interacts with A2058. Part of it overlaps with the normal location of the tip of L22, which must then move into a new location. In so doing , the tip of L22 blocks the tunnel.

The folding of the polypeptide in this tunnel has also been studied. With the knowledge of the exit tunnel, the conformation of the nascent peptide becomes an interesting topic. Is it straight and does it have a secondary structure? Can the ribosome allow a partial folding (become pregnant) with a growing polypeptide? Lim and Spirin (1986) made a stereochemical analysis and suggested that the nascent polypeptide forms an α-helix in the tunnel. Using labeled N-termini of the nascent polypeptides, the labels were observed to emerge with different lengths of nascent polypeptide for different proteins (Tsalkova *et al.*, 1998; Hardesty & Kramer, 2001). Between 44 and 72 amino acid residues could be hidden in the tunnel. An α-helix of 72 amino acids would have a length of 108 Å. It was found that the N-termini of the proteins with large amounts of helical conformation emerged later than proteins with large amounts of β-structure. The possibility of partial tertiary folding of some proteins within the exit tunnel remains an interesting topic (Gilbert *et al.*, 2004).

Certain peptides can get stuck in the exit tunnel. This is well characterized for the secretion monitor polypeptide (SecM; Nakatogawa & Ito, 2001; Tenson & Ehrenberg, 2002; Jenni & Ban, 2003). The polypeptide (FXXXXWIXXXXGIRAGP) is one of the peptides that get stalled (Nakatogawa & Ito, 2002). An attempt to characterize the interactions that lead to stalling has been made (Berisio *et al.*, 2003b). Mutations in the constriction part of the tunnel can prevent the arrest of translation (Nakatogawa & Ito, 2002). Peptides with a C-terminal prolyl residue seem to induce stalling of the ribosome (Gong & Yanofsky, 2002).

Two proteins in the archaeal ribosome are associated with the opening of the tunnel on the solvent side of the large subunit. They are L23 and L39e (Nissen *et al.*, 2000). In bacteria, there is only one protein, L23. L39e is a small and extended protein that in archaea replaces the tail of bacterial L23 (Harms *et al.*, 2001). L23 is part of the contact to the protein folding and transport machinery that docks at the opening of the exit site (Kramer *et al.*, 2002b; Pool *et al.*, 2002).

8.5 THE GTPASE BINDING SITE

One group of translation factors binds and hydrolyzes GTP (see Section 9.1). They are part of the large family of G-proteins (Bourne *et al.*, 1990; 1991). The translational GTPases (tGTPases) probably all bind to essentially the same site and their GTPase activity is probably induced in similar ways (see Chapters 9 and 11).

Elongation factor Tu (EF-Tu or EF1 in eukarya) binds to the ribosome as a ternary complex with an aminoacyl tRNA and GTP (Kaziro, 1979). Cryo-EM has given the best insight into how these factors bind to the ribosome and what conformational changes are induced. EF-Tu was shown to bind at the base of the L12 stalk of the large subunit, while the opposite end of the ternary complex, the anticodon of the tRNA binds to the decoding site of the small subunit (Plate 8.1; Moazed *et al.*, 1988; Stark *et al.*, 1997; 2002; Valle *et al.*, 2002, 2003a). The cryo-EM studies of the ternary complex were all done in the presence of kirromycin. This antibiotic binds to EF-Tu and allows the factor to hydrolyze its bound GTP

molecule to GDP, but prevents the factor from adopting the GDP conformation and dissociate from the ribosome (see Sections 9.2 and 10.7; Wolf
et al., 1977; Parmeggiani & Stewart, 1985; Vogeley *et al.*, 2001).

Cryo-EM studies of EF-G bound to the ribosome have also been performed (Agrawal *et al.*, 1998; 1999; Stark *et al.*, 2000; Valle *et al.*, 2003b).
The antibiotic fusidic acid (FA) or GDPNP was used to lock EF-G on the
ribosome (see Sections 9.3 and 10.7). The overall shape of EF-G is very
similar to that of the ternary complex of EF-Tu with tRNA (see Section 9.3;
Nissen *et al.*, 1995). Thus, one domain of EF-G (domain IV) corresponds
structurally to the ASL of the tRNA. Not surprisingly, when EF-G binds to
the ribosome, it overlaps with the binding site of the ternary complex of
EF-Tu and tRNA (Agrawal *et al.*, 1998; 1999). Domain IV of EF-G interacts
with the decoding region of the small subunit and the G-domain of EF-G
interacts with the GTPase associated region (GAR) of the large subunit as
the G-domain of EF-Tu (Agrawal *et al.*, 1998; 1999; Valle *et al.*, 2003b). All
information suggests that IF2 and RF3 interact in related manners.

The binding sites for EF-Tu and EF-G, which are partially overlapping, have important contributions from rRNA. The conserved domain II
of all tGTPases interacts with the small subunit in the region of helix h5
of the 16S RNA and protein S4 (Stark *et al.*, 1997; Agrawal *et al.*, 1988;
Wilson & Noller, 1998; Valle *et al.*, 2003a,b). EF-G makes a footprint in the
1067 region of the 23S RNA (Moazed *et al.*, 1988). This region that can be
called the tGTPase-binding site of the ribosome is built of two lobes, the
α-sarcin/ricin loop (SRL) and the GTPase associated region (GAR; Valle
et al., 2003a).

The lobe formed by the SRL of the 23S RNA contains residues
2653–2667 (Endo *et al.*, 1987). The loop got its name from the inhibiting
modifications that a number of enzymes, among others α-sarcin and ricin,
can have on this region in eukaryotes. SRL is built of H95 of 23S RNA. It
interacts with both elongation factors as seen from protection of the
rRNA to chemical reagents (Moazed *et al.*, 1988). SRL interacts with the
G-domains (Hausner *et al.*, 1987; Moazed *et al.*, 1988; La Teana *et al.*, 2001;
Valle *et al.*, 2002; 2003a). EF-G has been shown to bind a 12-nucleotide
fragment of the SRL (Munishkin & Wool, 1997). No structural information
about the interaction is available at atomic level (Wriggers *et al.*, 2000).

The lobe called GAR, at the base of the L12 stalk, is composed of H43,
H44 and ribosomal proteins L11 and the L10:L12$_4$ complex (Ban *et al.*, 2000;
Harms *et al.*, 2001). GAR interacts with EF-Tu and EF-G far from the

binding site for GTP. Nevertheless, the substitution of H43 (A1067U) inter-feres with the GTPase activity of EF-Tu and EF-G (Saarma *et al.*, 1997).

Thiostrepton (see Section 10.7) is an antibiotic that binds to GAR of the large subunit and is reported to prevent EF-G from binding (Bodley *et al.*, 1970; Cameron *et al.*, 2002). Thiostrepton binds to residues in the 1050–1110 region of domain II of the 23S RNA (Gale *et al.*, 1981). This part of the 23S RNA also binds protein L11 and the pentameric complex L10:L12$_4$ (Dijk *et al.*, 1979; Beauclerk *et al.*, 1984; Wimberley *et al.*, 1999). At an early stage, a crosslink was also identified between EF-G and nucleotide A1067 of the 23S RNA (Sköld, 1983).

8.6 THE RIBOSOMAL STATES

In the early studies of the ribosomes, different states have been discussed and reversible conformational changes have been suggested (Spirin, 1968; Spirin, 1985; Burma, *et al.*, 1985; Noller, 1991; Agrawal *et al.*, 1999b; Frank & Agrawal, 2000). Specific insights into these states are now gained through structural studies. The nomenclature of the ribosomal states is somewhat confusing. There may be fewer states than there are names for them.

Free Subunits and Translating Ribosomes, Initiation

The free subunits have been thoroughly studied at good resolution by crystallography, while the 70S ribosome has been studied only at moder-ate (Yusupov *et al.*, 2001) or low resolution (Valle *et al.*, 2003a,b). Some insights concerning the whole system will remain less detailed or prelim-inary until data at better resolution are available. Some conclusions drawn from studies of subunits may not be valid for the full ribosome.

The phase of initiation goes from separated subunits to 70S ribo-somes with an fMet-tRNA in the P-site. Naturally, the inter-subunit bridges are relevant for the association of the subunits, as is the initiator tRNA in the P-site. One of the main bridges, B2a (the universally conserved helix H69), has been observed to alter its conformation from the free 50S subunit to the 70S ribosome by as much as 13.5 Å at the tip of the stem-loop (Harms *et al.*, 2001). Since this helix is next to the P-site, the binding of the initiator-tRNA may induce the conformational change. The peptidyl-tRNA naturally is also a very important subunit bridge in the elongating ribosome (Zavialov & Ehrenberg, 2003). Similarly, the inter-subunit bridge B1a (the A-site finger or helix H38 of the 23S RNA) is not fully resolved in the subunit structures, probably due to flexibility (Ban *et al.*, 2000; Harms *et al.*, 2001). However, in the 70S ribosome, it is ordered (Yusupov *et al.*, 2001).

The Pre- and Post-translocation States

In the elongation cycle, the ribosome oscillates between two main states called the pre- and post-translocation states. In the pre-translocation state, a peptidyl-tRNA is bound to the A-site and a deacylated tRNA to the P-site. The post-translocation state is induced by elongation factor EF-G that, by binding to the ribosome in complex with GTP, causes a transloca-tion of the two tRNAs to the P- and E-sites, respectively, and a movement of the mRNA to expose the next codon in the now empty A-site. These two states can be identified in different ways. One classical method involves the use of puromycin (see Section 10.4). As long as a tRNA is in the A-site, puromycin cannot bind to the A-site part of the PTC and it reacts with and releases the nascent peptide. However, after translocation, which leads to the post-translocation state, this is possible. One simple feature that identi-fies the post-translocation ribosome is the peptidyl-tRNA in the P-site. The location of the peptidyl-tRNA also dictates which elongation factor should bind (Valle *et al.*, 2003b; Zavialov & Ehrenberg, 2003).

In going between the two main states, EF-G:GTP induces a ratchet-like motion in which the subunits move with regard to each other (Agrawal *et al.*, 1999b; Frank & Agrawal, 2000; 2001; Gao *et al.*, 2003; Valle *et al.*, 2003b). The states before, during and after the interaction with EF-G:GTP have been characterized by cryo-EM at resolutions approaching 10 Å (Valle *et al.*, 2003b). The action is a significant rotation of the two subunits relative to each other at the binding of EF-G with a GTP analogue (Fig. 8.4). After GTP hydrolysis (the ribosomal complex with EF-G:GDP and fusidic acid, FA), this rotation is reversed to about half. The subunits finally return to the normal orientation after the dissociation of EF-G.

The rotation is about 10°, which corresponds to movements of about 20 Å at the edge of the small subunit (Valle *et al.*, 2003b). The angle reported differs somewhat depending on the resolution and method of structure alignment. The binding site of EF-G explains the observed rota-tion. On the small subunit, it interacts between the body and the head

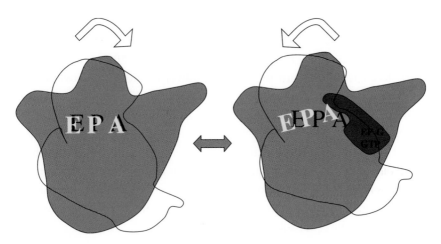

Fig. 8.4 The ratchet-like rotation induced by binding of EF-G:GTP is a counter clockwise rotation of the small subunit in relation to the large subunit of about 10° (Valle *et al.*, 2003b). After dissociation of EF-G:GDP, the ribosome returns to the subunit orientations seen to the left. Both pre- and post-translational ribosomes have essentially the same subunit orientation. During these conformational changes, the tRNAs in the A- and P-sites move into the P- and E-sites and the mRNA moves to expose the next codon in the A-site. Due to the rotation, the A-site tRNA on the small subunit moves closer to the P-site on the large subunit. The 50S subunit is shown in gray with black letters and the 30S subunit is transparent with white letters. EF-G:GTP is shown in dark gray binding at the subunit interface on the right-hand.

(Fig. 8.4). The head is pushed in an anti-clockwise direction in relation to the large subunit.

The largest relative movement occurs between the head of the small subunit and the central protuberance of the large subunit. The axis of rotation is more or less perpendicular to the subunit interface and passes through h44 between the bridges B3 and B5a just above h27 (Gabashvili *et al.*, 2000; Yusupov *et al.*, 2001; Gao *et al.*, 2003). In this movement, three of the inter-subunit bridges (B1a, B1b and B1c) change significantly. For bridges B1b and B1c, protein L5 finds alternate interactions with S13. H38 of bridge B1a moves from an interaction with protein S13 to one with protein S19 (Valle *et al.*, 2003b).

The Locked and Unlocked States

Locked and unlocked states of the ribosome were discussed at an early state (Spirin, 1968) and have gained a renewed interest. The ratchet-like rotation of the ribosomal subunits is normally induced by EF-G:GTP in the pre-translocation ribosomes (Frank & Agrawal, 2001). It can be induced even in empty ribosomes (Agrawal *et al.*, 1999) or when a deacylated tRNA is situated in the P-site (Valle *et al.*, 2003b). However, if a peptidyl-tRNA is located in the P-site, no ratchet-like movement can be induced. To this state EF-G:GDP but not EF-G:GTP is able to bind in the presence of fusidic acid (FA; Valle *et al.*, 2003b). This is apparently a locked state as far as EF-G actions are concerned (Zavialov & Ehrenberg, 2003; Valle *et al.*, 2003b). The locked state can be unlocked by binding an aminoacyl-tRNA to the A-site, which leads to continued peptide transfer and translocation; or by removing the peptide from the P-site tRNA with puromycin (Valle *et al.*, 2003b).

The Restrictive and Ram States

During translation, the ribosome also oscillates between two different states, which can be related to the fidelity of translation (see Section 11.4). In the restrictive state, the ribosome preferably stabilizes the binding of cognate tRNA. This is due to a relatively low affinity for near-cognate aminoacyl-tRNA (Pape *et al.*, 1998; 1999; 2000). In the ribosome ambiguity state (*ram*), the affinity for tRNA, cognate as well as non-cognate, in the A-site is higher. The fidelity of translation is affected if one state becomes stabilized over the other (Allen & Noller, 1989; Lodmell & Dahlberg, 1997).

These two states correspond to two conformations of the small subunit (Ogle *et al.*, 2003). When the A-site is empty, the subunit remains in an open or restrictive conformation (see Table 8.2). When a cognate ASL

Table 8.2 Induction of the *Ram* and Restrictive Conformations of the Small Subunit

Ribosomal State	*Ram*	Restrictive
Conformation of 30S subunit	closed	open
Empty A-site		X
Cognate tRNA in A-site	X	
Mutations in H27 of 16S RNA	X	or X
Paromomycin	X	
Streptomycin	X	
Str resistance, S12 mutants		X
Revertants from Str resistance	X	

of tRNA interacts with the mRNA in the decoding site, the small subunit adopts the closed or *ram* conformation (Plate 8.9, Ogle *et al.*, 2003). The conformational change, due to the strong cognate interaction, is composed of rotational movements of the head and shoulder of the small subunit toward the subunit center (Ogle *et al.*, 2002; 2003). One element of the closed state is the change in conformation and the participation of G530, A1492 and A1493 in the decoding mechanism of the codon-anticodon (Carter *et al.*, 2000; Ogle *et al.*, 2001; 2002; 2003). A near-cognate tRNA-ASL associates weakly with the decoding site and does not induce the closed form of the subunit (Ogle *et al.*, 2002).

The ribosome adopts the closed conformation in complex with error-inducing antibiotics (see Section 10.2). Thus, paramomycin makes the small subunit adopt its closed conformation even when a near-cognate tRNA is bound to the decoding site. In particular, the bases A1492 and A1493 are swung out of their normal positions as if they would identify a cognate interaction of mRNA and tRNA (Ogle *et al.*, 2002).

Streptomycin is another error-inducing antibiotic (see Section 10.2). When bound, it leads to the closed conformation by interaction with well-separated parts of the small subunit (Carter *et al.*, 2000; Ogle *et al.*, 2003). Resistance to streptomycin is primarily due to mutations in ribosomal protein S12 that is located on the shoulder of the small subunit near the A-site and near the binding site of streptomycin (Plate 8.9; Kurland *et al.*, 1996, Wimberley *et al.*, 2000; Carter *et al.*, 2000). The S12 mutants are restrictive. Many of them disfavor the closed conformation (Ogle *et al.*, 2003). Some S12 mutants are hyper-accurate. To get a balanced level of fidelity and rate of translation the bacterium can become streptomycin dependent (Bilgin *et al.*, 1992). Revertants from streptomycin dependence are primarily found as mutants of ribosomal proteins S4 and S5 (Kurland *et al.*, 1996). These are *ram* mutants and are located on the opposite side of the subunit from S12 and on each side of the interface between the shoulder and body of the subunit where the conformational alteration occurs (Wimberley *et al.*, 2000; Ogle *et al.*, 2003). The *ram* mutants decrease the number of bonds that need to be broken to allow closure of the subunit (Ogle *et al.*, 2003).

Another component close to S12 that has been identified as important is helix h27, where mutations can both increase and decrease fidelity (Lodmell & Dahlberg, 1997). It was suggested that this helix has two different modes of base-pairing and that mutants would affect the balance. It now appears that these mutations affect the interface of the shoulder and the environment around h44 (Ogle *et al.*, 2003).

The restrictive and *ram* conformations are related to the pre- and post-translocational conformations. The open or restrictive state that normally has no tRNA bound to the A-site corresponds to the post-translocational ribosome. The closed state, normally induced by the cognate tRNA, leads to a number of subsequent steps. To what extent it differs from the pre-translocational ribosome is not known at present.

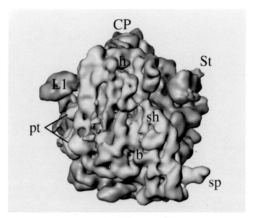

Plate 2.1 The structure of bacterial ribosomes studied by cryo electron microscopy using single particle reconstruction methods. The large subunit is shown in blue and the small subunit in yellow. CP means central protuberance; L1 is the L1 stalk feature; St shows the location of the L12 stalk. In the small subunit, h is the head, b is the body, sh is the shoulder, pt is the platform and sp is the spur. (Reprinted with permission from Gabashvili, *et al*. Solution structure of the *E. coli* 70S ribosome at 11.5 Å resolution, *Cell* **100**: 537–549. Copyright (2000) Elsevier.)

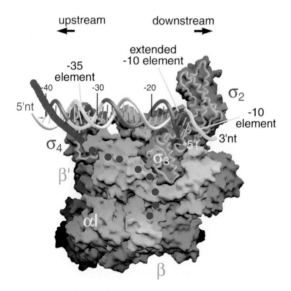

Plate 4.1 The structure of RNA polymerase from *T. aquaticus* showing the outline of the protein subunits and the DNA substrate (Murakami *et al.*, 2002). The likely path of the product RNA is shown as red dots. The process by which the DNA unwinds and the RNA is synthesized is described by Murakami & Darst (2003). (Reprinted with permission from Murakami *et al*. Structural basis of transcription initiation: an RNA polymerase holoenzyme-DNA complex, *Science* **296**:1285–1290. Copyright (2002) AAAS.)

A

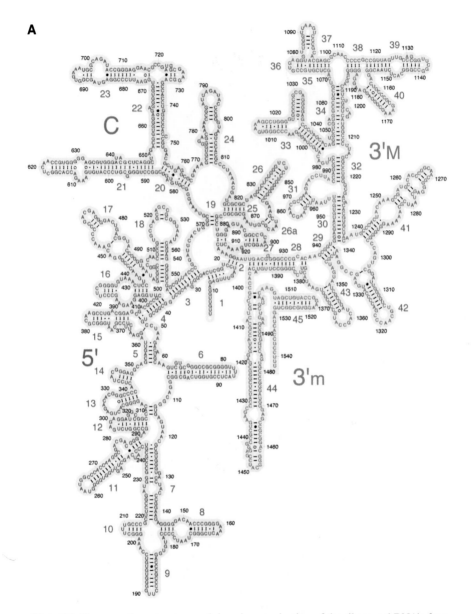

Plate 6.1 The secondary structure and domain organization of the ribosomal RNA's from *T. Thermophilus*: **A)** 16S RNA. The four domains are shown in different colors. The 5'-domain in blue, the central domain in purple, the 3'-major domain in red, and the 3'-minor domain in yellow. The helices are numbered from 1–45.

(Cont'd)

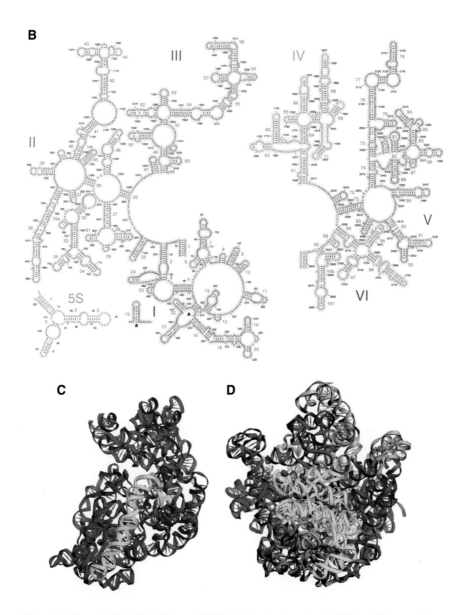

Plate 6.1 (Cont'd) **B)** 23S RNA and 5S RNA. The six domains (I–VI) in different colors. The helices are numbered 1–101. **C)** and **D)** The three-dimensional structures of the rRNAs of the small and large subunits, respectively. The coloring scheme is the same as in **A)** and **B)**. Notice in particular the long blue helix, the penultimate helix (h44) of the small subunit. While the RNA domains in the small subunit are separate folding units, the domains of the large subunit are intervowen. (Reprinted with permission from Yusupov *et al.* Crystal structure of the ribosome at 5.5Å resolution, *Science* **292**:883–896. Copyright (2000) AAAS.)

A

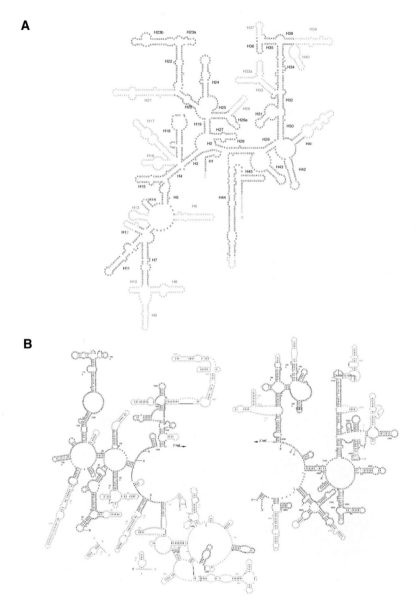

B

Plate 6.2 A) A comparison of the organization of 16S RNA from *T. thermophilus* with 12S of bovine mitochondria. **B)** A comparison of 23S RNA from *H. marismortui* with 16S RNA from human mitochondria. The pieces that are lacking in mitochondria are shown in red. (Reprinted with permission from Cavdar Koc, *et al*. The small subunit of the mammalian mitochondrial ribosome. Identification of the full complement of ribosomal proteins present, *J Biol Chem* **276**:19363–19374; and Cavdar Koc, *et al*. The large subunit of the mammalian mitochondrial ribosome. Analysis of the complement of ribosomal proteins present, *J Biol Chem* **276**:43958–43969. Copyright (2001) ASBMB.)

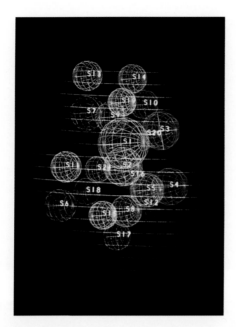

Plate 7.1 The distribution of proteins in the small subunit as observed from neutron scattering experiments (Capel *et al.*, 1987). (The illustration was kindly provided by Prof. P.B. Moore.)

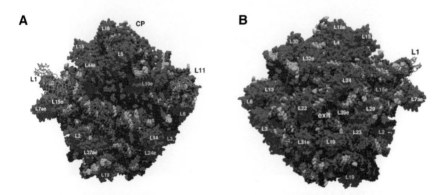

Plate 7.2 The 3-D structure of the large subunit (Nissen *et al.*, 2000). RNA is shown in red and white, whereas the proteins are blue. **A**) The subunit seen from the interface side. The deep canyon is the place for the acceptor ends for the tRNAs and at its center is the PTC situated. The proteins are dispersed towards the rim. **B**) The external side of the 50S subunit. The peptide exit channel is marked. The external side is richer in proteins than the interface side. (Reprinted with permission from Nissen *et al.* The structural basis of ribosome activity in peptide bond synthesis, *Science* **289**:920–930. Copyright (2000) AAAS.)

A

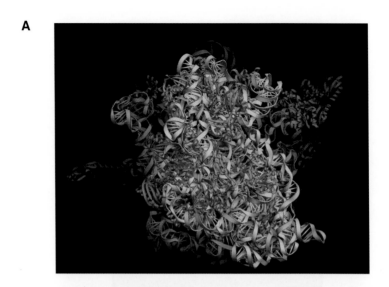

B

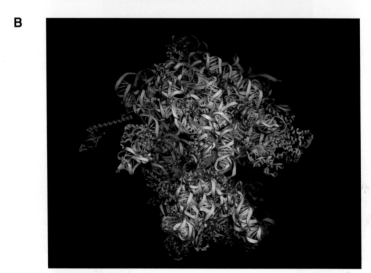

Plate 7.3 A) The 5.5Å-resolution structure of *T thermophilus* ribosomes studied using crystallography (Yusupov *et al.*, 2001). The three sites for tRNA molecules are shown from right to left (A-site yellow, P-site orange and E-site red). The RNA of the small subunit is shown in light blue and proteins in dark blue. The RNA of the large subunit is shown in light gray and the proteins are shown in purple. A. The same direction as Plate 2.1. **B**) The 70S ribosome seen from top. (Reprinted with permission from Yusupov *et al*. Crystal structure of the ribosome at 5.5Å resolution, *Science* **292**:883–896. Copyright (2000) AAAS.)

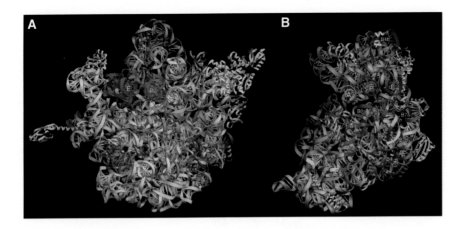

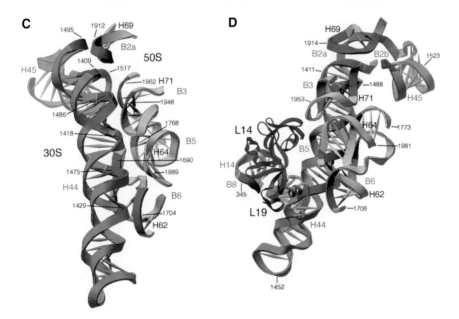

Plate 7.4 A) and **B)** The interface side of the two subunits with the inter-subunit bridges indicated. **C)** and **D)** Part of the interface between the two ribosomal subunits illustrating the multitude of contacts with helix 44 (h44) of the 16S RNA. (Reprinted with permission from Yusupov *et al.* Crystal structure of the ribosome at 5.5Å resolution, *Science* **292**: 883–896. Copyright (2000) AAAS.)

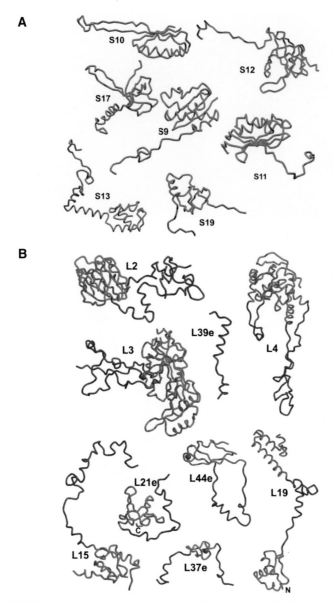

Plate 7.5 Many ribosomal proteins have a globular unit (green) with unusually extended tails or loops (red). Some (L39e) lack a globular part entirely. **A)** The small subunit proteins are from *T. thermophilus*. (Reprinted with permission from Ramakrishnan, V. and Moore, P.B. 2001. Atomic structure at last: the ribosome in 2000, *Curr Op Struct Biol* **11**: 144–154. Copyright (2001) Elsevier.) **B)** The large subunit proteins are from *H. marismurtui*. (Reprinted with permission from Ban *et al*. The complete atomic structure of the large ribosomal subunit at 2.4 Å resolution, *Science* **289**:905–920. Copyright (2000) AAAS.)

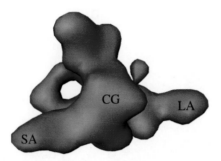

Plate 7.6 The structure of ribosomal protein S1 as seen by cryo-EM at 11.5 Å resolution. The LA extension from the central globular region may represent the N-terminal region and SA, the C-terminal part due to their interactions in the ribosome (Sillers & Moore, 1981). (Reprinted with permission from Sengupta *et al.* Visualization of protein S1 within the 30S ribosomal subunit and its interaction with mRNA, *Proc Natl Sci Acad USA* **98**: 11991–11996. Copyright (2001) National Academy of Sciences, USA.)

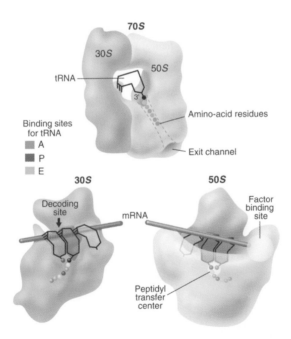

Plate 8.1 The functional sites of the ribosome. The small subunit (30S) is shown in purple and the large subunit (50S) in gray. The functional sites of the ribosome are built around two perpendicular channels. One, between the subunits, is the channel where mRNA and tRNAs move through and the other goes through the large subunit and is the exit channel for the nascent peptide. At the subunit interface side of the exit channel is the peptidyl transfer site situated. (Reprinted with permission from Liljas. Function is structure, *Science* **285**:2077–2078. Copyright (1999) AAAS.)

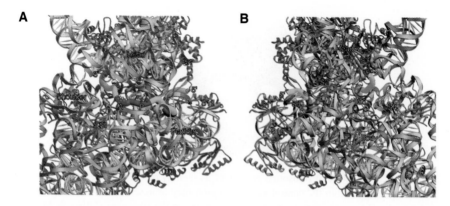

Plate 8.2 The binding site of the mRNA shown as an orange (A-site codon), red (P-site codon) and yellow ribbon on the small subunit. The groove on the subunit makes a tight interaction with the mRNA along its whole length. **A)** The narrow neck region of the small subunit shown from the interface side with the head up and the body down. The platform is on the right hand side and the shoulder, on the left hand side. Only nucleotides −1 to +7 of the mRNA, counting from the first nucleotide of the P-site codon, are exposed. **B)** The opposite side of the small subunit. The Shine-Dalgarno interaction between the mRNA and the 16S RNA is seen as the helical feature on the left hand side. (Reprinted with permission from Yusupova *et al*. The path of messenger RNA through the ribosome, *Cell* **106**:233–241. Copyright (2001) Elsevier.)

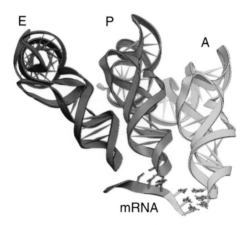

Plate 8.3 The three sites for tRNA molecules are shown from right to left (A-site yellow, P-site orange and E-site red) interacting with the mRNA as seen in the 5.5Å-resolution crystal structure of *T thermophilus* ribosomes. (Reprinted with permission from Yusupov *et al*. Crystal structure of the ribosome at 5.5Å resolution, *Science* **292**:883–896. Copyright (2000) AAAS.)

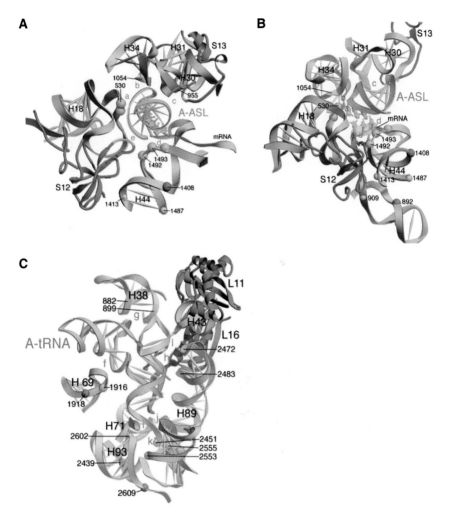

Plate 8.4 The interactions around the A-site tRNA. The 16S RNA is shown in cyan and proteins from the small subunit are shown in blue, whereas the 23S RNA is gray and large subunit proteins are purple. **A**) and **B**) The anticodon part and its proximity to h44. **C**) The acceptor interactions with the peptidyl transfer region. Two of the important intersubunit bridges, the A-site finger (H38) and H69 are seen. (Reprinted with permission from Yusupov *et al.* Crystal structure of the ribosome at 5.5Å resolution, *Science* **292**:883–896. Copyright (2000) AAAS.)

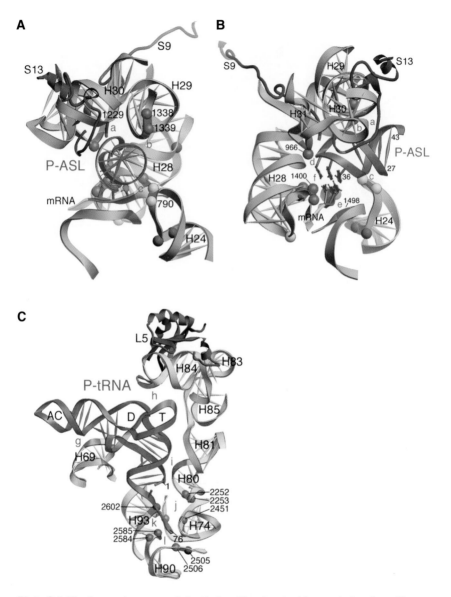

Plate 8.5 The interactions around the P-site. (Reprinted with permission from Yusupov *et al.* Crystal structure of the ribosome at 5.5Å resolution, *Science* **292**:883–896. Copyright (2000) AAAS.)

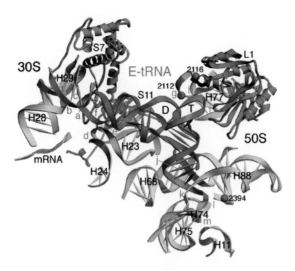

Plate 8.6 The interactions around the E-site tRNA. (Reprinted with permission from Yusupov *et al.* Crystal structure of the ribosome at 5.5Å resolution, *Science* **292**: 883–896. Copyright (2000) AAAS.)

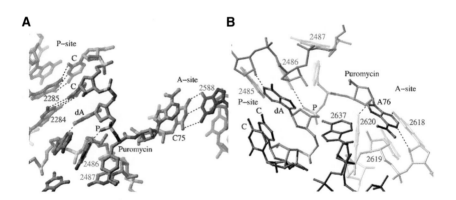

Plate 8.7 Two views of the binding of the CcdA-p-Puro to the peptidyl transfer site of *H. marismortui* large subunits. The ligand mimics the acceptor ends of both the A-site and P-site tRNAs at the stage it forms a tetrahedral intermediate like the one formed during peptidyl transfer. **A**) The interactions of C75 in the A-site with the A-loop (G2588) are seen on the right hand side. The interactions between the CCA and the P-loop (G2284 and G2285) are seen on the left hand side. **B**) The most important residues around the site of peptidyl transfer. (Reprinted with permission from Nissen *et al.* The structural basis of ribosome activity in peptide bond synthesis, *Science* **289**:920–930. Copyright (2000) AAAS.)

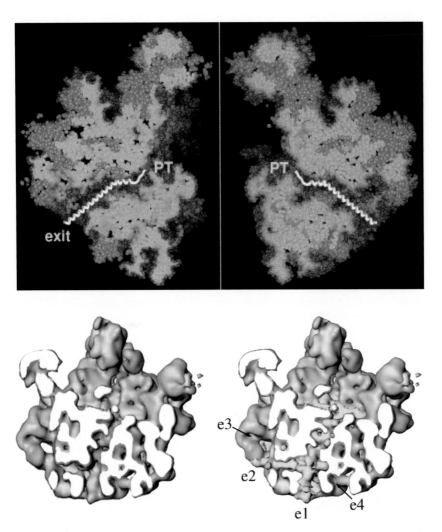

Plate 8.8 *Top*. The polypeptide exit tunnel of the large subunit seen by crystallography. (Reprinted with permission from Nissen *et al*. The structural basis of ribosome activity in peptide bond synthesis, *Science* **289**:920–930. Copyright (2000) AAAS.) The 23S RNA is shown in gray, whereas the proteins are green. The PTC is indicated in the upper picture as PT and a red marker. *Bottom*. The tunnel is shown as seen by cryo-EM. (Reprinted with permission from Gabashvili *et al*. The polypeptide tunnel system in the ribosome and its gating in erythromycin resistance mutants of L4 and L22, *Mol Cell* **8**:181–188. Copyright (2001) Elsevier.)

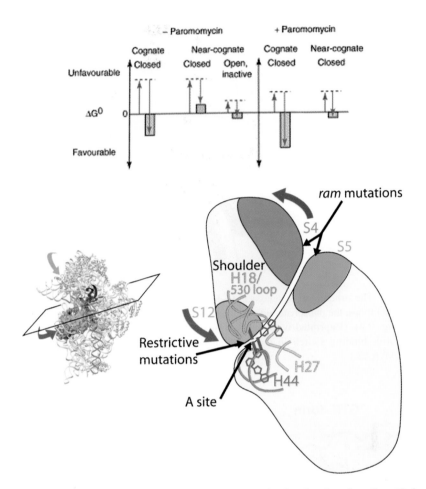

Plate 8.9 The conformational changes of the small subunit related to decoding (Ogle *et al.*, 2003). The shoulder moves in relation to the rest of the subunit. The section (*left*) through the subunit shows the two mobile parts (*right*). The decoding site is located on the left hand opening of the divide of the subunit. A cognate interaction of codon and anti-codon causes the reorientation and closing of the subunit as indicated by the red arrows. The binding of streptomycin at this site causes a closing of the subunit, leading to errors in decoding. Restrictive mutations in S12 (yellow) shift the balance to a more open subunit and fewer decoding errors. *Ram* mutations in S4 and S5 on the external side of the subunit shift the balance towards a closed conformation of the subunit and a lower fidelity of translation. (Reprinted with permission from Ogle *et al.* Insights into the decoding mechanism from recent ribosome structures, *Trends Biochem Sci* **28**:259–266. Copyright (2003) Elsevier.)

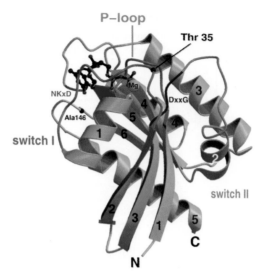

Plate 9.1 The structure of the classical G-domain with a bound GTP analogue. The three important loops, the phosphate biding loop (P-loop) and Switch I and II are indicated (see also Fig. 9.4). (Reprinted with permission from Vetter & Wittinghofer. The guanine nucleotide-binding switch in three dimensions, *Science* **294**:1299–1304. Copyright (2001) AAAS.)

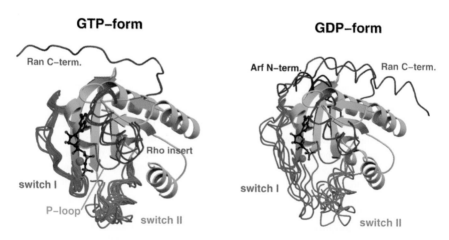

Plate 9.2 Comparison of the structure of G-proteins in the GTP and GDP conformations. In the GTP form, the Switch I and Switch II loops from all proteins are seen to have very similar conformations, whereas in the GDP form, they differ substantially. (Reprinted with permission from Vetter & Wittinghofer. The guanine nucleotide-binding switch in three dimensions, *Science* **294**:1299–1304. Copyright (2001) AAAS.)

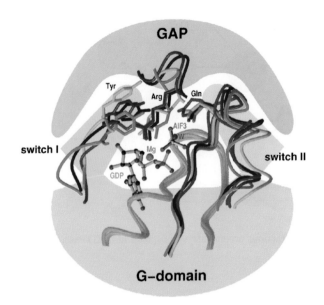

Plate 9.3 Several examples of interactions with the GAP-s (GTPase activating protein). Two crucial residues for GTP hydrolysis is a Gln from switch II and an Arg called the Arg-finger that comes either from the GAP (*trans*) or from switch I (*cis*). (Reprinted with permission from Vetter & Wittinghofer. The guanine nucleotide-binding switch in three dimensions, *Science* **294**:1299–1304. Copyright (2001) AAAS.)

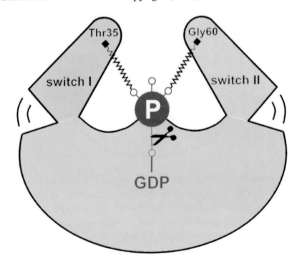

Plate 9.4 A key role of Switch I and II of GTPases is to identify the γ-phosphate of the GTP. They have defined positions in complex with GTP, but alter their conformations when the GTP is hydrolyzed. This conformational change is transmitted to the whole protein. (Reprinted with permission from Vetter & Wittinghofer. The guanine nucleotide-binding switch in three dimensions, *Science* **294**:1299–1304. Copyright (2001) AAAS.)

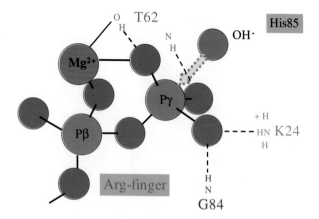

Plate 9.5 The interactions around the γ-phosphate in a tGTPase. The example is taken from EF-Tu.

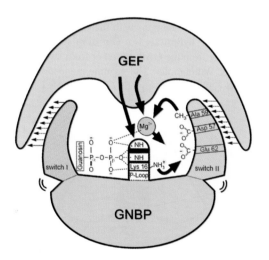

Plate 9.6 The interaction of switch I and II with the GEF (G nucleotide exchange factor). The binding of the GDP is destabilized by specific interactions releasing the magnesium ion and changing the conformation of the P-loop. (Reprinted with permission from Vetter & Wittinghofer. The guanine nucleotide-binding switch in three dimensions, *Science* **294**: 1299–1304. Copyright (2001) AAAS.)

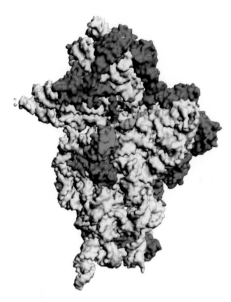

Plate 9.7 The structure of IF1 when bound to the 30S subunit seen from the side of the ribosomal interface (Carter *et al.*, 2001). The 16S RNA is shown in beige and the ribosomal proteins in blue. IF1 is shown in brown and a part of the mRNA is shown in green. (The drawing was kindly produced by Dr. M. Fodje using PyMOL: DeLano, W.L. The PyMOL Molecular Graphics System (2002) http://www.pymol.org.)

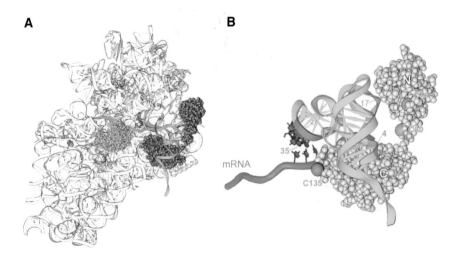

Plate 9.8 The binding of IF3, IF1 and the initiator tRNA on the small subunit. The factors occupy the E- and A-sites, respectively, surrounding and guiding the initiator tRNA to the P-site. (Reprinted with permission from Dallas & Noller. Interaction of translation initiation factor 3 with the 30S ribosomal subunit, *Mol Cell* **4**:855–864. Copyright (2001) Elsevier.)

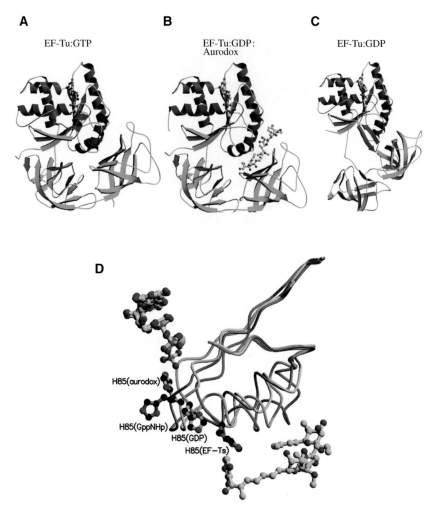

Plate 9.9 Three conformations of EF-Tu: **A**) with GTP; **B**) with GDP and the antibiotic aurodox; and **C**) with GDP. **D**) The dynamics of His84 in different conformations of EF-Tu. This residue is suggested to activate the water that attacks the γ-phosphate. (Reprinted with permission from Vogeley *et al.* Conformational change of elongation factor Tu (EF-Tu) induced by antibiotic binding. Crystal structure of the complex between EF-Tu:GDP and aurodox, *J Biol Chem* **276**:17149–17155. Copyright (2001) ASBMB.)

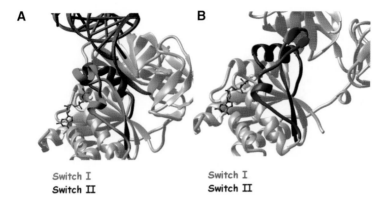

Switch I
Switch II

Switch I
Switch II

Plate 9.10 **A**) The conformation of Switch I and II in EF-Tu in the GTP conformation. **B**) The conformation of Switch I and II in EF-Tu in the GDP conformation. (The illustration was kindly provided by M. Laurberg (2002).)

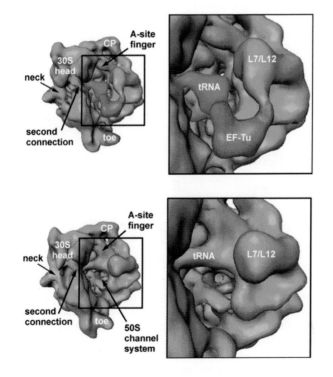

Plate 9.11 The ribosome with the ternary complex of EF-Tu:GDPNP:aa-tRNA bound. The G-domain interacts with the large subunit in particular with the SRL and the base of the L12 stalk. (Reprinted with permission from Stark *et al.* Visualization of elongation factor Tu on the Escherichia coli ribosome, *Nature* **389**:403–406. Copyright (1997) Nature.)

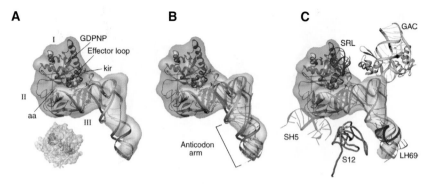

Plate 9.12 The structure of the kirromycin inhibited ternary complex when bound to the ribosome. **A**) The crystallographic structure of the ternary complex fitted into the cryo-EM density. **B**) The RNA structure is adjusted to fit the density. **C**) Different ribosomal interactions are shown. (Reprinted with permission from Valle *et al.* Incorporation of aminoacyl-tRNA into the ribosome as seen by cryo-electron microscopy, *Nat Struct Biol* **10**:899–906. Copyright (2003) Nature.)

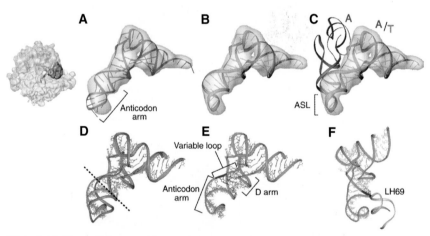

Plate 9.13 The structure and interactions of the tRNA in the A/T site and the A-site. **A**) The anticodon region fits poorly. **B**) The kink between the anticodon and D-stems is made to adjust to the density. **C**) The difference in structure between the A/T- and A-site tRNAs is shown. **F**) The interaction between the tRNA and H69 may be involved in the induction of the GTP hydrolysis by EF-Tu. (Reprinted with permission from Valle *et al.* Incorporation of aminoacyl-tRNA into the ribosome as seen by cryo-electron microscopy, *Nat Struct Biol* **10**:899–906. Copyright (2003) Nature.)

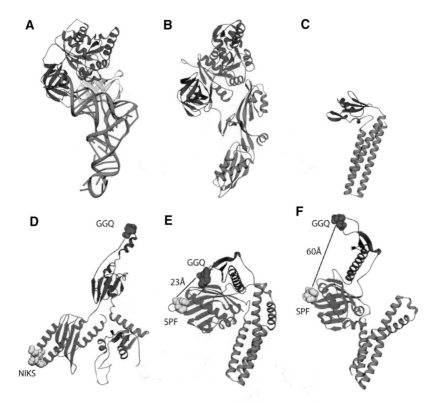

Plate 9.14 Some examples of tRNA mimicry. Not all of them correspond with a predicted functional relationship. **A**) The ternary complex of EF-Tu:GTP and aminoacyl-tRNA. **B**) EF-G. **C**) RRF. **D**) eRF1. **E**) RF2 as observed in the crystal structure. **F**) RF2 as interpreted from the cryo-EM. (Reprinted with permission from Brodersen & Ramakrishnan. Shape can be seductive, *Nature Struct Biol* **10**:78–80. Copyright (2003) Nature.)

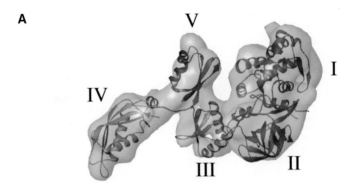

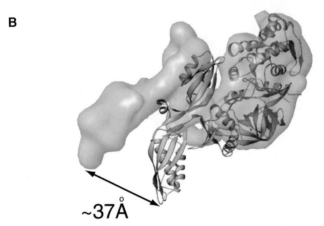

Plate 9.15 Two conformations observed for EF-G. The density shows EF-G when bound to the ribosome with GDPNP. **A**) The crystal structure of the domains is fitted into the density. **B**) The discrepancy is shown between the cryo-EM density and the crystal structure of a mutant presented by Laurberg *et al.* (2000). (Reprinted with permission from Valle *et al.* Locking and unlocking of ribosomal motions, *Cell* **114**:123–134. Copyright (2003) Elsevier.)

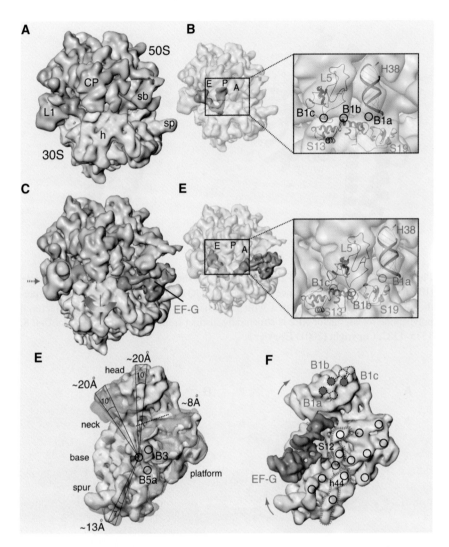

Plate 9.16 The ratchet-like movement induced by the binding of EF-G*GDPNP. **A**) and **B**) show the situation before binding, including the inter-subunit bridges and **C**) and **D**) the situation when EF-G has bound. The rotation of the small subunit is shown in **E**) and the location of EF-G in relation to the 30S subunit is shown in **F**). The inter-subunit bridges that have moved are shown in red in **F**). Their original positions are shown as dashed black circles. (Reprinted with permission from Valle *et al.* Locking and unlocking of ribosomal motions, *Cell* **114**:123–134. Copyright (2003) Elsevier.)

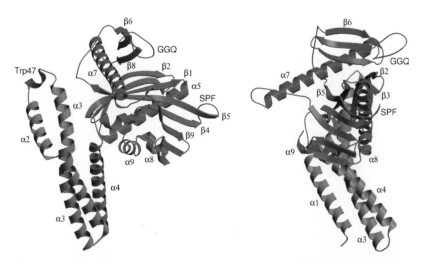

Plate 9.17 The structure of RF2. The domains are shown in brown (I), blue (II), purple (III) and green (IV). The distance between the GGQ and SPF motifs is shorter than predicted from function. (Reprinted with permission from Vestergaard *et al.* Bacterial polypeptide release factor RF2 is structurally distinct from eucaryotic eRF1, *Mol Cell* **8**: 1375–1382. Copyright (2001) Elsevier.)

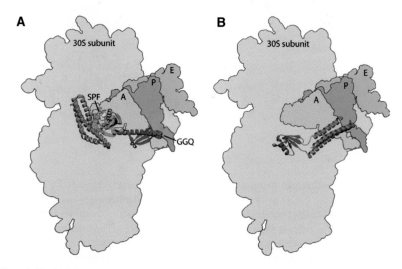

Plate 9.18 A) The structure of RF2 as seen when bound to the small subunit (Brodersen & Ramakrishnan, 2003). The color code is the same as in Plate 9.14. Domain I is close to the factor binding site while the GGQ region reaches into the PTC area. **B**) The binding of RRF as determined through labeling (Lancaster *et al.*, 2002). RRF also binds in the same general manner. (Reprinted with permission from Brodersen & Ramakrishnan. Shape can be seductive, *Nature Struct Biol* **10**:78–80. Copyright (2003) Nature.)

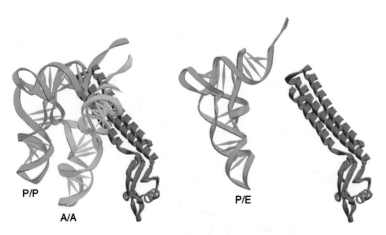

P/P

A/A

P/E

Plate 9.19 The binding site for RRF on the ribosome in relation to the tRNA sites as identified by chemical labeling. (Reprinted with permission from Lancaster *et al.* Orientation of ribosome recycling factor in the ribosome from directed hydroxyl radical probing, *Cell* **111**:129–140. Copyright (2003) Elsevier.)

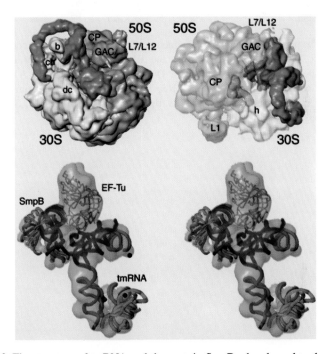

Plate 9.20 The structure of tmRNA and the protein SmpB when bound to the ribosome. (Reprinted with permission from Valle *et al.* Visualizing tmRNA entry into a stalled ribosome, *Science* **300**:127–130. Copyright (2003) AAAS.)

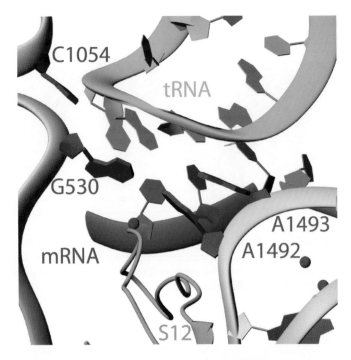

Plate 11.1 The participation by the ribosome in differentiating the correct codon from the nearly correct. The mRNA (polyU, purple), the tRNA^{Phe} (yellow), parts of the 16S RNA (gray) and ribosomal protein S12 (brown). The bases G530, A1492 and A1493 interact with the codon anti-codon base-pairs. (Reprinted with permission from Ogle *et al.* Recognition of cognate transfer RNA by the 30S ribosomal subunit, *Science* **292**: 897–902. Copyright (2001) AAAS.)

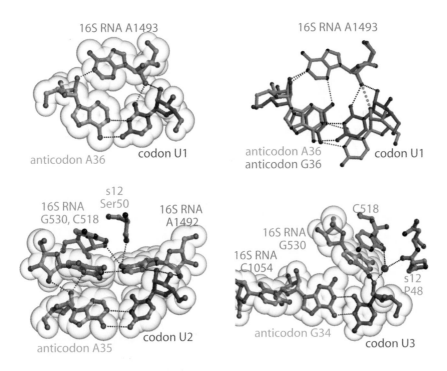

Plate 11.2 The participation of the small subunit in the decoding of a cognate from a near-cognate tRNA (Ogle *et al.*, 2001; 2002; 2003). (Upper right reprinted with permission from Ogle *et al*. Selection of tRNA by the ribosome requires a transition from an open to a closed form. *Cell* **111**:721–732. Copyright (2002) Elsevier. The others reprinted with permission from Ogle *et al*. Recognition of cognate transfer RNA by the 30S ribosomal subunit. *Science* **292**:897–902. Copyright (2001) AAAS.)

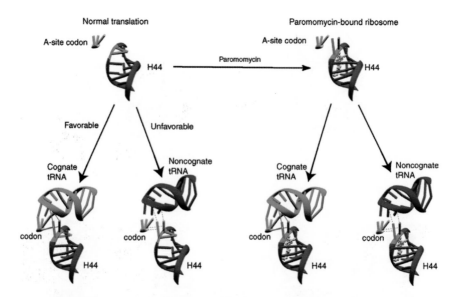

Plate 11.3 A1492 and A1493 participate in the decoding of the mRNA. In case of a correct match they can interact with the Watson-Crick base-pair; in case of a noncognate base-pair they do not interact. When paromomycin is bound, these bases adopt the conformation they normally have for a cognate codon anticodon interaction. This leads to a stabilization of the tRNA interaction and possibly the signal for GTP hydrolysis. (Reprinted with permission from Ramakrishnan, V. and Moore, P.B. 2001. Atomic structure at last: the ribosome in 2000, *Curr Op Struct Biol* **11**:144–154. Copyright (2001) Elsevier.)

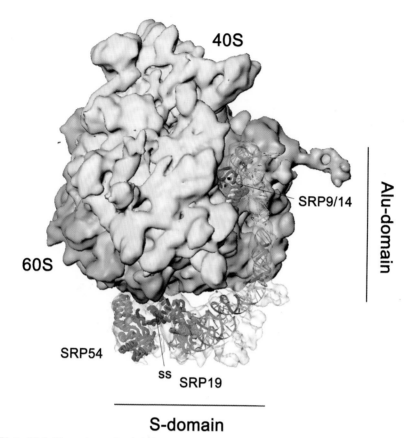

40S

60S

SRP9/14

Alu-domain

SRP54

ss **SRP19**

S-domain

Plate 12.1 The eukaryotic signal recognition particle bound to the 80S ribosome (Halic *et al.*, 2004). In bacteria, the 4.5S RNA corresponds to the S-domain and the single protein Ffh corresponds to SRP54. The eukaryotic Alu-domain interacts with the binding site for tGTPases. (The illustration was a kind gift from Dr. R. Beckmann.)

9

The Catalysts — Translation Factors

In vivo, protein synthesis is catalyzed by a number of translation factors most of which bind transiently to the ribosome during the different phases of translation (see reviews by Lipmann, 1969; Kaziro, 1978). It is an important issue whether protein synthesis is a spontaneous process or whether an extra input of energy is needed. Experiments have been reported where protein synthesis was performed without translation factors (see review by Spirin, 1978). Slight impurities of the translation factors could have given the low rates of synthesis observed, but since different agents inhibit normal translation and factor-free translation, this possibility could be excluded. GTP was not needed for factor-free protein synthesis and no GTP hydrolysis was observed. Furthermore, non-cleavable GTP analogues did not inhibit the process. Thus, it seems that protein synthesis is a spontaneous process.

Most translation factors from bacteria have been extensively studied and this has led to a proposed detailed mechanism for bacterial translation. However, the archaeal and eukaryotic translation factors are numerous and have only been partly investigated, and therefore a detailed description of protein synthesis in these organisms is lacking. An attempt

Table 9.1 Bacterial Translation Factors

Protein	Universally Conserved*	GTPase	Role
Initiation			
IF-1	X		Assists IF-2 in initiation
IF-2	X	X	Binds 50S subunit to initiation complex
IF-3			Assists in dissociation of subunits
Elongation			
EF-P	X		Homologous to eIF5A. Role unknown
EF-Tu	X	X	Binds aatRNA to A-site
SelB		X	Binds SeCys-tRNA to A-site
EF-Ts			Nucleotide exchange factor for EF-Tu
EF-G	X	X	Catalyzes translocation of peptidyl tRNA from A- to P-site
Termination			
RF-1,2			Recognizes termination codons and releases peptide from P-site tRNA
RF-3		X	Releases RF-1,2 from ribosome
Recycling			
RRF			Dissociates the terminated ribosomes

*Pandit & Srinivasan, 2003.

will be made here to discuss several of these factors. Table 9.1 gives a summary of the key bacterial translation factors. In addition, there are a number of other proteins, different GTPases (see Section 9.7) and ribosome rescue proteins (see Section 9.8) that interact with the ribosome. Furthermore, there are chaperones and signal recognition proteins that participate in the folding and export of different proteins that also interact with the ribosome (see Chapter 12).

9.1 THE GTPASES

Several of the translation factors bind and hydrolyze GTP. Thus, they belong to the family of GTPases or G-proteins with several sequence motifs (Bourne *et al.*, 1990; 1991; Vetter & Wittinghofer, 2001). This family of proteins is structurally and functionally related to a large family of ATP hydrolyzing enzymes (Leipe *et al.*, 2002). The translation factors that belong to this family are IF-2, EF-Tu, SelB, EF-G and RF-3. As a group, they can be called the translation GTPases (tGTPases). Some organisms do not have all the genes for these tGTPases (Table 9.2; Pandit & Srinivasan, 2003). Thus, RF3 seems to be dispensable. For other tGTPases, there are sometimes two or three genes.

The GTP-binding domain of the GTPases is called the G-domain (Plate 9.1). The G-domain normally has 160–200 amino acid residues and is a special version of the Rossmann fold with a large central parallel β-sheet surrounded by α-helices (Fig. 9.1; Kjeldgaard & Nyborg, 1992).

The core of the G-domain is well conserved in all G-proteins, but variations in the domain are frequent. A number of consensus elements or motifs are characteristic (Walker *et al.*, 1982; Saraste *et al.*, 1990; Bourne 1990; 1991; Table 9.3). Amino acid residues of the consensus sequences are primarily involved in the binding of the nucleotide and a magnesium ion. In addition, they can sense whether the bound nucleotide is a di- or triphosphate.

There are G-proteins composed only of one domain, but many are multi-domain proteins (Plate 9.1; Vetter & Wittinghofer, 2001). All tGTPases are multi-domain proteins (Fig. 9.2). They all have at least two domains in common, the G-domain and the subsequent domain, usually called domain II (Ævarsson, 1995). Domain II is a unique β-barrel. There are a number of G-proteins that are not translation factors but nevertheless may interact with the ribosome in different states (see Section 9.7; Leipe *et al.*, 2002). Most of these additional G-proteins that interact with the ribosome lack domain II. Two exceptions are Tet(M) and LepA. Tet(M) is a member of a protein family that causes resistance against the antibiotic tetracycline (see Sections 9.7 and 10.6; Burdett, 1996, Levy *et al.*, 1999). The role of LepA remains unclear.

Table 9.2 Number of Genes for tGTPases in Archaea and Bacteria (Pandit and Srinivasan, 2003*)

	IF2#	EF-Tu	SelB	EF-G	RF3	LepA	TypA/BipA
Archaea	2	1–2	0(1)	1	—	—	—
Bacteria	1	1–2	0–1	1–3	0–1	1	0–1

*42 genomes were searched.
#In archaea, there is one protein of the eIF2γ type and one of the IF2 type.

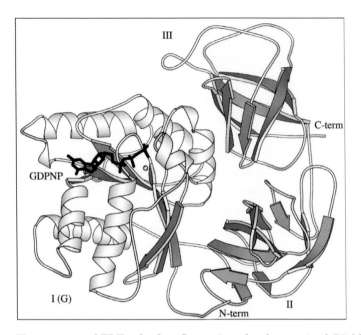

Fig. 9.1 The structure of EF-Tu, the first G-protein to be characterized (Kjeldgaard *et al.*, 1993). Here it is in complex with a GTP analogue, GDPNP, and thus in the closed form. The domains are denoted I (the G-domain), II and III. A magnesium ion is seen as a small grey circle between the β- and γ-phosphates. The figure has kindly been provided by Dr. M. Fodje using the program MOLSCRIPT (Kraulis, 1991).

Table 9.3 Consensus Elements of the tGTPases and LepA (all sequence numbers correspond to the *E. coli* proteins)

Element	PO_4-loop	Switch I	Switch II		
Alternative name*	G1	G2	G3	G4	G5
Sequence	GXXXXGKS/TS/T	RGITI	DXXGH	NKXD	GSAL/K
Role	Interactions with α- and β-phosphates	Binding of γ-phosphate and Mg^{2+}	Indirect Mg^{2+} binding	Recognition of the G nucleotide	Binding of nucleotide
IF2	398–406	422–426	444–449	498–501	533–536
EF-Tu	18–26	58–62	80–85	135–138	172–175
SelB	7–15	34–38	57–62	112–115	194–197?
EF-G	19–27	61–65	83–88	137–140	261–264
Tet(O)	12–20	52–56	80–85	128–131	165–168?
RF-3	20–28	66–70	88–93	142–145	257–260
LepA	11–19	50–54	77–82	131–134	161–164?

*Bourne (1990).

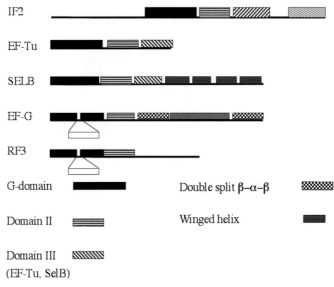

Fig. 9.2 The domain arrangement of the tGTPases. The common domains are the G-domain and domain II. Normally these domains occur at the N-terminus, but in IF2 some species have a long N-terminal addition of unknown structure. These versions of IF2 are called IF2α. IF2 without the N-terminal extension is called IF2β. The G-domain of EF-G has an insert called G'. The structure of RF3 is not known, except for the recognition of the sequence relationship of domains G, G' and II to other tGTPases.

A central event for the tGTPases is obviously the GTP hydrolysis on the ribosome. All G-proteins undergo conformational changes associated with GTP hydrolysis (Plate 9.2). The conformational changes of the tGTPases catalyze different steps of protein synthesis, primarily through their interactions with tRNA on the ribosome.

Most G-proteins act as molecular switches. They have an active ON state in complex with GTP (Bourne *et al.*, 1990; Wittinghofer & Pai, 1991; Vetter & Wittinghofer, 2001). In this state, they bind to their receptor. After hydrolysis of the GTP to GDP and inorganic phosphate (Pi), they adopt the OFF state and fall off the receptor. The steps catalyzed by the G-proteins are irreversible due to the GTP hydrolysis. The normal functional cycle for tGTPases is shown in Fig. 9.3.

G-proteins are incomplete enzymes. They have a low intrinsic GTPase activity (Kaziro, 1978). Therefore, they need to interact with the appropriate components of the cell to become activated. Usually proteins called

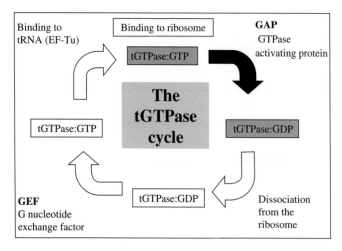

Fig. 9.3 The common elements of the functional cycle for the tGTPases. Two of the four steps are on the ribosome (*dark background*) and two are off the ribosome (*white background*).

GAP (GTPase Activating Protein) activate the GTP hydrolysis of G-proteins. Obviously, there are multiple possibilities to control this activity (Bourne *et al.*, 1990; 1991). In the presence of ribosomes but without other components of the system, some of the tGTPases are able to hydrolyze GTP in a manner that is uncoupled from protein synthesis (Arai & Kaziro, 1975).

When the G-proteins with a bound GTP molecule are in the ON state, they can bind to a receptor or effector (Bourne *et al.*, 1990; Vetter & Wittinghofer, 2001; Fig. 9.3). This interaction may lead to an interaction with GAP that induces the G-protein to hydrolyze its bound GTP molecule (Bourne *et al.*, 1990; 1991). After the GTP hydrolysis, the conformation of the G-protein changes in such a way, that it dissociates from the effector.

A significant number of the related ATPases have mechanochemical functions or are the so-called motor proteins. The energy from the ATP hydrolysis is used to drive different molecular processes (Goody & Hoffmann-Goody, 2002). Obviously, this involves conformational changes of the ATPase. Do some of the tGTPases function as motor proteins as well (Cross, 1997)? A characteristic difference between a motor protein and a molecular switch is in what state the nucleotide is hydrolyzed. Thus, a classical motor protein hydrolyzes its ATP molecule before the conformational change and before the work is done. However, a molecular switch induces a process by binding to its receptor. When this process is completed, the nucleotide is hydrolyzed, leading to a loss of affinity of the molecular switch for the receptor (Spirin, 2002).

The Consensus Elements. Nucleotide and Mg^{2+} Binding

Five loops in the G-domains containing highly conserved sequences, the so-called consensus elements, primarily connect the C-terminal side of the β-strands to the α-helices and constitute the binding site for the nucleotide and a magnesium ion (Table 9.3; Figs. 9.4–9.6). Sometimes these consensus elements are called G1 to G5 (Bourne, 1990). The role of some of the individual residues can also be identified from Plate 9.3 (Vetter & Wittinghofer, 2001). Generally, the three first consensus elements, the PO_4-loop (normally called the P-loop, but since this term is also used to denote the interaction of the CCA-end of the P-site tRNA, we use PO_4-loop), switches I and II, interact with the GTP/GDP phosphates and the magnesium ion and the two last consensus elements provide the selection of G-nucleotides (Table 9.3; Fig. 9.4). The magnesium ion is an essential cofactor for GTP hydrolysis (Kjeldgaard *et al.*, 1996).

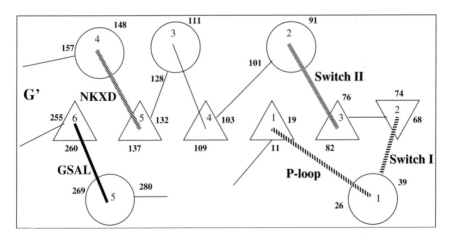

Fig. 9.4 The organization of the G-domain (EF-G). The circles represent α-helices and the triangles, β-strands. The numbers on the secondary elements indicate the order number of helices and strands, respectively. The numbers on the side represent the amino acid sequence number entering and exiting from the secondary structure element. The five loops with the consensus elements are highlighted. The PO_4-loop (consensus **G/AHVDAGKT/ST**), switch I (**RGITI**) and switch II (**DXPGHXDF/Y**) are essential for interaction with phosphates and recognition of γ-phosphate. The **NKXD** and **GSAL** loops identify the G-nucleotides.

The first loop (G1) is called the phosphate binding or PO_4-loop (residues 19–27 in EF-G). It folds around the α- and β-phosphate moieties of the G nucleotide. Residues 22–27 of EF-G all make hydrogen bonds to the GDP phosphates (Fig. 9.5; Al-Karadaghi et al., 1996). Switch I (G2), or the effector region (residues 38–68 in EF-G) and switch II (G3, residues 83–101 in EF-G) are loops that primarily interact with the β- and γ-phosphates of the nucleotide through the magnesium ion (Fig. 9.6). The conformations of these loops respond to whether a GTP or a GDP molecule is bound (Plate 9.4), or whether the nucleotide-binding site is empty. The effector loop (switch I) is involved in receptor binding and switches conformation drastically between the GDP and GTP states (Plate 9.2). Switch II relays the state of the bound nucleotide to the conformation of the multidomain protein.

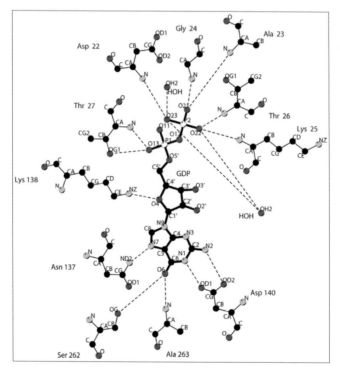

Fig. 9.5 The interactions between the GDP and *T. thermophilus* EF-G. (Reprinted with permission from Al-Karadaghi *et al.*) The structure of elongation factor G in complex with GDP: conformational flexibility and nucleotide exchange. *Structure* 4, 555–565. Copyright (1996) Elsevier.

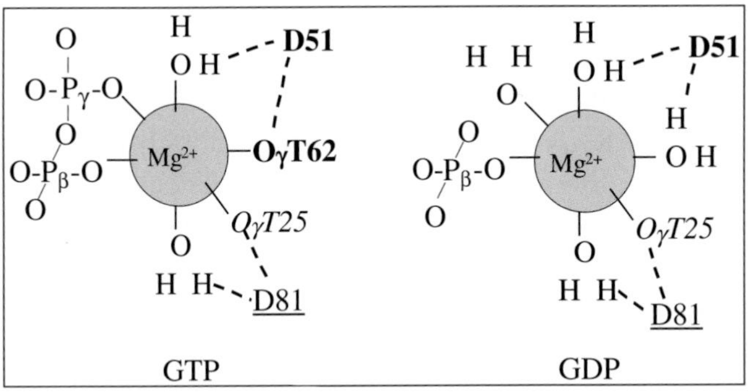

Fig. 9.6 The coordination around the magnesium ion in the GTP and GDP states of EF-Tu. The γ- and β-phosphates of the nucleotide bind to the magnesium ion. Thr25 of the PO_4-loop is a constant ligand to the magnesium ion and Asp51 of switch I and Asp81 of switch II constantly interact with magnesium ligands. Thr62 of switch I is an added ligand to the magnesium in the GTP state.

Ribosome Binding

The tGTPases bind to overlapping binding sites on the ribosome (Heimark *et al.*, 1976; see Section 8.5). The two domains that are common between these proteins (G and II) are expected to interact similarly with the ribosome. A number of studies have provided a gross picture as well as some detailed insights. The interaction is best characterized for EF-Tu and EF-G by cryo-EM (Stark *et al.*, 1997; 2000; 2002; Agrawal *et al.*, 1998; 1999; Valle *et al.*, 2002, 2003a,b). These studies show that the G-domains interact primarily with the SRL and the GAR of the large subunit at the base of the L12 stalk (see Section 8.5; Fig. 8.4; Plate 8.1). Domain II interacts with the small subunit in the general area of the shoulder and h5 (Wilson & Noller, 1998). At the same time, the ASL of the tRNA in the ternary complex, EF-Tu:GTP:aa-tRNA or domain IV of EF-G interact with the decoding region of the small subunit.

Which tGTPase Should Bind?

With one binding site for tGTPases on the ribosome, what makes the right tGTPase bind at the right moment? Without a control mechanism it would lead to an idling of the GTPases and a significant waste of GTP. Recent investigations show that the state of the ribosome and particularly the presence and position of the peptidyl-tRNA controls what GTPase binds (Valle *et al.*, 2003b; Zavialov & Ehrenberg, 2003). The two main conformations, pre- and post-translocation, provide useful information as to which factor should bind.

If a peptidyl-tRNA is bound to the P-site, only EF-Tu can bind to the ribosome while the other tGTPases are prevented from binding (Zavialov & Ehrenberg, 2003). When, however, the P-site has an fMet-tRNA bound, IF2 binds (Table 9.4). The initiation factors must dissociate to allow EF-Tu to bind its aminoacyl-tRNA to the A-site. The binding of EF-Tu does not depend on whether there is a deacylated tRNA in the P-site (Zavialov & Ehrenberg, 2003). The binding is only controlled by the codon in the A-site.

EF-G is prevented from repeatedly binding to ribosomes and hydrolyzing its GTP by the peptidyl tRNA it translocates to the P-site. It is not understood how EF-G can discriminate between a peptidyl-tRNA and a deacylated tRNA in the P-site and bind only to the later (Zavialov & Ehrenberg, 2003; Valle *et al.*, 2003b). Finally, RF3 binds and hydrolyzes its GTP if RF1/2 is bound at a stop codon and the nascent peptide is hydrolyzed and released from the tRNA in the P-site (Table 9.4).

The Induction of GTP Hydrolysis — The Ribosomal GAP

The tGTPases do not normally hydrolyze their bound GTP molecules, but do so when bound to the ribosome and in response to specific states of the ribosome. Generally, as part of inducing the GTP hydrolysis, an Arg-finger of the GAP (trans) interacts with the phosphates to stabilize the transition state (Plate 9.3; Vetter & Wittinghofer, 2001). The question is whether the ribosome supplies this Arg-finger or whether it, by other means, induces an active GTPase conformation in the factor. Studies of

Table 9.4 Ribosomal States Required for the Binding and GTP-hydrolysis by the tGTPases (from Zavialov & Ehrenberg, 2003; Valle *et al.*, 2003b)

Factor	IF2	EF-Tu	EF-G	RF3
Ribosomal state	30S	Post-translocation	Pre-translocation	Pre-translocation
A-site	IF1	—	pp-tRNA	RF1/2
P-site	fMet-tRNA	pp-tRNA	tRNA	tRNA
E-site	IF3-NTD	tRNA	—	?

EF-Tu suggest that all residues needed for catalysis are available in the tGTPases themselves. EF-Tu is able to hydrolyze its bound GTP in the presence of kirromycin but without ribosomes (see Section 10.7; Wolf *et al.*, 1977). Probably kirromycin induces a conformation in EF-Tu similar to the one caused by the ribosome. The ribosomal GAP may not contain an Arg-finger but induce an internal Arg-finger (*cis*) to interact with the phosphates. There is no reason to suspect that the other tGTPases function in a different manner than EF-Tu.

The nature of the ribosomal GAP has been investigated but without a final conclusion. One would assume that all tGTPases are activated by the same GAP. SRL is close to the nucleotide-binding site for bound factors, but Valle *et al.* (2003a; see Section 8.5) have identified a more dynamic relationship of the GAR to the bound factor. GAR is the base of the L12 stalk of the large subunit (Valle *et al.*, 2003a). The protein L12 is a ribosomal component closely associated with GTP hydrolysis. Removal of L12 leads to a very significantly reduced GTPase activity (Kischa *et al.*, 1971; Fakunding *et al.*, 1973; Wahl & Möller, 2002; Mohr *et al.*, 2002). The removal of the whole protein may not only lead to a loss of the functional components, but also induce conformational changes in the GAR (see Section 7.4). Furthermore, there are observations that the isolated protein can activate a low level of GTPase activity in the elongation factors (Donner *et al.*, 1977; Savelsbergh *et al.*, 2000b). The GTPase transition state can be studied by the induction of fluorescence in a GTP derivative by AlF_4. This works when L12 is present in the ribosome, but not when L12 is removed (Mohr *et al.*, 2002). Thus, L12 stabilizes the transition state (Mohr *et al.*, 2002) and may tentatively be called the ribosomal GAP.

The ribosome contains two dimers of L12 (see Section 7.4). The protein is highly flexible as has been revealed by NMR (Morrison *et al.*, 1977; Tritton, 1980; Gudkov *et al.*, 1982; Cowgill *et al.*, 1984; Bushuev *et al.*, 1989; Bocharov *et al.*, 2004; Mulder *et al.*, 2004) and mild proteolysis (Liljas *et al.*, 1978; Wahl *et al.*, 2000). The number of flexible L12 monomers varies, depending on whether the ribosomes are free from factors or whether EF-Tu or EF-G is bound to the ribosome with GDPNP or with the antibiotics kirromycin and fusidic acid, respectively (see Section 11.4; Gudkov *et al.*, 1982; Gudkov & Bubunenko, 1989; Gudkov, 1997). A variation in stalk structure is also observable from cryo-EM studies at lower resolution of ribosomes in various states (Stark *et al.*, 1997; Agrawal *et al.*, 1998; 1999). The flexibility of L12 depends on the flexible hinge between its

N- and C-terminal domains (Bushuev *et al.*, 1989). The length of the hinge can be decreased to 9 amino acid residues without the ribosomes loosing significant amounts of activity (Gudkov *et al.*, 1991; Bubunenko *et al.*, 1992). No surface of interaction of the L12 CTD and elongation factors has been identified despite investigations of a number of mutations of surface residues. L12 has only one arginyl residue, Arg74. Mutations of this residue have no convincing effect on the induction of the GTPase activity of EF-G (Savelsbergh *et al.*, 2000b).

Mutations of Lys70 of L12 lead to a slow release of the inorganic phosphate from the GTP hydrolysis by EF-G (Savelsbergh *et al.*, 2003). Probably this does not correlate with the observation of a bound sulphate ion to the crystal structure of the C-terminal domain of L12 (Leijonmarck *et al.*, 1980). The sulphate ion was found to interact with Lys65 and some main chain atoms.

Even though the isolated L10:L12$_4$ complex is quite stable, observations suggest that L12 dissociates easily from the ribosome (Subramanian & van Duin, 1977). It remains to be explored if this is of relevance for the L12 as the possible ribosomal GAP. In the classical extraction, ammonium chloride and ethanol can be used to remove L12 exclusively (Hamel *et al.*, 1972). Furthermore, L12 could most easily be made to dissociate from the ribosome in the electro-spray MS investigations (Hanson *et al.*, 2003).

In conclusion, as for the other G-proteins, the ribosomal GAP may induce a conformational change of the tGTPase that stabilizes the transition state. It remains possible that there is no Arg-finger involved with the tGTPases. No arginine has been identified to be essential for the GTPase in EF-Tu (Knudsen & Clark, 1995; Zeidler *et al.*, 1995). For EF-G, Arg29 is essential for GTP hydrolysis (Mohr *et al.*, 2000). However, this residue is far from the phosphates of the GDP molecule in the known conformations and does not correspond to the location of essential residues in the other GTPases.

GTP Hydrolysis by tGTPases

For many of the G-proteins the mechanism of GTP hydrolysis is thoroughly studied (Vetter & Wittinghofer, 2001; Kosloff & Selinger, 2003). The role of the GAP is to induce a conformational change in the protein that makes it an active GTPase. The tGTPases bind to the same binding site of the ribosome and depend on very similar interactions. The GTPase induction is probably similar for all tGTPases.

A water molecule, suitably placed near the γ-phosphate, on the opposite side of the β-phosphate, is needed for the GTP hydrolysis (Plate 9.5). Such water molecules have frequently been observed in GTPases (Vetter & Wittinghofer, 2001). Two requirements for the activation of a GTPase are the stabilization of the transition state and the activation of the water molecule by removing a proton so that the water molecule can make an associative in-line sn^2 attack on the γ-phosphate and hydrolyze the GTP to GDP and inorganic phosphate.

Some of the critical residues inducing GTP hydrolysis have been identified (Vetter & Wittinghofer, 2001). Normally, metal ions or histidyl residues are involved in activating the water molecules. However, for many of the GTPases a glutamine residue is found to be essential for the water activation (Plate 9.3; see Vetter & Wittinghofer, 2001 and references therein). In addition, the Arg-finger (see above) is generally found to interact with the phosphates to stabilize the transition state for GTP hydrolysis (Vetter & Wittinghofer, 2001). As mentioned, this can be part of the GTPase itself (*cis*) or of the GAP (*trans*; Plate 9.3).

There is no conserved Gln in switch II of the tGTPases. In the corresponding position, there is a conserved histidyl residue (His87 in EF-G; His84 in EF-Tu; residue numbers correspond to *T. thermophilus*). This would be a better base for the water activation than glutamine. The histidyl residue (His84) is far from the phosphate-binding site in most of the states of EF-Tu. This would contradict the view that it is involved in the water activation. However, in the crystallographic structure of the complex of EF-Tu:GDP:aurodox (aurodox is N-methyl-kirromycin) His84 is close to the γ-phosphate (Vogeley *et al.*, 2001). This complex is active in GTP hydrolysis of the ribosome.

One would expect all GTPases to operate according to the same catalytic mechanism and to use the corresponding groups for the critical steps in the process. Thus, some GTPases using histidine for water activation and some using glutamine would sound unlikely for several reasons. An alternative mode of water activation is by the γ-phosphate of GTP, through the so-called substrate induced catalysis (Schweins *et al.*, 1995). In this case, the glutamyl/histidyl residue may stabilize the transition state and assist the γ-phosphate in placing the water molecule suitably for the activation. Thus, the prime suggestion of a mechanism is activation by the γ-phosphate.

With a mechanism of GTP hydrolysis that is the same for all GTPases, including the tGTPases, the most critical residue for GTP hydrolysis by EF-Tu would seem to be His84. His84 could be involved in stabilizing the water molecule close to the γ-phosphate for activation. The corresponding His of the other members of the tGTPase family would be expected to have the same role (Table 9.3).

The tGTPases have two common domains: the G-domain and domain II. It is interesting that the homologous domains in EF-Tu, eIF5B and EF-G, behave differently. In the conformational changes of the factors, domains G and II of EF-G seem to retain their interaction (Laurberg *et al.*, 2000), whereas for EF-Tu (Kjeldgaard *et al.*, 1993) and eIF5B (Roll–Mecak *et al.*, 2000), domains II and III move in relation to the G-domain. This may be related to the induction of conformational changes these factors induce in the ribosomal subunits (see Section 8.5). A detailed analysis is not yet available.

Nucleotide Exchange

After GTP hydrolysis, the G-proteins need to release the GDP molecule and be recharged with GTP. This is normally done through the interaction with a G-nucleotide exchange factor (GEF, Bourne *et al.*, 1990; 1991). The GEF proteins have frequently been found to work by removing the magnesium ion bound at the phosphates of the nucleotide (Vetter & Wittinghofer, 2001; Plate 9.6). The complexes between a number of G-proteins and GEFs have been characterized (Vetter & Wittinghofer, 2001).

The affinities for GTP and GDP differ greatly between the translational GTPases (Table 9.5). Depending on the relative affinities for GTP and GDP, the GTPases may require G-nucleotide exchange factors (GEFs). The best-known GEF is EF-Ts catalyzing the exchange of GDP for GTP in EF-Tu. For IF-2 and EF-G, exchange factors are not known and do not seem to be needed. RF3 uses the ribosome for the exchange of GDP for GTP (Zavialov *et al.*, 2001; Zavialov & Ehrenberg, 2003). Further details will be discussed below for the individual factor. A general observation is that the GEFs interact with switches I and II. In particular, they reduce the affinity of the Mg^{2+} ion and the PO_4-loop for the α- and β-phosphates (Plate 9.6; Vetter & Wittinghofer, 2001).

Table 9.5 Functional Properties of the tGTPases (after Selmer, 2002)

	IF2	EF-Tu	SelB	EF-G	RF3
K_dGTP	143 µM	0.36 µM	0.74 µM	14 µM	2.5 µM
K_dGDP	12.5 µM	4.9 nM	13.4 µM	11 µM	5.5 nM
Affinity ratio GTP/ GDP	0.091	0.014	18	0.77	0.0022
References	Pon *et al.*, 1985	Kaziro, 1978	Thanbichler *et al.*, 2000	Kaziro, 1978	Zavialov *et al.*, 2001
GEF	—	EF-Ts	—	—	Ribosome
GTPase Trigger	50S binding	Codon recognition	Codon recognization	Ribosome binding	Ribosome binding
Role of GTP Hydrolysis	Fast association of subunits	Conf. change of EF-Tu leads to tRNA release into the PTC	Conf. change of SelB leads to tRNA release into the PTC	Accelerated translocation	Factor recycling

9.2 INITIATION FACTORS

Initiation of protein synthesis is performed on the small subunit. To be able to bind the mRNA, the small and large subunits need to be separated since the mRNA passes between them and is wrapped around the neck of the small subunit (Yusupova *et al.*, 2001). The correct methionine codon (AUG) for the start of translation in bacteria is positioned with the aid of the SD interaction (Shine & Dalgarno, 1974), whereby an A- and G-rich sequence of the mRNA binds through base-pairing to a region of the 3'-terminal sequence of the 16S RNA (Fig. 8.1A). Through this procedure, the initiator AUG codon is placed in the P-site. No translation factor is required for these interactions in bacteria. Three bacterial initiation factors, IF1, IF2, and IF3 assist the placement of the fMet-tRNA at the AUG start codon in the P-site and IF2 catalyzes the association of the two subunits.

Eukaryotic initiation is performed with the aid of at least 12 initiation factors, including 27 protein subunits (see reviews: Pestova & Hellen, 2000; Nyborg *et al.*, 2003). For some time, it was thought that the initiation factors of bacteria, archaea and eukaryotes were unrelated. However, the analysis of complete genomes has shown that bacterial initiation factors IF1 and IF2 have homologues in the other domains. Thus, at least some common elements of initiation were present at the universal ancestor stage of evolution (Kyrpides & Woese, 1998a).

One factor, eIF2, has no counterpart in bacteria. eIF2 binds the initiator tRNA and carries it to the small subunit and eIF2 is released after GTP hydrolysis (see review by Hinnebusch, 2000). eIF2 is a three-subunit protein where one of the subunits is related to EF-Tu or even more closely to SelB (Keeling & Doolittle, 1995; Keeling *et al.*, 1998). The close relationship was confirmed by the determination of the structure of the subunit e/aIF2γ (Schmitt *et al.*, 2002). The dissociation of eIF2 from the initiator tRNA and the ribosome is associated with GTP hydrolysis (Lee *et al.*, 2002). This GTPase is induced differently than the regular tGTPases, since it cannot interact with the GTPase associated region (GAR) of the large subunit, which binds to the small subunit at a later stage. However, another eukaryotic initiation factor, eIF5, induces the GTPase activity (Das & Maitra, 2000). Thus, there are two GTP-dependent steps in archaeal and eukaryotic initiation (Lee *et al.*, 2002).

A separate tGTPase, eIF5B, catalyzes the joining of the two subunits. eIF5B is an orthologue of IF2 in bacteria and its structure and function are well characterized. The universally conserved protein EF-P has a homologue called eIF5A in archaea and eukarya. EF-P may not be an initiation factor (see below).

IF1

IF1 has been identified as a factor that stimulates the dissociation of the ribosomal subunits and the binding of IF2 (Grunberg-Manago *et al.*, 1975). Proteins related to IF1 are found universally and are called eIF1A in eukaryotes (Kyrpides & Woese, 1998a). The structure of IF1 as well as of eIF1A is known (Sette *et al.*, 1997; Battiste *et al.*, 2000). This small globular protein has the OB-fold (Fig. 9.7). It binds to the decoding part of the A-site as has been shown in different ways, including a crystal structure of the complex with the small subunit (Moazed *et al.*, 1995; Dahlquist & Puglisi, 2000; Carter *et al.*, 2001; Plate 9.7). By this binding, it prevents tRNA molecules from binding here, in particular initiator tRNA. In this way, it assists in directing the initiator tRNA into the P-site. IF1 interacts with IF2 on the ribosome, which may explain why the two proteins are conserved throughout evolution (Stringer *et al.*, 1977; Choi *et al.*, 2000).

IF2

The protein of archaea and eukaryotes that is homologous to bacterial IF2 is called eIF5B. This protein contains most of the features of IF2.

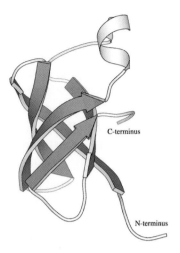

Fig. 9.7 The structure of IF1 (Sette *et al.*, 1997; Battiste *et al.*, 2000). The protein has a version of the OB fold. The figure has kindly been provided by Dr. M. Fodje using the program MOLSCRIPT (Kraulis, 1991).

IF2/eIF5B is found in all organisms (Lee *et al.*, 1999; Choi *et al.*, 2000). The function of IF2/eIF5B is to stabilize the bound initiator tRNA and catalyze the joining of the ribosomal subunits in a GTP-dependent manner (Pestova *et al.*, 2000; Shin *et al.*, 2002; Antoun *et al.*, 2003). IF2 is one of the largest of the bacterial translation factors and a GTPase. Like other tGTPases, it interacts with the large subunit to hydrolyze its GTP. The interaction of IF2 with the ribosomal subunits can be described as four step process:

1. Binding of IF2 to the pre-initiation complex.
2. Association of the pre-initiation complex with the large subunit.
3. GTP hydrolysis.
4. Dissociation of IF2 from the initiated 70S ribosome.

   IF2 binds equally well to the pre-initiation complex with GTP or GDPNP and slightly less well with GDP. The association with the large subunit with GDP is considerably slower. In the case of GDPNP, the factor will be stuck on the ribosome and the 70S initiation complex can form only after the IF2-GDPNP has dissociated. In the case of GDP, there is a slow production of peptide bonds compared to when GTP is used (Antoun *et al.*, 2003). Evidently, IF2/eIF5B depends on GTP hydrolysis to be released from the ribosome. The possibility that IF2/eIF5B would have a mechanochemical role (Tomsic *et al.*, 2000) seems to be excluded, since GTP hydrolysis was not required for translation but primarily for release of the factor (Shin *et al.*, 2002).

The domain organization of IF2/eIF5B varies considerably. In some species, both bacterial and eukaryotic, there is an N-terminal extension of variable length (Sacerdot *et al.*, 1992; Tiennault-Desbordes *et al.*, 2001). In *E. coli*, the two forms are called IF2α and IF2β. The latter is produced from a downstream in-frame GUG codon to yield IF2β that lacks the N-terminal domain (Fig. 9.2; Plumbridge *et al.*, 1985). The N-terminal domain participates in the binding of IF2 to the small subunit (Moreno *et al.*, 1998).

The structure of a bacterial IF2 is not known, but the structure of an archaeal eIF5B has been determined for the empty form, with GDP and with GDPNP (Roll-Mecak *et al.*, 2000). The protein is very much extended, about 110 Å, and has the shape of a chalice (Fig. 9.8). The G-domain is situated at the N-terminus followed by the classical domain II as in all tGTPases (Ævarsson, 1995). Domain III has a α/β-fold reminiscent of

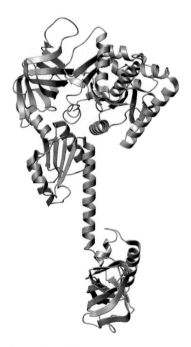

Fig. 9.8 The structure of eIF5B (Roll-Mecak *et al.*, 2000). The G-domain is seen on the upper right followed by domain II (*upper left*), domain III (*middle left*) followed by the long α-helix and the domain IV. The illustration was taken from Selmer (2002) with kind permission.

a four-stranded Rossmann-fold protein. Domain IV has the same anti-parallel β-barrel fold as has domain II in all tGTPases. The protein is remarkable in that domain IV, the base of the chalice, is situated far away from the main body of the molecule due to a 40 Å long α-helix that extends from the three N-terminal domains. One part of the helix that is solvent-exposed on all sides has the unusual sequence EEEKKKK. Switch I is disorganized in all crystal structures. Switch II is situated at the heart of IF2/eIF5B, between domains I, II and III, a situation reminiscent of EF-Tu and EF-G (see Section 9.3). The differences between the three forms of the protein primarily concern switch II that undergoes distinct conformational changes, but not as dramatic as for EF-Tu (Roll-Mecak *et al.*, 2000). In the conformational changes from the GDP to the GTP state, domains II and III move as a block as for EF-Tu, but not for EF-G. The small changes induced by the nucleotide, transmitted through Switch II to the inter-domain helix, become a more significant movement of domain IV.

If eIF5B is bound to the ribosome in the same manner as the elongation factors are known to bind, domain IV would be situated close to the P-site, where the initiator tRNA has previously been placed (Roll-Mecak *et al.*, 2001). Thus, when IF1, IF3, mRNA and initiator tRNA are bound to the small ribosomal subunit, IF2 can bind with GTP. The initiation phase is completed when the large subunit associates and the GTP molecule bound to IF-2 or eIF5B is hydrolyzed (see Section 11.2).

A number of mutants in switches I and II have been analyzed in relation to the structures of eIF5B. Mutations of the conserved threonyl residue in switch I (that co-ordinates the magnesium ion) to alanine, and of the catalytically important histidyl residue of switch II to glutamate both block GTP hydrolysis (Shin *et al.*, 2002). Nevertheless, these mutations stimulated the joining of the subunits and stabilization of initiator tRNA binding but could not function in translation since the factor remained bound to the ribosome. A suppressor mutation, also in the G-domain of eIF5B, caused weaker binding to the ribosome and permitted release of the factor even though the GTP was not hydrolyzed (Shin *et al.*, 2002).

IF3

IF3 has multiple functions during initiation. It prevents the association between the two ribosomal subunits before proper initiation has been

Fig. 9.9 The dumbbell structure of IF3 (Biou *et al.*, 1995). The N-terminal domain and the connecting α-helix (*left*) are followed by the C-terminal domain (*right*). The illustration was taken from Selmer (2002) with kind permission.

achieved (Kaempfer, 1972) and promotes dissociation of the 70S ribosomes (Subramanian & Davis, 1970). In addition, it directs the initiator tRNA to the P-site and influences the kinetics and fidelity of codon-anticodon recognition of the fMet-tRNA (Meinnel *et al.*, 1999; O'Connor *et al.*, 2001).

The structure of IF3 has been investigated (Biou *et al.*, 1995; Garcia *et al.*, 1995a,b). The protein has a dumbbell shape with two domains separated by an α-helix (Fig. 9.9). The location of the two domains on the ribosome has been investigated by crystallography, cryo-EM and labeling studies (Moazed *et al.*, 1995; McCutcheon *et al.*, 1999; Pioletti *et al.*, 2001; Dallas & Noller, 2001). The results are partly divergent. Cryo-EM has identified a location of the C-terminal domain of IF3 on the small subunit at the interface side of the platform (McCutcheon *et al.*, 1999). This is the site of interaction of the inter-subunit bridge B2b or H69 from the large subunit. This site agrees well with the labeling studies and would readily explain the anti-association character of IF3. The location of the N-terminal domain initially suggested from cryo-EM has subsequently been identified as a movement of the platform of the small subunit. The labeling experiments (Dallas & Noller, 2001) suggest that the N-terminal domain of IF3 is bound at the 30S part of the E-site (Plate 9.8). This site was not accessible for binding due to packing reasons in the crystals (Pioletti *et al.*, 2001). The question whether the crystallographically observed location of the N-terminal domain is an artifact or it could be an alternate binding site remains to be firmly established.

The functional properties, at least *in vitro*, seem all to be associated with the C-terminal domain (Petrelli *et al.*, 2003).

Additional Initiation Factors

With an increasing number of complete genomes having been sequenced, it is clear that there are proteins identified as initiation factors in archaea and eukarya that are also present in bacteria (Kyrpides & Woese, 1998a,b). A factor, which has long been discussed as a possible elongation factor, is called EF-P (Glick *et al.*, 1979). In archaea and eukarya, it is called aIF5A and eIF5A, respectively (Kyrpides & Woese, 1998a). The protein is universal (Harris *et al.*, 2003) and is essential for the viability of bacteria (Aoki *et al.*, 1997). Mutations in yeast cause growth defects (Valentini *et al.*, 2002). The real function may not be in initiation but more likely in multiple steps of RNA metabolism (Valentini *et al.*, 2002). Whether eIF5A is involved in the decay of mRNA or regulation of ribosome synthesis, two of several possibilities, remain unclear (Valentini *et al.*, 2002).

The structure of EF-P/eIF5A has been determined both for a eukaryotic version (Kim *et al.*, 1998) and a bacterial version of the protein. The latter is formed by three domains with the shape of an L, reminiscent of the shape of a tRNA (Benson *et al.*, 2000). The N-terminal domain is unique in structure, whereas the two domains towards the C-terminus have similar folds. The archaeal and eukaryotic factors lack the third domain. With somewhat different structures, the functions may also differ.

All eukaryotic and archaeal genomes contain a sequence encoding a factor called eIF1 or SUI1. This sequence is also found in some, but far from all, bacterial genomes (Kyrpides & Woese, 1998a). Evidently, the understanding of bacterial initiation of translation may not yet be complete.

9.3 ELONGATION FACTORS

EF-Tu

The role of the elongation factor Tu (EF-Tu) is to bind charged and cognate tRNAs to the ribosomal A-site (see review by Krab & Parmeggiani, 1998). It does not bind uncharged tRNAs but binds all aminoacyl-tRNAs. The tRNAs charged with amino acids by the synthetases, need to be protected from hydrolysis that would easily occur in the cytoplasm. EF-Tu is very abundant in the bacterial cell and can thus protect all charged tRNA from hydrolysis. In eukaryotes and archaea, it is called elongation factor 1 (EF1) and is composed of several subunits. EF-Tu is a tGTPase.

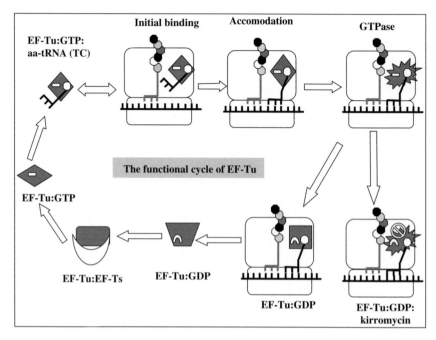

Fig. 9.10 The functional cycle of EF-Tu starts at the middle left where EF-Tu in complex with GTP is in the ON state and can bind any aminoacyl-tRNA. The ternary complex (TC) formed binds to the ribosome without contact between codon and anticodon. In the accommodation step, a kink in the tRNA allows the anticodon to be checked against the codon. In case of a cognate fit, the GTP will be hydrolyzed to GDP. This changes the conformation of EF-Tu to the OFF state, and it dissociates from the tRNA and the ribosome. Subsequently, EF-Tu will be recharged with a new GTP molecule by EF-Ts.

When EF-Tu is activated with a GTP molecule, it can bind the aminoacyl-tRNA (Fig. 9.10). The complex between EF-Tu, GTP and aminoacyl-tRNA is called the ternary complex (TC). This complex binds to the ribosome. If the anticodon of the aminoacyl-tRNA bound to EF-Tu matches the codon of the mRNA in the decoding part of the A-site on the small subunit, EF-Tu is induced to hydrolyze its GTP to GDP. This has the effect that EF-Tu undergoes a large conformational change that leads to its dissociation from the tRNA and the ribosome. To be able to engage in a new elongation cycle, the GDP needs to be exchanged for GTP. This exchange is catalyzed by EF-Ts.

EF-Tu recognizes the aminoacyl ester moiety of tRNAs but does not discriminate between different species of amino acid or tRNA, with some exceptions. It does not bind fMet-tRNA or Se-Cys-tRNA^{Sec} (see below). Obviously, specific features of these special amino acids and tRNAs are recognized.

Structure

EF-Tu was the first G-protein to be characterized (Miller & Weissbach, 1977). Thus, the structure of the protein has had significant interest for comparisons with numerous other GTPases (Vetter & Wittinghofer, 2001).

Initially, the structure of the factor was characterized in complex with GDP (Plate 9.9C; Kjeldgaard & Nyborg, 1992). The effector loop or switch I was nicked in such a way that 14 residues were removed. The factor is composed of three domains: the G-domain at the N-terminus and domains II and III (see Fig 9.2). The conformation of this GDP-complex is an open structure, where a hole is generated between the three domains. Subsequently, other crystal forms have allowed the examination of the complete EF-Tu, including the effector loop in complex with GDP, GDPNP, GDPNP and aminoacyl-tRNA (Berchtold *et al.*, 1993; Kjeldgaard *et al.*, 1993, Nissen *et al.*, 1995). EF-Tu in complex with the G-nucleotide exchange factor is also described below.

After the dissociation of GDP assisted by EF-Ts, EF-Tu is able to bind a molecule of GTP. EF-Tu in complex with GTP analogues has a significantly different and more closed structure (Plate 9.9A). GTP analogues such as GDPNP are used for structural studies, since GTP would be hydrolyzed during the time of the experiment. A large conformational change occurs upon binding of GTP (Berchtold *et al.*, 1993; Kjeldgaard *et al.*, 1993). It was found that domains II and III move as a block with regard to the G-domain. The largest movement of atoms on the surface is about 40 Å and can be described as a rotation of the G-domain in relation to domains I and II by 90° (Plate 9.9). The loss of the γ-phosphate after hydrolysis of the GTP is transmitted through the PO_4-loop to the effector loop and Switch II loop to affect the whole molecule. In particular, Switch I or the effector loop undergoes a dramatic change in location but also with regard to its secondary structure (Polekhina *et al.*, 1996). A β-ribbon in the GDP form, including residues 56–66 (*T. thermophilus* numbering), is converted to a short α-helix and a region without specific secondary structure in the GTP form (Plate 9.10).

Ternary Complexes

Crystallographic structures of the complexes between EF-Tu and tRNA^{Phe} and tRNA^{Cys} have been determined (Nissen *et al.*, 1995; 1999; Valle *et al.*, 2003a; Nielsen *et al.*, 2004). GTP analogues were also used in these experiments. The structures of the complexes are highly similar. The conformation of the protein is marginally changed from the GTP-binding closed conformation. Likewise, the tRNA has the normal L-shaped structure. In the case of the complex with tRNA^{Cys}, the angle of the L is somewhat larger than normal, about 100° (Nissen *et al.*, 1999).

All three domains of EF-Tu are engaged in the binding of tRNA but only the aminoacylated-CCA-end, the acceptor stem and the T-stem and loop of the tRNA participate in the binding (Fig. 9.11C). Thus, the CCA-end interacts with specific pockets on domain II of EF-Tu, while the amino acid binds in a large cavity between domains I and II (Nissen *et al.*, 1995; 1999). The conserved residue Glu 271 is stacked over A76 and makes a hydrogen bond with the 2'-OH of the ribose. The amino group forms hydrogen bonds to the main chain carbonyl of Asn285 and to His273, which is not conserved. The side chain of the aminoacyl moiety (Phe or Cys) is in van der Waals' contact with the side chains of Asn285 and His67. The acceptor stem also makes contacts with switches I and II. The phosphorylated 5'-end is bound to a positively charged pocket between all the three domains of EF-Tu and interacts with the conserved residues Lys90 and Arg300. The T-stem has a large interface with domain III. This interaction may be an important part of the selection of tRNAs (Nissen *et al.*, 1995).

EF-Tu should not form ternary complexes with initiator tRNA or tRNA^{Sec}. The formyl-group of fMet alone will prevent the specific recognition of the amino group by EF-Tu. Also, the missing base pair at positions 1–72 of tRNA^{fMet} could be important for the discrimination. Three base pairs inducing a special structure at the joining of the acceptor and T-stems may be responsible for the discrimination of tRNA^{Sec} (Rudinger *et al.*, 1996). A C-terminal segment of EF-Tu (domain III) interacts with these nucleotides. Normally, specific hydrogen bonds are formed between G63 and G64 of the tRNA with Glu390 and Gly391 of EF-Tu (Nissen *et al.*, 1999).

The ternary complex is in the conformation that binds to the ribosome. An overview of the conformations of EF-Tu is shown in Fig. 9.11.

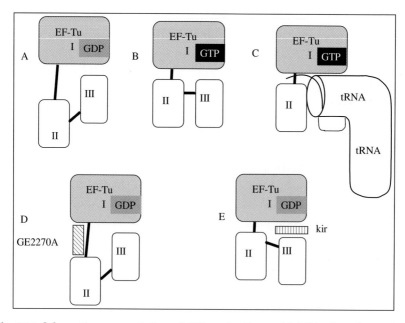

Fig. 9.11 Schematic representation of different active and inhibited conformations of EF-Tu. (**A**) The GDP conformation is open without contact between domains I and II. (**B**) The GTP conformation has a closed arrangement of the domains. (**C**) A tRNA bound to the closed GTP conformation of EF-Tu. The tRNA binds in the interface region. In particular, the CCA-aminoacyl part binds between domains I and II. (**D**) When inhibited by GE2270A, EF-Tu is in the GDP conformation with the inhibitor bound between domains I and II (Heffron & Jurnak, 2000). In particular, the binding sites for the terminal A and the aminoacyl moiety are occupied, preventing the formation of the ternary complex. (**E**) The kirromycin-inhibited form of EF-Tu is close to the GTP conformation (Vogeley *et al.*, 2001). The inhibitor is bound between domains I and III, locking a closed conformation that does not dissociate from the ribosome even though GTP hydrolysis has taken place. This conformation can also bind aminoacyl-tRNA.

Antibiotics Targeting EF-Tu

A number of antibiotics inhibit the function of EF-Tu (see Krab & Parmeggiani, 1998, for a review). They fall into two functional groups: one blocks the formation of the ternary complex and the other one blocks the release of EF-Tu from the ribosome after GTP hydrolysis (Fig. 9.11). The best characterized inhibitor is kirromycin (Parmeggiani & Swart, 1985). It can bind to EF-Tu:GTP, to the ternary complex or to this complex

when bound to the ribosome. Kirromycin inhibits the dissociation of EF-Tu:GDP from the ribosome (Wolf *et al.*, 1977). An interesting aspect of kirromycin is that it can induce EF-Tu to hydrolyze its GTP even off the ribosome (Wolf *et al.*, 1974). This may suggest that tGTPases in general contain all the residues that are needed for GTP hydrolysis.

The structure of the complex of EF-Tu:GDP with the kirromycin derivative aurodox (N-methyl-kirromycin) has been investigated (Vogeley *et al.*, 2001). Here EF-Tu is not open like the GDP form, but closer to the GTP conformation (Fig. 9.11). This explains why the complex is retained on the ribosome despite the fact that the GTP molecule is hydrolyzed. The reason why the conformation is retained is that the aurodox molecule acts like glue between the G-domain and domain III. To allow the antibiotic to bind, the tight interaction between these domains is opened up somewhat (Plates 9.9A and B). A crystallographic structure of a ternary complex with kirromycin has also been determined (Valle *et al.*, 2003a).

Mutants of EF-Tu resistant to kirromycin have been characterized. From this, a possible binding site for the antibiotic was suggested (Abdulkarim *et al.*, 1994; Mesters *et al.*, 1994). The predicted site generally agrees with the experimentally determined one. Most of the mutations occur close to the binding site for aurodox (Vogeley *et al.*, 2001). However, a number of mutations are located further away from the binding site. Q341H and E390K affect the binding of aa-tRNA, thus decreasing the likelihood that EF-Tu with aurodox will be firmly bound to the ribosome and preventing the reading of an mRNA (Vogeley *et al.*, 2001).

Other types of antibiotics are pulvomycin and GE2270A. They inhibit the formation of the ternary complex. GE2270A also inhibits the transition from the inactive GDP conformation to the active one with GTP. The structure of the complex between EF-Tu and GE2270A is known (Heffron & Jurnak, 2000). The inhibitor is bound at the interface between domains I and II and the pocket where the terminal A of the acceptor end of the tRNA binds is occupied by the inhibitor (Fig. 9.11). This, then, explains the mode of inhibition of the antibiotic. The structure of a complex with pulvomycin has not been determined.

Binding of EF-Tu to the Ribosome

EF-Tu binds as a ternary complex to the GTPase binding site of the ribosome (see Section 8.5). The complex between the ternary complex and the

ribosome has been analyzed by cryo-EM techniques (Plate 9.11; Stark *et al.*, 1997; 2002; Valle *et al.*, 2002; 2003a). To make a stable complex with the ribosome that is amenable to study, kirromycin was used. Kirromycin locks the ternary complex on the ribosome despite the fact that GTP hydrolysis has occurred. The best resolved cryo-EM densities are at about 9 Å resolution (Valle *et al.*, 2003a).

The P- and E-sites are both filled with tRNAs in one of the structures (Valle *et al.*, 2002), while in the other there is no tRNA in the E-site (Stark *et al.*, 2002). Thus, somewhat different states of the ribosome seem to be observed. Furthermore, no density is observed for the A-site finger (the B1a inter-subunit bridge or helix H38 of the 23S RNA), indicating flexibility of this part of the structure (Stark *et al.*, 2002).

The 9 Å structure of the ternary complex when bound to the ribosome shows a movement of the elbow of the tRNA by about 7 Å (Plates 9.12A and B; Valle *et al.*, 2003a). This is paralleled by a similar movement of helix H43 around the nucleotide A1067 that is in contact with the tRNA of about the same distance. This region of the base of the L12 stalk (GAR) also seems to go between an open and a closed state. With an empty or filled A-site the open state is adopted, while it is closed when the ternary complex is bound to the T-site (Valle *et al.*, 2003a). How these structural changes are induced remains unknown.

Even though EF-Tu binds the tRNA somewhat differently and the conformation of the tRNA is different, the structure of EF-Tu is essentially as in the crystal structures of the ternary complex (Valle *et al.*, 2003a; Nissen *et al.*, 1995; 1999; Nielsen *et al.*, 2004).

EF-Tu interacts with the SRL of the 23S RNA near the base of the L12 stalk (Plate 9.12C; Stark *et al.*, 2002; Valle *et al.*, 2002; 2003a). SRL interacts with the nucleotide-binding site and the effector loop or switch I of EF-Tu (Stark *et al.*, 2002). Valle *et al.* (2003a) made the observation that the effector loop of EF-Tu is invisible, probably due to multiple conformations. No contact with the L12 stalk could be observed in this complex, even though such contacts had been seen at lower resolution (Stark *et al.*, 1997). It remains an unresolved problem how the stalk protein L12 participates in the interplay with the tGTPases.

Domain II of EF-Tu interacts with helix h5 of the 30S subunit (Plate 9.12C; Valle *et al.*, 2003a). The tRNA interacts with the GAR at the base of the stalk including helices H43, H44 and protein L11 and stretches across to the decoding site on the small subunit as seen by cryo-EM (Stark *et al.*,

1997; 2002; Valle *et al.*, 2002; 2003a). If the anticodon of the tRNA of the complex is cognate to the codon in the A-site, the tRNA will bind in the A/T state. Here, the ASL of the tRNA is bound at the decoding part of the A-site, while the rest of the tRNA is bound to EF-Tu (Plate 9.13; Valle *et al.*, 2003a). Since EF-Tu, that binds the acceptor stem of the tRNA, is at the factor-binding site of the large subunit, the aminoacyl moiety is far from the PTC, where it may be incorporated into the nascent peptide at a later state (Plate 9.11; see Section 11.4).

GTP Hydrolysis by EF-Tu

Since the tGTPases are dormant enzymes, they need to be activated. There is no final conclusion with regard to the mechanism of the GTP hydrolysis. However, the most likely mechanism is the substrate-induced catalysis (see Section 9.1).

EF-Tu may not be able to hydrolyze its bound GTP even when bound to the ribosome. What is required for activation is a cognate interaction between the codon of the mRNA and the anticodon of a tRNA. Here H69 may have an important role, that is, to communicate that the anticodon is cognate and that GTP should be hydrolyzed. The signal needs to be passed from the decoding site on the small subunit to the GAR on the large subunit (Yusupov *et al.*, 2001; Stark *et al.*, 2002; Bashan *et al.*, 2003; Valle *et al.*, 2003a). Residues important for the tGTPase are then induced to adopt a different conformation to stimulate GTP hydrolysis. The two residues that are most important in general for GTP hydrolysis are the Arg-finger and the Gln residue in switch II. In the tGTPases the Gln corresponds to a His residue (see Section 9.1).

In EF-Tu, His84 corresponds to the essential Gln residue in other G-proteins. In the GTP conformation of EF-Tu, it is at a distance of 5 Å from the water molecule, that is, it is too far away to have any significant role in activating it (Kjeldgaard *et al.*, 1993). The histidyl residue is shielded from close interaction with the water molecule by two hydrophobic residues, Val20 and Ile61. These residues have been called the hydrophobic gate (Berchtold *et al.*, 1993). The hydrophobic gate has been analyzed by mutations. V20G and I61A did not lead to an increase of intrinsic or ribosome-induced GTP hydrolysis (Jacquet & Parmeggiani, 1988; Krab & Parmeggiani, 1999). Evidently, it is not enough to remove the hydrophobic gate to induce GTP hydrolysis. However, the structural effects of these mutations are unknown.

Mutations of His84 reduce the GTPase activity of EF-Tu significantly (Cool & Parmeggiani, 1991; Zeidler *et al.*, 1995). Daviter *et al.* (2003) have done the most thorough investigation. In one of the steps involving His84, it plays the very important role of stabilizing the transition state. A variation of pH leads to the conclusion that the histidyl residue does not act as a general base. Kirromycin and aurodox induce EF-Tu to hydrolyze GTP in the absence of ribosomes (Wolf *et al.*, 1974). The structure of the complex of EF-Tu with aurodox illustrates that His84 is capable of approaching the water molecule at the γ-phosphate (Vogeley *et al.*, 2001). Apparently, the hydrophobic gate is destabilized and His84 swings into close proximity of the water molecule at the γ-phosphate (Plates 9.9B and D). However, in the ternary complex with kirromycin, His84 is not close to the γ-phosphate (Nielsen *et al.*, 2004). Some unknown conformational change due to the binding to the ribosome may open the gate to allow His84 to interact with the water molecule and induce GTP hydrolysis.

The consensus sequence of switch I (RGITI) has several important roles in conformational changes associated with GTP-binding, ribosome binding and GTP hydrolysis. Thr62 binds to the magnesium ion at the phosphates (Berchtold *et al.*, 1993; Kjeldgaard *et al.*, 1993). Ile61, together with Val20, forms the hydrophobic gate preventing His84 getting close to the water molecule (Berchtold *et al.*, 1993). The Arg-finger, frequently the arginyl residue of the switch I consensus sequence, stabilizes the transition state by electrostatic interactions. Mutations of Arg58 of EF-Tu have little effect on GTP hydrolysis (Knudsen & Clark, 1995; Zeidler *et al.*, 1996). Arg58 is 8 Å away from the phosphates in the GTP conformation or in the ternary complexes (Berchtold *et al.*, 1993; Nissen *et al.*, 1995). In the structure of the EF-Tu complex with aurodox, Arg58 is flexible and invisible (Vogeley *et al.*, 2001).

The proper interaction with the ribosome induces the GTPase conformation in EF-Tu. The best structure of the ribosome in complex with EF-Tu has been obtained with cryo-EM studies of the ternary complex locked on the ribosome with kirromycin at 9 Å resolution (Valle *et al.*, 2003a). The structure does not permit detailed examination of the orientation of different residues, but whereas the conformation does not differ significantly from EF-Tu in the ternary complex, there is no density for the switch I, or effector, loop. Evidently, in this state, these residues do not have a defined state but can adopt different conformations. As discussed above, the GTP hydrolysis is most likely due to the substrate-induced catalysis mechanism (Schweins *et al.*, 1995).

GTP hydrolysis induces a large conformational change in EF-Tu. In this conformational change, EF-Tu reverts to the GDP conformation where domains II and III move as a unit with regard to domain I or the G-domain. In this way, the interaction between the G-domain and domain II is broken.

EF-Ts

EF-Ts is the GEF for EF-Tu. When EF-Tu is released from the ribosome in complex with GDP, EF-Ts binds to the complex to release the GDP molecule. The EF-Ts structures are known from *T. thermophilus* (Jiang *et al.*, 1996), and in complex with EF-Tu from *E. coli* and *T. thermophilus* (Fig. 9.12; Kawashima *et al.*, 1996; Wang *et al.*, 1997). The structure of the

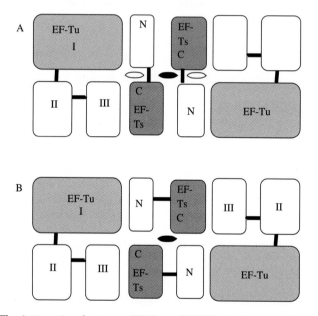

Fig. 9.12 The interaction between EF-Tu and EF-Ts in the crystal structures of: (**A**) *E. coli* (Kawashima *et al.*, 1996); (**B**) *T. thermophilus*. The region called N in *T. thermophilus* EF-Ts interacts with the G-domain of EF-Tu. It is N-terminal to the region (C) that interacts with domain III of EF-Tu (Wang *et al.*, 1997). The filled symbol between the subunits signifies a molecular two-fold axis and the open symbol represents a local pseudo two-fold symmetry axis. In the two cases, the structures of EF-Ts that interact with EF-Tu are related, even though they belong to one molecule in the case of *E. coli* and to two molecules in the case of *T. thermophilus*.

corresponding complex from yeast, eEF1A:eEF1Bα has also been determined *et al.*, 2000). Even though the structures of the complexes are quite different, the functional principles are the same.

The two bacterial EF-Tu:EF-Ts complexes are organized quite differently. In the crystal structure, EF-Ts from *E. coli* forms a tight dimer, where each of the EF-Ts monomers binds one EF-Tu molecule (Kawashima *et al.*, 1996). *E. coli* EF-Ts is an elongated protein with four structural modules. The core of the protein has a structural repeat, where the N-terminal region is related to the C-terminal region through an approximate two-fold symmetry axis (Fig. 9.12a; Kawashima *et al.*, 1996). The N-terminal region interacts with the G-domain of EF-Tu, while the C-terminal region interacts with domain III of the same molecule of EF-Tu (Fig. 9.12a). In *E. coli*, there is thus an interaction between one molecule of EF-Ts with one molecule of EF-Tu (Kawashima *et al.*, 1996).

EF-Ts from *T. thermophilus* forms a dimeric structure (Jiang *et al.*, 1996; Wang *et al.*, 1997). Here, both monomers of EF-Ts interact with one molecule of EF-Tu (Fig. 9.12b; Wang *et al.*, 1997). One monomer of EF-Ts interacts with the G-domain through its N-terminal region, and the other monomer interacts with domain III of the same molecule of EF-Tu through its C-terminal region (Fig. 9.12b).

The two molecules of EF-Ts are very differently organized. The two-fold axis observed between the two hetero-dimers of *E. coli*, EF-Ts:EF-Tu (Kawashima *et al.*, 1996) corresponds to a two-fold axis of the hetero-tetramer of *T. thermophilus*, EF-Ts:EF-Tu (Fig. 9.12; Wang *et al.*, 1997). Despite the differences in organization, the structures and amino acid sequences of EF-Ts interacting with EF-Tu are very similar (Wang *et al.*, 1997). Evidently, the two different structures of EF-Ts are related and have an interesting evolutionary background. The conformation of EF-Tu in both complexes is closest to the GDP conformation even though no nucleotide is present. EF-Ts can also remove GTP from EF-Tu but since the conformation required for complex formation differs much from the one with GTP, the kinetics is unfavorable (Wang *et al.*, 1997).

The removal of the magnesium ion from EF-Tu by EDTA leads to a loss in affinity of the nucleotide (Arai *et al.*, 1972). Thus, it was expected that EF-Ts would act in a similar way. The crystal structures show, that the most important interactions of EF-Ts are with the PO_4-loop and switch II of EF-Tu. Asp80 and Phe81 of the conserved sequence element TDFV in EF-Ts induce structural changes in EF-Tu, leading to a loss of the magnesium and GDP. Asp80 causes a shift in position of helix B of switch II in such a way that the residues involved in the binding of the magnesium ion are moved from their binding positions (Kawashima *et al.*, 1996; Wang *et al.*, 1997). Phe81 of EF-Ts binds to a hydrophobic pocket of EF-Tu, causing a series of conformational changes through the side chains of residues His119, Gln115 and His19 of EF-Tu, that results in a flip of the peptide between Val20 and Asp21 (Wang *et al.*, 1997). This flip leads to a loss of the hydrogen bond donor, the peptide nitrogen, to the GDP β-phosphate and a replacement with a hydrogen bond acceptor, the peptide carbonyl oxygen (see Fig. 9.5). This becomes highly unfavorable for the interaction with the GDP. Additional distortions of the binding site of the ribose and guanine base are induced. A kinetic analysis shows that all these changes and distortions are needed to explain the efficiency of the release of the GDP (Wieden *et al.*, 2001).

The structures of GEFs are not generally related to each other (Vetter & Wittinghofer, 2001). This is also true for the structure and interaction of the yeast factor eEF1Bα, that functions as a GEF for the factor corresponding to EF-Tu in eukarya, eEF1A (Andersen *et al.*, 2000). eEF1Bα is not structurally related to the bacterial EF-Ts molecules and does not bind in a corresponding region. However, the main effect of EF-Ts and eEF1Bα, as of all GEFs (Vetter & Wittinghofer, 2001), is to remove the magnesium ion from its position near the GDP phosphates in EF-Tu and to induce a peptide flip in the PO_4-loop of EF-Tu. This reduces the binding affinity for GDP. Since the concentration of GTP in the cytoplasm is much higher than the concentration of GDP, GTP will preferably bind to an empty nucleotide site in EF-Tu.

SelB

Some proteins in the organism that are involved in oxidation/reduction need selenium in the form of seleno-cystein (Se-Cys) for functioning (Forchhammer *et al.*, 1989). This residue is not among the 20 normal amino acid residues for which there are genetic code words. Se-Cys is rather incorporated through the translation of a stop codon, UGA. A hairpin loop of the mRNA, following the UGA, distinguishes a true stop codon from an UGA codon that should be translated as Se-Cys (Fig. 9.13). The distance between the UGA codon and the hairpin loop is about 20 nucleotides.

SelB is essential for incorporating this 21st amino acid, seleno-cystein, into a number of proteins (Forchhammer *et al.*, 1989). The protein functions

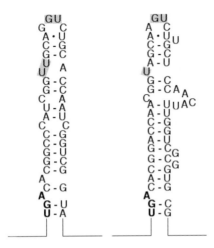

Fig. 9.13 mRNA hairpin structures of *E. coli fdhF* and *fdnG* directing the insertion of Se-Met into corresponding proteins. The UGA codon is shown at the base of the hairpin and the nucleotides protected against chemical modification by SelB are shaded (Hüttenhofer *et al.*, 1996). With kind permission from Selmer (2002).

as an EF-Tu for aminoacylated SeCys-tRNA (Baron *et al.*, 1993). SelB recognizes only tRNA^{Sec} in contrast to EF-Tu (Suppmann *et al.*, 1999).

SelB:GTP binds tRNA^{Sec}, which EF-Tu cannot. In addition, SelB recognizes and binds to the hairpin loop of the mRNA. When the UGA codon reaches the A-site, the likelihood that RF2 will bind and terminate protein synthesis is decreased and instead SeCys is incorporated into the nascent peptide due to the ternary complex of SeCys-tRNA, SelB and GTP that is bound to the mRNA hairpin (Suppmann *et al.*, 1999).

The N-terminal half of SelB corresponds to the three domains of EF-Tu (Fig. 9.2; Kromayer *et al.*, 1996). The structure of the C-terminal half, the mRNA-binding part of SelB, was found to be composed of four closely similar domains arranged in the form of an "L" (Fig. 9.14; Selmer & Su, 2002). From the location of the conserved residues and from studies of functional mutants, it was evident that the main interactions with the stem-loop of the mRNA are due to the seventh or the C-terminal domain (Kromayer *et al.*, 1999; Li *et al.*, 2000). The ternary complex of SeCys-tRNA, GTP and SelB obviously can interact with the two points on the mRNA. The anticodon of SeCys-tRNA interacts with the UGA codon and the stem-loop structure of the mRNA interacts with the C-terminal domain of SelB. Since the EF-Tu related parts of SelB bind to the T-site in

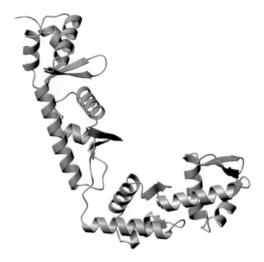

Fig. 9.14 The structure of the C-terminal half of SelB (Selmer & Su, 2002). The four structurally similar domains form the shape of an "L". The very last domain contains highly conserved residues. Mutations in this domain can lead to a changed specificity for the hairpin loop of the mRNA. (With kind permission from Selmer (2002).)

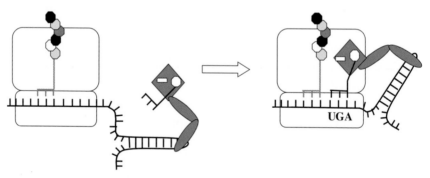

Fig. 9.15 SelB in complex with Se-Cys-tRNA^{Sec} binds to the specific stem loop structure of the mRNA and travels with the mRNA until the UGA stop codon is exposed in the decoding site when the anticodon of tRNA^{Sec} interacts with the codon. The final result is that Se-Cys is incorporated in response to a stop codon.

analogy with EF-Tu, and since the anticodon of the tRNA extends far from the factor protein, it is evident that an elongated structure of the C-terminal domains of SelB is needed (Fig. 9.15) to reach the stem-loop structure of the mRNA (Selmer & Su, 2002). The approximate position of the stem-loop was estimated from structural examination of the path of the mRNA on the 30S subunit (Yusupova *et al.*, 2001).

EF-G

Elongation factor G (EF-G) is the translocase of translation and is a large tGTPase with a molecular weight of around 80 kDa (Ovchinnikov *et al.*, 1982). The archaeal and eukaryotic protein corresponding to EF-G is called EF2. After peptidyl transfer, a peptidyl tRNA is located in the A-site and a deacylated tRNA in the P-site. They need to be translocated to the P- and E-sites, respectively. The mRNA is also moved with the peptidyl-tRNA so that a new codon is exposed in the A-site. The deacylated tRNA in the E-site will subsequently fall off from the ribosome. This process is catalyzed by EF-G (Kaziro, 1978; Spirin, 1985). When EF-G has dissociated, the ribosome gets ready for a new cycle of elongation.

The functional cycles of EF-G and EF2 have a number of different states (Fig. 9.16). How these states differ from each other is only partly

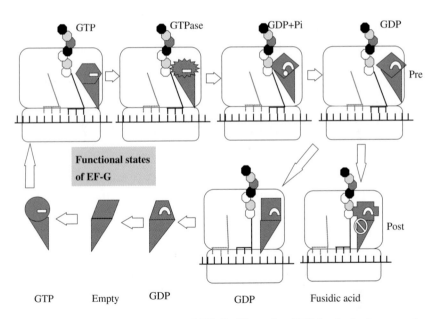

Fig. 9.16 The functional states of EF-G. Here the GTP hydrolysis precedes translocation (Rodnina *et al.*, 1997). In the upper row, the ribosome is in the pre-translocation state and, in the lower it is in the post-translocation state. The white little strip symbolizes the GTP molecule and the little half circle represents GDP. Fusidic acid (FA) inhibits the dissociation of EF-G from the ribosome after translocation. EF-G binds to ribosomes where the peptidyl-tRNA is in the A-site (Valle *et al.*, 2003b; Zavialov & Ehrenberg, 2003).

known. The bacterial system has been most thoroughly investigated, but additional insights are gained from eukaryotic systems. The order of the different steps is discussed. The classical view is that translocation precedes GTP hydrolysis (Inoue-Yokosawa *et al.*, 1974; Belitsina *et al.*, 1975; Modollel *et al.*, 1975; Spirin, 1985; 2002). However, from the rates of the different steps of translocation, it seems as if GTP hydrolysis must precede translocation (Rodnina *et al.*, 1997). In this way, the function of EF-G may be more closely related to the motor proteins than to the molecular switches (Cross, 1997).

Release of Inorganic Phosphate, Pi

One of the functional steps after GTP hydrolysis that has not always been considered important is the release of the inorganic phosphate (Pi). Kinetic analysis of the Pi release using a fluorescent-labeled phosphate binding protein (Brune *et al.*, 1994) has shown that it is not concomitant with GTP hydrolysis, but significantly delayed (Wintermeyer *et al.*, 2001; Savelsbergh *et al.*, 2002). Even though the release is intrinsically rapid, a step of rearrangement of the complex of ribosomes with EF-G:GDP:Pi has to precede the release of the phosphate. No structural insights have yet been gained.

Antibiotics Targeting EF-G and EF2

One antibiotic that affects the function of EF-G is fusidic acid (FA; see Sections 9.1 and 10.7). FA traps EF-G on the ribosome after GTP hydrolysis and translocation (Willie *et al.*, 1975). This inhibition of protein synthesis is a parallel to the inhibition by kirromycin that locks EF-Tu on the ribosome. Contrary to EF-Tu, which can bind kirromycin off the ribosome, FA binds only when EF-G is bound with GTP to the ribosome (Baca *et al.*, 1976). Thus, FA cannot induce EF-G to hydrolyze its GTP off the ribosome as the EF-Tu:kirromycin complex does. In the case of EF2, the antibiotic sordarin has similar effects as FA on protein synthesis. EF2:GDP in complex with sordarin remains bound to the ribosome after translocation (Capa *et al.*, 1998; Justice *et al.*, 1998).

Resistance to FA is obtained by mutations of the amino acid residues in several regions of EF-G (Johansson & Hughes, 1994). These mutations occur primarily in the interfaces between domains G, III and V. It is evident that only some of these mutations, if any, can be at the FA-binding site of EF-G (Johansson *et al.*, 1996). At least three mechanisms could be involved in obtaining FA resistance. In addition to interfering with the binding site, a mutation could lower the energy barrier that prevents the factor from changing its conformation to allow EF-G:GDP to dissociate from the ribosome. Furthermore, a different group of mutations could lower the affinity of EF-G:GDP:FA for the ribosome, thus avoiding the blockage of protein synthesis that would occur if EF-G remained bound on the ribosome.

Mutant EF-Gs that are more sensitive to FA than wild-type EF-G are also known (Martemyanov *et al.*, 2001). These FA-sensitive mutants are revertants from FA resistance where the original mutation has been removed (Johanson *et al.*, 1996). The sensitivity to FA was coupled to a higher affinity for GTP (Martemyanov *et al.*, 2001). FA resistance, on the other hand, was related to a lower affinity for GTP. A high affinity for GTP may be the primary fact that leads to a higher or lower affinity for the ribosome and thus would increase the chances for FA to bind to EF-G on the ribosome.

Structure

When the amino acid sequences became available, it was obvious that the N-terminal region of EF-G is homologous to the N-terminus of EF-Tu (Laursen *et al.*, 1981). Part of the G-domain was identified this way.

The structure of EF-G (Fig. 9.17) is highly elongated, about 120 Å, and composed of six domains, where a large insertion in the G-domain, called G', can be considered as a separate domain (Czworkowski *et al.*, 1994; Ævarsson *et al.*, 1994). In the initial structures, neither domain III or Switch I could be observed. From the structures, it was evident that the whole G-domain and domain II of EF-Tu and EF-G are homologous.

The G-domain is quite normal except for the G' insertion. EF2 from yeast has a different structure in this part of the protein. A corresponding domain extends from a neighboring loop (Fig. 9.18; Ævarsson *et al.*, 1994, Jørgensen *et al.*, 2003). Archaeal EF2 has smaller extensions in both these loops (Fig. 9.18; Ævarsson *et al.*, 1994).

The initial structural reports describe two closely related conformations, one with a bound GDP molecule (Czworkowski *et al.*, 1994) and one without any bound nucleotide (Ævarsson *et al.*, 1994). In both of these structures, domain III was only partly interpretable due to its flexibility.

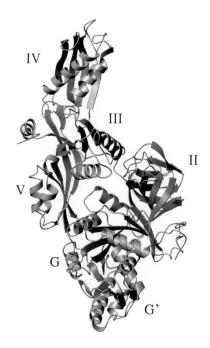

Fig. 9.17 The structure of EF-G from *T. thermophilus* (Ævarsson *et al.*, 1994; Czworkowski *et al.*, 1994; Al-Karadaghi *et al.*, 1996; Laurberg *et al.*, 2000). The five domains I (G), II, III, IV, V and the subdomain G' form an elongated molecule. Helix A is the N-terminal helix. (With kind permission from Selmer (2002).)

A mutant, H573A, has given a picture of a different GDP structure, in this case with a bound magnesium ion (Laurberg *et al.*, 2000). This structure has a somewhat different arrangement of the domains and the entire domain III becomes visible.

When the structure of the ternary complex of EF-Tu with aminoacyl-tRNA and the GTP-analogue GDPNP was determined, it was obvious that EF-G mimics the whole ternary complex (Nissen *et al.*, 1995). Domains III, IV and V of EF-G are the tRNA mimicking parts. EF-G in complex with GDP or without a bound nucleotide is highly similar to the GTP conformation of EF-Tu with bound tRNA (Plates 9.14A and B). Thus, the OFF (GDP) conformation of EF-G mimics the ON (GTP) conformation of the ternary complex. This may make good sense since EF-G·GDP disso-ciates from a space on the ribosome, leaving an imprint that will next bind the ternary complex (Liljas, 1996).

Two main states of EF-G have been studied by crystallography, with and without bound GDP (Ævarsson *et al.*, 1994; Czworkowski *et al.*, 1994;

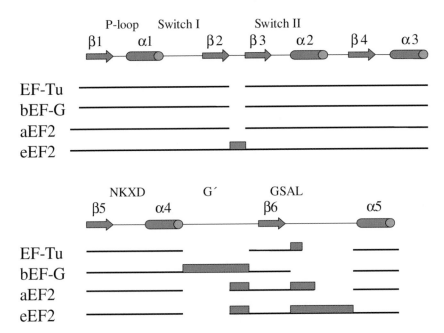

Fig. 9.18 The different inserts into the G-domain of EF-G and EF2 from the different domains of life compared to EF-Tu. While bacterial EF-G has the long insert called G′ before β-strand 6, eukaryal EF2 has a short insert here and a longer one after β6. Archaeal EF2 has short inserts in both places (Ævarsson, 1995).

Al-Karadaghi *et al.*, 1996). In addition, two different conformations with bound GDP have been observed. The structure of the mutant H573A is more bent than the wild-type protein (Laurberg *et al.*, 2000). The difference observed is probably due to the binding of a magnesium ion at the β-phosphate. However, the structures of the empty EF-G and the one with GDP without magnesium are more extended and relatively similar compared with the structure of a mutant EF-G with GDP and magnesium. The bending occurs between the G-domain and domain II that form one block, against which the tRNA-mimicking domains III, IV and V are rotated by about 10°. This leads to a displacement of the tip of domain IV by about 10 Å (Laurberg *et al.*, 2000). The bending of EF-G also leads to a full visibility of domain III in the electron density maps.

The gross conformational changes are related to a chain of residues undergoing conformational changes starting at the nucleotide binding site and extending across the interface between the two blocks that are mobile with regard to each other (Laurberg *et al.*, 2000). This chain of residues was partly identified by comparing the different conformations so far characterized for EF-G. Thus, the bound nucleotide affects residues of the PO_4-loop. They undergo conformational changes that in turn affect residues of Switch II to adopt different conformations like a domino effect (Laurberg *et al.*, 2000). In this way, the state of the nucleotide binding site in the G-domain is communicated to domain III, which is a highly flexible domain and probably critical for the different states of EF-G. Two residues that seem central for this process are Phe90 and Leu457 (Laurberg *et al.*, 2000). Mutations of essentially all the residues along the "domino" path can cause FA resistance (Laurberg *et al.*, 2000).

Three different types of crystals have been obtained from EF-G without nucleotide or with GDP (Laurberg, 2002). The classical form, the rhombic crystals, has been most thoroughly examined (Chirgadze *et al.*, 1983). In addition, plate-like and needle-formed crystals can be obtained. The plate crystals have two molecules per asymmetric unit (Laurberg, 2002). However, the main crystal contact is the same in all the crystal forms of EF-G. The β-sheet of domain IV connects with the β-sheet of the G'-domain of a neighboring molecule (Laurberg, 2002). This β-β interaction limits the range of flexibility of EF-G in the present crystal forms and may prevent access to additional functional states (Laurberg, 2002).

So far, there is no crystal structure of EF-G in the GTP conformation despite the fact that the complex readily forms in solution (Kaziro, 1978). Addition of the GTP analogue GDPNP inhibits crystal formation (Laurberg, 2002). The mutant G16V has an elevated affinity for GTP (Martemyanov *et al.*, 2001). When crystallized in the presence of a high concentration of GDPNP and alkaline phosphatase that degrades any GDP present, the resulting crystal structure was devoid of any bound nucleotide (Laurberg *et al.*, 2004). This could mean that the intramolecular β-sheet in the crystal packing prevents the GTP conformation of EF-G. The GTP conformation may be an even more bent form of the factor than the one observed for the mutant H573A (Laurberg *et al.*, 2000).

Due to its flexibility, the effector loop of EF-G has not been seen in any crystal structure. The amino acid sequences of this region of EF-Tu and EF-G are highly similar. For these two molecules that have clearly related structures and bind to overlapping binding sites on the ribosome, one could expect that the effector loops undergo related conformational changes (Laurberg *et al.*, 2000). Kolesnikov and Gudkov (2002) tested this possibility and concluded that the effector loops are designed for different functions or ribosomal states since a hybrid of EF-G with the effector loop of EF-Tu does not function properly in relation to the ribosome.

The conformational changes of EF-G (Fig. 9.16), which are central to the role of the molecule, depend on communication between the nucleotide-binding site and the domain interfaces and relay information between the factor and the ribosome. One experiment that illustrates this is the introduction of an intramolecular disulphide crosslink between the G-domain and domain V (Peske *et al.*, 2000). The resulting EF-G retains some capacity of single round GTP hydrolysis but cannot translocate and cannot be released from the ribosome. If the disulphide is reduced, the mutant EF-G functions normally. The GTP hydrolysis is not much disturbed by the crosslink, but evidently translocation cannot be performed without movement of domain V in relation to the G-domain, as is also evident from the cryo-EM studies (Plate 9.15; Valle *et al.*, 2003b). These observations further support the concept that GTP hydrolysis precedes translocation (Peske *et al.*, 2000).

The structure of EF2 in complex with as well as without sordarin has been determined (Jørgensen *et al.*, 2003). The structures are significantly different with regard to the domain orientations. As in EF-G, the G-domain and domain II form one block and domains IV and V another. The mobile domain III is not part of the two blocks but has its own mobility and acts as a spacer and a flexible linker. The conformation of EF2 with sordarin may illustrate the conformation after translocation has occurred and before EF2 dissociates from the ribosome. Sordarin is bound at the interface between domains III, IV and V. This binding site coincides with a number of mutations conferring sordarin resistance (Shastry *et al.*, 2001). The location between domains suggests that the antibiotic inhibits the structural transition that is needed for the factor to fall off from the ribosome. A location for FA on EF-G between the G-domain and domain III has been suggested from the location of a number of mutations causing resistance to high levels of FA (Laurberg *et al.*, 2000). No experimental verification of this suggestion is available. Whether FA and sordarin have related binding sites remains to be seen.

EF-G Bound to the Ribosome

The studies of EF-G bound to the ribosome have been performed both with FA and with GDPNP or GDPCP (Agrawal *et al.*, 1998; 1999; Frank &

Agrawal, 2000; 2001; Stark *et al.*, 2000; Valle *et al.*, 2003b). EF-G binds to the ribosome much in the same way as the ternary complex of EF-Tu:GTP and aa-tRNA. These results agree with studies by chemical methods (Wilson & Noller, 1998). When the crystal structure of EF-G with GDP is superimposed on the EM density originating from the bound EF-G:GDPNP, it is evident that the factor on the ribosome has undergone significant conformational changes. The investigations of around 11 Å resolution give the best picture (Plates 9.15 and 9.16; Valle *et al.*, 2003b). EF-G:GDPNP is seen bound between the two subunits, stretching from the base of the L12 stalk into the decoding part of the A-site (Table 9.6). Domain IV is located in the decoding site with the tip of the domain close to the top of h44 of the small subunit. The conformation of the factor is distinctly different from that of the isolated factor. The protein is more extended, and while domains G and II retain their normal interactions with the 50S and 30S subunits respectively, domains III, IV and V are stretched out in such a way that the tip of domain IV is about 37 Å from its position in the crystal structure of H573A in relation to the G-domain (Laurberg *et al.*, 2000). The difference is somewhat less in relation to the wild-type factor (Al-Karadaghi *et al.*, 1996). The structures of EF-G bound to the ribosome with GDPNP or GDP and FA are overall similar, but the maps with GDP and FA are less well defined for domains III, IV and V (Valle *et al.*, 2003b). The cryo-EM experiments of EF-G bound to the ribosome may not be fully representative for normal translocation. A P-site with a peptidyl-tRNA prevents the binding of EF-G:GTP, as seen by biochemical experiments (Zavialov & Ehrenberg, 2003) and the very low

Table 9.6 Interaction of EF-G with the Ribosome (Wilson & Noller, 1998; Frank & Agrawal, 2001; Valle *et al.*, 2003b)

Region of EF-G	30S	50S
G-domain		GAR at base of L12 stalk
GTP binding site		SRL (of H95)
Switch I		
G'-domain		L11 NTD
Domain II	shoulder; h4, h5, h15, S4 region	
Domain III	S12	
Domain IV	decoding site, h44	H69
Domain V		H43, H49; Thiostrepton loop, L11

occupancy of the factor (Agrawal *et al.*, 1999). Therefore, the peptide was removed with puromycin, leaving EF-G:GDPNP or EF-G:GDP:FA bound to 70S ribosomes with a deacylated tRNA in the P-site (Valle *et al.*, 2003b). The fact that the ratchet-like movement was observed in the former case suggests that the complexes are anyway functional. Since the complex was already pretranslocated, domain IV is located in the decoding part of the A-site.

GTP Hydrolysis and Translocation

As has already been described for EF-G, GTP hydrolysis precedes translocation but the details of the enzymatic mechanism are not known. As for all the GTPases, one would expect there to be an Arg-finger, possibly in Switch I. However, Arg59 is not the Arg-finger since mutations of it have a limited effect (Mohr *et al.*, 2000). The conserved Arg29 has a more dramatic effect on the GTP hydrolysis and seems essential (Mohr *et al.*, 2000). Arg29 is at a distance of about 11 Å from the β-phosphate of GTP. Its role in GTP hydrolysis may be direct or indirect.

In studies of translocation with different nucleotides, Zavialov and Ehrenberg (2003) came to conclusions different from Rodnina *et al.* (1997). The former investigators used puromycin or RF2 to analyze to what extent translocation has occurred with EF-G in complex with GDP, GDPNP and GTP. They made the observation that translocation with GDP is insignificant. Using EF-G with GDPNP, the nascent peptide is reactive with puromycin but not RF2. Evidently, the peptidyl tRNA is no more in the A-site part of PTC. They suggest that EF-G:GDPNP may drive the A- and P-site tRNAs to intermediate states, the A/P and P/E hybrid states. This may coincide with an initial part of the ratchet-like movement of the subunits (Valle *et al.*, 2003b). This movement may be related to the placement of domain IV of EF-G at the decoding part of the A-site (Valle *et al.*, 2003b). Whether the ribosome is in the pretranslocation state, an intermediate state or the post-translocational state, as long as peptidyl-tRNA or EF-G is bound to the A-site, it is not accessible for RF2. Since RF2 needs to bind to the A-site to hydrolyze the nascent peptide from the P-site tRNA, EF-G has to dissociate before this can happen. This does not prevent puromycin from reacting at the PTC on the large subunit. As soon as the CCA-end of the A-site tRNA is moved into the A/P-site, puromycin can bind and react (Sharma *et al.*, 2004). A slow increase of the accessibility of both puromycin and RF2 to react with the nascent peptide is interpreted as a slow dissociation of EF-G that would lead to completion of translocation.

EF-G cannot rebind or hydrolyze GTP on the post-translocated ribosome with a peptidyl-tRNA in the P-site, since this is the locked state (see Section 8.6; Zavialov & Ehrenberg, 2003). When EF-G is trapped on the ribosome with GDPNP or FA, further protein synthesis is inhibited because the A- and T-sites are blocked.

EF-G Mutants

The domains, functional groups and sites of EF-G have been thoroughly investigated by mutational analysis (Table 9.7). The functional properties

Table 9.7 Some Mutants in EF-G and their Functional Effects (residue numbers refer to *T. thermophilus*). For references, see the text.

Mutation	GTPase	Trans-location[1]	Puromycin Reactivity[2]	Reference
Domain deletions				
ΔI	−	−	Low	Borowski *et al.*, 1996
ΔIII	−	−	−	Martemyanov & Gudkov, 2000
ΔIV	+	Low	Low	Rodnina *et al.*, 1997
	+	−	−	Martemyanov & Gudkov, 1999
ΔIV + V	+		Low	Savelsbergh *et al.*, 2000a
ΔV	+		Low	Savelsbergh *et al.*, 2000a
Local mutations				
G502D	?	Low		Hou *et al.*, 1994
R504T	+			Kolesnikov & Gudkov, 2003
501–504 GSGT	+	Low	Low	Kolesnikov & Gudkov, 2003
H573K/R	+	−		Savelsbergh *et al.*, 2000a
H573A	+	+		Martemyanov & Gudkov, 1998
573–578 GTGSVD	+	Low	−	Martemyanov & Gudkov, 1998
G162C–T649C	+	−		Peske *et al.*, 2000

[1]Translocation was identified by the release of deacylated tRNA from the E-site.
[2]Puromycin reactivity with peptidyl tRNA after treatment with the mutant factor.

investigated were the ability to perform GTP hydrolysis and translocation and the coupling between these two properties. In addition, several mutants were investigated as to whether they, after interaction with the ribosome, would permit puromycin to react with the nascent peptide as an early step of translocation.

In some experiments, whole domains were deleted. Here it was seen that a deletion of domain IV and/or domain V had no effect on GTP hydrolysis, but that such truncated factors could not translocate (Rodnina *et al.*, 1997; Martemyanov & Gudkov, 1999; Savelsbergh *et al.*, 2000a). This would agree with the movement of domain IV into the decoding site during translocation. EF-G, where domain III was deleted, could neither hydrolyze GTP nor translocate (Martemyanov & Gudkov, 2000). However, this truncated factor bound quite well to ribosomes in the presence of GDP and FA. We do not understand EF-G well enough to be able to explain why this truncated factor cannot hydrolyze GTP, but it is evident that domain III is a central element in the factor.

With the realization that domain IV at translocation enters into the decoding site (Frank & Agrawal, 2000; 2001; Valle *et al.*, 2003b), a focus on mutations in the tip of domain IV seemed natural. Domain IV has three loops at the tip of the domain 501–504, 530–534 and 573–578. Martemyanov *et al.* (1998) initially focused on the loop with a conserved histidyl residue His583 (573 in *T. thermophilus*). Both replacements and insertions were tried without and with effect on translation. Kolesnikov & Gudkov (2003) analyzed the effect of mutations in all the three loops and came to the conclusion that an intact conformation of two loops (501–504 and 573–578), is needed for efficient translocation. On the other hand, mutations of the loop 530–534 on the tip side of domain IV or mutations of the loops at the base of the domain (517–519 and 554–556) had no effects. The mutation that is locked by a disulphide bridge cannot translocate, but can hydrolyze GTP even though the identification of the residue proximity comes from the GDP conformation (Peske *et al.*, 2000).

9.4 RELEASE FACTORS

The termination or release factors RF1 and RF2 hydrolyze and release the completed polypeptide from the P-site tRNA in response to a stop codon (Ganoza, 1966; Kisselev *et al.*, 2003). RF1 and RF2 are homologues (Craigen *et al.*, 1985). RF1 recognizes the stop codons UAA and UAG, whereas RF2 responds to UAA and UGA (Scolnick *et al.*, 1968). The expression of RF2 in most bacteria needs a shift of the reading frame (see Chapter 4). In eukaryotes, there is only one factor, eRF1, which recognizes all three stop codons (Zouravleva *et al.*, 1995). These release factors are of class 1. RF3, which is a class 2 release factor (Zouravleva *et al.*, 1995) is a tGTPase. It catalyses the removal of release factors RF1 and RF2 from the ribosome (Caskey *et al.*, 1969; Capecchi & Klein, 1969; Freistroffer *et al.*, 1997). RF3 is not found in some bacterial species, among them *T. thermophilus*. Thus, RF3 could not be essential. A corresponding molecule in eukaryotes is called eRF3 (for a review, see Inge-Vechtomov *et al.*, 2003). This type of factor is not found in archaea (Kisselev & Buckingham, 2000). Whether a different protein substitutes for RF3 in the species that lack it is not known.

The termination factors compete with the tRNAs for the decoding of the mRNA. Obviously, there is normally no tRNA for the stop codons. Two exceptions are the tRNAs coding for Se-Cys and pyrrolysine (see above Section 4.4).

A genetic search by swapping conserved regions between RF1 and RF2 identified a region that switches the recognition of stop codons (Ito *et al.*, 2000). The so-called anticodon mimic was found to be the tripeptides PAT for RF1 and SPF for RF2 (Ito *et al.*, 2000; Nakamura & Ito, 2003). Mutations of these sites changed the codon specificity. It may seem surprising that a peptide as short as three residues is able to identify and discriminate between the stop codons. The side chains of three consecutive amino acid residues rarely point in the same direction. It is also surprising that nonpolar residues like alanyl, prolyl and phenylalanyl would be used for selective interactions. Further analysis is needed to understand this mechanism. A conserved sequence in eRF1 that could have the corresponding function is NIKS (Frolova *et al.*, 2002).

Class 1 release factors hydrolyze or trigger the hydrolysis of the ester bond between the completed peptide and the P-site tRNA. The sequence GGQ is universally conserved (Frolova *et al.*, 1999). Mutants of class 1 release factors where the conserved motif changed are slow in hydrolyzing the peptide (Frolova *et al.*, 1999; Seit-Nebi *et al.*, 2001; Zavialov *et al.*, 2002).

For a long time, it was thought that release factors of class 1 would mimic tRNA (Smrt *et al.*, 1970; Brown & Tate, 1994; Ito *et al.*, 2000). Thus, the decoding and hydrolyzing sites would be expected to be about 75 Å apart. The first experimental insight into this was obtained when Song *et al.* (2000) determined the structure of human eRF1. The protein is composed of three domains with the shape of a Y. Some spatial relationship to tRNA could be identified. Here the peptide, GGQ, is located at one extreme of the molecule. The part that probably decodes the mRNA is located at the end of another domain. Whether the conformation of eRF1 is the same when bound to the ribosome as in the crystal structure remains to be elucidated.

The structure of *E. coli* RF2 has also been determined (Plate 9.17; Vestergaard *et al.*, 2001). As was expected, from sequence comparisons, the eukaryotic and bacterial class 1 release factors have no structural similarity, except possibly a distant tRNA similarity. Only the tri-peptide GGQ is common to bacteria and eukarya (Frolova *et al.*, 1999). In the crystal structure of *E. coli* RF2, the two functional regions, the GGQ and SPF sequences, were about 23 Å apart, a much shorter distance than that between the decoding site on the small subunit and the PTC on the large subunit (Vestergaard *et al.*, 2002). From a comparison of cryo-EM images of RF2 bound to ribosomes and the crystal structure of the protein, a large conformational change in the factor could explain how the two functional sites of the molecule can reach the distant sites (Klaholz *et al.*, 2003; Rawat *et al.*, 2003). Such a drastic conformational change, where the structures of the domains are also rearranged, is quite rare and the inter- and intramolecular interactions allowing this flexibility are of great interest. When bound to the ribosome, domain I of RF2 is close to the factor-binding site, while the GGQ region reaches into the PTC (Plate 9.18A).

RF3 is a tGTPase that binds to terminated ribosomes, where RF1/2 is bound to the A-site in addition to the deacylated tRNA in the P-site (Freistroffer *et al.*, 1997). The structure of RF3 is not known except that it has domains G and II as in all tGTPases (Ævarsson, 1995). RF3 is unusual among the tGTPases in that it needs the terminated ribosome in complex with RF1/2 for nucleotide exchange (Zavialov *et al.*, 2001; 2002). RF3 has been studied when bound to the ribosome by cryo-EM at moderate resolution (Klaholz *et al.*, 2004). Two different conformations were observed: one in a prestate, and the other in a state where RF3 is bound at the GTPase center and associated with a conformational change of the ribosome like the one induced by EF-G.

9.5 RIBOSOME RECYCLING FACTOR (RRF)

Once the release factors have terminated the synthesis of a protein, the ribosome is in a post-termination state. The mRNA remains bound to the

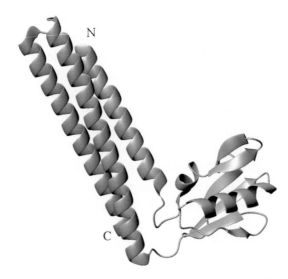

N

C

Fig. 9.19 The structure of RRF (Selmer *et al.*, 1999). (With kind permission from Selmer (2002).)

70S ribosome as well as a deacylated tRNA in the P-site. The ribosome may be part of a polysome. Observations of random reinitiation of protein synthesis have been made (Janosi *et al.*, 1998). The ribosome recycling factor (RRF) prevents such random reinitiation by recycling the components that are bound to the mRNA (Janosi *et al.*, 1996; Kaji *et al.*, 2001). RRF functions together with EF-G and is a conserved protein in bacteria and chloroplasts (Hirashima & Kaji, 1973).

There are different views on what RRF does. One line of observation is that RRF dissociates polysomes into monosomes and releases the mRNA and tRNA (Hirashima & Kaji, 1972). The other observation is that RRF only separates the two ribosomal subunits without releasing the mRNA or the deacylated tRNA (Karimi *et al.*, 1999).

The structure of RRF has been determined for a number of species (Selmer *et al.*, 1999; Kim *et al.*, 2000; Toyoda *et al.*, 2000; Yoshida *et al.*, 2001). The protein has two domains with the shape of an L (Fig. 9.19; Plate 9.14). Domain I is composed of three long α-helices and domain II is composed of β-strands as well as some shorter helices (Selmer *et al.*, 1999). The hinge between the two domains allows a fair amount of flexibility as is seen when the structures are compared.

The structure fits exceedingly nicely within the envelope of a tRNA molecule. This led to the hypothesis that RRF can initially bind to the ribosomal A-site and be translocated to the P-site by EF-G. This leads to the subsequent dissociation of the termination complex, whereby the deacylated tRNA and the mRNA are released (Selmer *et al.*, 1999). This hypothesis has been tested using chemical labeling of RRF when bound to the ribosome (Lancaster *et al.*, 2002) as well as a cryo-EM study (Agrawal *et al.*, 2004). The site of binding is more complex than anticipated (Plates 9.18B and 9.19). Domain I of RRF overlaps with the site of the acceptor ends of both the P- and A-site tRNAs. Domain II of RRF overlaps with the binding site for domain IV of EF-G (Agrawal *et al.*, 2004). After termination, there is a deacylated tRNA in the P-site. For the binding of RRF, this tRNA must move into the P/E hybrid site (Lancaster *et al.*, 2002).

9.6 tRNA MIMICRY

Many proteins that participate in protein biosynthesis interact with the ribosomal sites for tRNA (Table 9.8). The structures of several of these have been unraveled (see Nyborg *et al.*, 2000; Kristensen *et al.*, 2001; Brodersen & Ramakrishnan, 2003 for reviews).

The first clear observation of tRNA mimicry was for EF-G, where domains III, IV and V mimic the tRNA part of the ternary complex EF-Tu:GDPNP:aminoacyl-tRNA (Nissen *et al.*, 1995). During translocation, the tRNA mimicry part binds to the decoding part of the A-site (Agrawal *et al.*, 1997; Valle *et al.*, 2003b). As expected, the structure of yeast EF2 has essentially the same tRNA mimicking structure. However, domain IV is considerably broader (Jørgensen *et al.*, 2003).

Table 9.8 Binding of tRNA and Proteins to the tRNA Sites during Translation

	GTPase	A	P	E
Initiation	IF-2	IF1	f-Met-tRNA	IF3-NTD
Elongation				
aa binding	EF-Tu	aa-tRNA	pp-tRNA	tRNA
translocation	EF-G	EF-G	pp-tRNA	tRNA
Termination		RF1/2	pp-tRNA	
	RF3		tRNA	
Recycling			RRF	tRNA
	EF-G	EF-G	RRF?	tRNA?
Proteins that interact with the A-site				
IF1, EF-G, RF1/2, RRF, stringent factor, RaiA.				

RRF is an excellent mimic of a tRNA, judging from its structure (Selmer *et al.*, 1999). However, it does not bind to the A-site in a tRNA-like manner after termination (Lancaster *et al.*, 2002; Agrawal *et al.* 2004). In addition, the eukaryotic termination factor 1 (eRF1) and bacterial RF2, that cause the hydrolysis of the peptide when a stop codon is encountered in the A-site, have structures that, to some extent, mimic tRNA (Song *et al.*, 2000; Vestergaard *et al.*, 2002). In the case of the bacterial factor, a large conformational change is necessary to bring it from the state in the crystals to the state when binding to ribosomes. Here the potential tRNA mimicry is not used at all (Rawat *et al.*, 2003; Klaholz *et al.*, 2003). As Brodersen & Ramakrishnan (2003) have phrased it: "Shape can be seductive"! (Plate 9.14).

Several other proteins that participate in protein synthesis also mimic tRNA in different manners. Thus, EF-P has a structure that has the shape of an L with the same dimensions as a tRNA molecule (Benson *et al.*, 2000). It is not known if, where and when EF-P binds to the ribosome and it thus remains to be established if this mimicry is coincidental or not (see Section 9.2). The four C-terminal domains of SelB also have a tRNA mimicry structure (Selmer and Su, 2002). In this case, the function is better understood. It has nothing to do with the tRNA binding sites. A ribosomal protein from *T. thermophilus*, TL5, also has the classical tRNA shape (see Section 7.4; Fedorov *et al.*, 2001). This protein is also a general stress protein, called CTC, that is expressed in high quantities in *B. subtilis* if the bacteria are exposed to stress conditions (Völker *et al.*, 1994). It is not known whether the receptor for this tRNA-like stress protein requires a tRNA-like shape.

The factor, IF1, is too small to mimic a tRNA. However, it binds to the decoding part of the A-site on the small subunit (Carter *et al.*, 2001). The N-terminal domain of IF-3 binds to the part of the E-site that is located on the small subunit (see Section 9.2; Moazed *et al.*, 1995; Dallas & Noller, 2001). Even if these proteins do not mimic a tRNA in shape, they bind to sites for tRNA on the ribosome. In some way, they also mimic tRNA. A summary of the proteins that mimic tRNA or bind to tRNA sites is provided in Table 9.9. It is remarkable that in all steps of protein synthesis, except during tRNA binding, specific proteins occupy the A-site (Kristensen *et al.*, 2001). Furthermore, the bacterial translation factors can simply be divided into two main categories (Kristensen *et al.*, 2001). EF-G belongs to both categories:

1. Factors that bind to the tRNA-binding sites
2. tGTPases.

Table 9.9 GTPases and tRNA Mimics in the Translation System

Factor	GTPase	tRNA Shape	tRNA Binding Site	Comments
IF-1			A	Anticodon stem loop
IF-2	+			Binds 50S to preinitiation complex
IF-3			E	Anti association
EF-P		+	?	Homologous to eIF5A
EF-Tu	+			Binds aa-tRNA to ribosome
SelB	+			mRNA binding part has L-shape
EF-G	+	+	Decoding site	Translocase
RF-1,2		+?	A	Peptide release
RF-3	+			RF-1/2 release
RRF		+	A + P	Recycling of ribosomal subunits
TL5		+		Ribosomal protein of variable size, shock factor

9.7 ADDITIONAL RIBOSOME-BINDING GTPASES

Bacteria have a core of 11 universally conserved GTPases. In many species, additional GTPases are important for cell survival. They interact primarily with RNA. Some GTPases are shown to bind to the ribosome (Table 9.10). The analyses are not as conclusive as for the classical factors, but in several cases, the functions as well as the binding site have been established. Caldon *et al.* (2001) make a summary of the current knowledge. One group, the FtsY/Ffh family, is involved in the transport of the nascent protein to a different compartment. They are part of the signal recognition particle with its associated proteins (see Chapter 12). According to the sequence relationship, LepA also belongs to the family of tGTPases (Ævarsson, 1995), but its function and structure remain unknown.

The function of the proteins of the Era family is poorly known. The structure of Era from *E. coli* has been determined (Chen *et al.*, 1999). The C-terminal domain has an RNA binding character. EngA also belongs to this family and is unusual in that it has two closely related G-domains (van Doorn *et al.*, 1997). TrmE of the Era family is involved in tRNA modification (Cabedo *et al.*, 1999). In yeast mitochondria, it binds to the small subunit 15S RNA (Decoster *et al.*, 1993).

Table 9.10 Additional Ribosome Binding GTPases (Caldon *et al.*, 2001)

Family/Protein	Binding Site	Comments	References
tGTPases			
LepA		Closely related to the tGTPases.	Caldon *et al.*, 2001
Era family			
Era	16S rRNA	Ribosomal maturation?	Chen *et al.*, 1999 Hang *et al.*, 2001 Inoue *et al.*, 2003
EngA		Double G-domain; Compensates for RrmJ deletion	Tan *et al.* 2002
TrmE/MnmE		C-term G-domain; involved in tRNA modification; binds to 16S rRNA.	Cabedo *et al.*, 1999
FtsY/Ffh family			
FtsY/Ffh	4.5S RNA	Signal recognition particle. Interacts with the ribosome.	Herskovits & Bibi, 2000
Obg family			
Obg	L13	Compensates for RrmJ deletion. Interacts with r-protein L13. Required for stress activation of transcription factor σ^B. Three-domain protein where the G-domain is central.	Tan *et al.*, 2002 Scott *et al.*, 2000 Buglino *et al.*, 2002 Kukimoto-Niino *et al.*, 2004
TypA family			
TypA/BipA		Tyrosine phosphorylated protein. Similar to EF-G. Regulation of expression of target proteins.	Pfennig & Flowers, 2001

The Era and Obg family may be sensors of the status of the ribosome (Caldon *et al.*, 2001). Obg binds specifically to protein L13 in the ribosome (Scott *et al.*, 2000). L13 is located below the tGTPase binding site (Ban *et al.*, 2000). The structure of a fragment of Obg from *B. subtilis* (Buglino *et al.*, 2002) and the complete protein have been determined for

T. thermophilus (Kukimoto-Niino *et al.*, 2004). Obg is a highly elongated protein composed of three domains, where the G-domain is the central domain. Flexible hinges connect the domains and large conformational changes have been identified.

9.8 RIBOSOME RESCUE FACTORS

A multitude of proteins that interact with ribosomes under different conditions have been identified (Table 9.11). These are more or less well characterized in their binding and function. A number of them are clearly ribosome rescue factors with a role at different physiological situations and problems. A selection of such proteins is briefly described here.

Tet(O)/Tet(M)

A number of mechanisms lead to resistance to tetracyclines. One of these includes a range of ribosome rescue proteins, the best characterized of which may be Tet(M), and Tet(O) (Burdett, 1996; Trieber *et al.*, 1998; Dantley *et al.*, 1998). These proteins eliminate the antibiotic inhibitor tetracycline from the ribosome in a GTP-dependent manner (Burdett, 1996; Trieber *et al.*, 1998). The primary binding site for tetracycline is located on the small subunit in the region between the shoulder and head near the decoding part of the A-site (see Section 10.3).

The tetracycline resistance proteins have a mass of 70 kDa and about 50% sequence similarity to EF-G (Sanchez-Pescador *et al.*, 1988). All six

Table 9.11 Ribosome Rescue Factors

Protein	Role	Binding Site
Tet(O)/Tet(M)	Tetracycline removal	tGTPase
RelA/SpoT	Regulation at amino acid starvation	A-site
RelE	Inhibitor of translation at nutritional stress.	?
tmRNA	Recovery of stalled ribosomes with a fragmented mRNA	A/T-site
RaiA	Prevents ribosome dissociation during environmental stress	30S subunit
RMF	Formation of inactive 100S ribosome dimers during environmental stress	50S subunit

domains of EF-G have corresponding parts in Tet(O). With this similarity, one would expect a similarity with EF-G in binding to the ribosome. Spahn *et al.* (2001) report a cryo-EM investigation of the binding of Tet(O) in complex with GTP-γS to the ribosome. Indeed, the protein binds essentially like EF-G with one clear difference: domain IV, which mimics the anticodon part of the tRNA in a ternary complex, does not extend into the decoding part of the A-site, but rather interacts with the junction between the head and the shoulder of the small subunit where the binding site for tetracycline is located (Brodersen *et al.*, 2000; Pioletti *et al.*, 2001). There is yet another important difference. While EF-G, when binding to the ribosome, induces a rotational rearrangement of the subunits with regard to each other and conformational changes within the small subunit (Agrawal *et al.*, 1999; Frank & Agrawal, 2000), Tet(O) does not induce these changes (Spahn *et al.*, 2001). While EF-G with a noncleavable GTP analog interacts through its domain IV with the largest subunit bridge (B2a; H69 of the 50S subunit), Tet(O) does not reach this bridge. The conformational changes induced by EF-G are probably important for translocation. However, Tet(M) or Tet(O) that do not induce these changes cannot substitute for EF-G (Burdett, 1991; 1996).

The mechanism of binding and dissociation of Tet(O) seems to be just like that of EF-G. The protein binds to the ribosome with GTP and after tetracycline removal and GTP hydrolysis, it dissociates from the ribosome (Connell *et al.*, 2003). It is unlikely that Tet(O) makes a steric clash with tetracycline. It is more probable that it induces conformational changes in h34 and h18 of the 16S RNA that leads to the dissociation of tetracycline (Spahn *et al.*, 2001).

When Tet(O) is bound to the ribosome, its domain III is close to protein S12 (Spahn *et al.*, 2001). Mutations in S12, whether causing streptomycin resistance or streptomycin dependence, result in decreased Tet(O) activity. One mutation (K42Q) abolishes the Tet(O) activity totally (Taylor *et al.*, 1998).

Stringent Response

When *E. coli* experiences starvation of an amino acid or other nutrients, the level of uncharged tRNA goes up significantly (Yegian *et al.*, 1966). This can lead to a situation where a deacylated tRNA occupies the A-site. This induces what is called stringent control of transcription, i.e. transcription is immediately inhibited (Stent & Brenner, 1961). A protein called stringent factor, ppGpp synthase I (PSI) or RelA is expressed by the gene *relA*. It binds to ribosomes with deacylated tRNA in the A-site to produce what is called magic spots I and II (MSI and MSII), guanine nucleotide tetra- and penta-phosphates (ppGpp and pppGpp). These nucleotides are produced from ATP and GTP or GDP (Fig. 9.20; Haseltine & Block 1973; Sy & Lipman, 1973). Stringent control is relieved by mutations of RelA or a protein produced by a gene called *relC*. The product of *relC* is identical to ribosomal protein L11 (Friesen *et al.*, 1974; Parker *et al.*, 1976). Ribosomes devoid of L11 are inactive in the synthesis of (p)ppGpp (Wendrich *et al.*, 2002). However, L11-deficient ribosomes still bind RelA. The synthesis of ppGpp by RelA does not lead to dissociation of deacylated tRNA from the A-site but causes RelA to dissociate from the ribosome (Wendrich *et al.*, 2002). When there is sufficient amounts of the lacking amino acid, the tRNAs will be charged and the ternary complex formed will compete with the bound deacylated tRNA for the A-site (Schilling-Bartetzko *et al.*, 1992; Wendrich *et al.*, 2002). ppGpp production inhibits transcription of genes involved in the translational apparatus (Lazzarine & Dahlberg, 1971; Dennis & Nomura, 1974) but upregulates genes encoding enzymes involved in amino acid biosynthesis (Cashel *et al.*, 1996; Zhou & Jin, 1998).

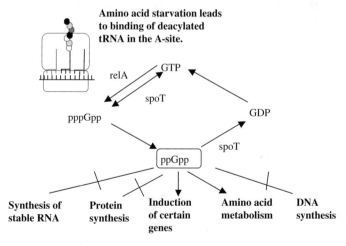

Fig. 9.20 The synthesis, degradation and role of (p)ppGpp dependent on amino acid starvation (after Izutsu *et al.*, 2001). RelA and SpoT control the synthesis and breakdown of the nucleotide, which can inhibit the synthesis of DNA, rRNA, tRNA and proteins but induce some genes related to amino acid metabolism.

In gram-negative bacteria, there is a second protein related to RelA. This protein produced by the *spoT* gene is called SpoT or PSII (Laffler & Gallant, 1974). While RelA and SpoT are guanosine 3′,5′-bis(diphosphate) synthetases, SpoT is also a guanosine 3′,5′-bis(diphosphate) 3′-phospho-hydrolase (ppGppase; Heinemeyer & Richter, 1977; Sy, 1977). Gram-positive bacteria have only one protein called Rel, which like SpoT, has both synthetase and hydrolase activities (Wendrich & Marahiel, 1997; Mittenhuber, 2001). In a strain where the PSII activity was deleted, the contribution of RelA to (p)ppGpp synthesis was analyzed. It was found that RelA was responsible for the initial burst of (p)ppGpp synthesis during glucose and amino acid starvation, but that RelA was inactive when amino acids were available (Gentry & Cashel, 1996).

An examination of the *E. coli* spoT protein of 702 amino acid residues showed that the ppGppase activity is contained in the first 203 residues and that the PSII activity is contained in the region 67–374 (Gentry & Cashel, 1996). A further study indicated that Asp 293 is indispensable for ppGpp synthesis in *E. coli* SpoT (Fujita *et al.*, 2002).

Structural information on the RelA/SpoT enzymes has recently become available. The structure of the N-terminal half of the Rel protein from *Streptococcus dysgalactiae equisimilis* has been determined (Hogg *et al.*, 2004). This is a gram-positive bacterium and the Rel protein contains both hydrolase and synthetase activities. Residues 5–159 contain the hydrolase domain and residues 176–371 contain the synthetase domain. An overlapping central 3-helix bundle connects the two domains.

The hydrolase domain is α-helical and is structurally homologous with the catalytic domain of cyclic nucleotide phosphodiesterases (PDEs; Xu *et al.*, 2000; Huai *et al.*, 2003). PDEs all have a conserved His-Asp sequence and are metallophosphohydrolases (Aravind & Koonin, 1998). In the case of Rel, a manganese ion is bound at these conserved residues. The synthetase domain is a mixed α/β-protein with a fold related to the palm domain of mammalian DNA polymerase β (Sawaya *et al.*, 1997). The two active sites are located about 30 Å apart (Hogg *et al.*, 2004).

The crystals of Rel contain two conformations of the N-terminal half of the protein. In both, there is a GDP molecule bound to the synthetase active site. In one molecule, the synthetase active site seems to be in an active conformation. Here, the hydrolase is in an inactive conformation. In the other molecule with the inactive conformation of the synthetase active site, there is an unusual GDP derivative seen at the hydrolase active site.

Table 9.12 Proteins Involved in Synthesis, Conversion of pppGpp to ppGpp and Breakdown of (p)ppGpp

Bacteria	Synthesis	Conversion	Degradation
Gram-negative	RelA, (SpoT)	GPP	SpoT
Gram-positive	Rel		Rel

This nucleotide is best interpreted as a guanosine 5′-diphosphate-2′: 3′-cyclic monophosphate (ppG2′:3′p; Hogg *et al.*, 2004). It seems evident that the two sites with opposing activities control each other. Most likely, the C-terminal, ribosome binding, domain also affects the activities of the two active sites.

The management of suitable levels of (p)ppGpp in bacteria (Table 9.12) involves one more protein, called GPP. It is produced from the *gppA* gene, and is important for the conversion of the penta-phosphate to the tetra-phosphate (Keasling *et al.*, 1993). The structure was recently determined (Kristensen *et al.*, 2004).

RelBE

Two additional proteins related to nutritional stress are RelB and RelE. These proteins form a toxin/antitoxin pair (Gotfredsen & Gerdes, 1998; Gronlund & Gerdes, 1999). RelE is a toxin by being a global inhibitor of translation, while RelB can neutralize its effect, probably by forming a protein-protein complex. RelE most likely binds to the ribosome (Galvani *et al.*, 2001), but no structural details are known. RelB is an autorepressor of RelBE transcription and RelE is a corepressor. Normally the proteins are not expressed. During amino acid starvation, the level of RelB decreases and, as a consequence, the expression of RelBE increases dramatically (Christensen *et al.*, 2001). The protease Lon specifically degrades RelB and thereby activates RelE as an inhibitor of translation (Christensen *et al.*, 2001). Evidently, this is an interesting set of proteins for structural and functional studies.

tmRNA

Bacterial ribosomes can get stalled at the 3′-end of a degraded or defective mRNA. The factor that recovers the ribosomes is an RNA molecule of about 300 nucleotides called 10Sa RNA, SsrA or tmRNA (see review by Kazai *et al.*, 2000). As the third name indicates, this RNA molecule functions both as messenger and tRNA in a process called trans-translation (Keiler *et al.*, 1996).

The tRNA-like domain (TLD) of tmRNA can be charged at its 3'-CCA-end with alanine by AlaRS and delivered to the stalled ribosome in complex with EF-Tu and GTP (Kazai *et al.*, 2000). Once the alanyl residue is added to the nascent peptide and the TLD is translocated to the ribosomal P-site, an open reading frame of the tmRNA will be translated. Ten amino acids are incorporated up to the stop codon of the tmRNA. In this way, the ribosome is rescued and the released peptide can be recognized and eliminated by the degrading systems of the cell. A small protein, SmpB, is associated with the TLD of tmRNA (Kazai *et al.*, 2000).

The structure of the tmRNA when bound to the ribosome in complex with SmpB, EF-Tu:GDP and kirromycin has been characterized (Valle *et al.*, 2003c). The antibiotic inhibits the release of EF-Tu and the tmRNA cannot extend the nascent peptide. The complex is, therefore, seen in the state when it has just bound to the ribosome. Since the secondary structure of tmRNA is well established and since EF-Tu has been seen when bound to the ribosome, the density can be interpreted with good confidence (Plate 9.20). Ribosomal protein S1 also seems to have a role in tmRNA function (Valle *et al.*, 2003c).

The crystal structure of the TLD in complex with SmpB has also been determined (Gutmann *et al.*, 2003). Several distinct differences are observed and compared with the classical tRNA structure. The TLD has an open L-shape and the T-arm of the tRNA is rotated about 90°. SmpB is a protein with the OB-fold binding to the elbow of the TLD on the side of the D-loop. This orientation of the T-arm causes SmpB to be placed in the decoding center of the ribosome. Many parts of the mechanism for the trans-translation remain unknown, but an interesting start is made.

Ribosome Modulation Factor — RMF

In the stationary phase, *E. coli* becomes highly resistant to a number of environmental stresses. This change occurs simultaneously with the expression of a large number of genes. Among these genes are a number of ribosome-associated genes encoding proteins such as RMF (ribosome modulation factor); SRA (stationary phase-induced ribosome associated protein); protein Y (RaiA); and protein G (YhbH; see Izutsu *et al.*, 2001 and references therein). RMF binds to the large subunit, which has the effect that 70S ribosomes dimerize and form inactive 100S particles (Wada *et al.*, 1990). In this way, RMF regulates translation. A mutant where RMF is deleted was unable to form 100S particles in the stationary phase (Yamagishi *et al.*, 1993). RMF transcription is induced by the global regulator ppGpp (Izutsu *et al.*, 2001). So far, there has been no information about the structure of RMF or how it binds to the ribosome.

Proteins that Respond to Temperature Stress

The state of the ribosome can induce cold-shock or heat-shock proteins (VanBogelen & Neidhardt, 1990). Some such proteins can be co-purified with the ribosome and seem to have ribosome-related functions.

Protein Y — pY — RaiA

Protein Y is a small protein found in many bacteria and also in chloroplasts. It is produced by the *yfiA* gene (Agafonov *et al.*, 1999; Maki *et al.*, 2000; Agafonov *et al.*, 2001; Zhou & Mache, 1989; Schmidt *et al.*, 1993; Johnson *et al.*, 1990). Sometimes it is also called protein F. Agafonov *et al.* (2001) suggest the name ribosome associated inhibitor A (RaiA). It binds to the small ribosomal subunit and prevents ribosomes from dissociating when bacteria are exposed to environmental stress, particularly at low temperatures or at the stationary phase with high cell density (Maki *et al.*, 2000; Agafonov *et al.*, 2001). Binding is probably to the 30S part of the A-site in the ribosomal interface where it inhibits translation (Agafonov *et al.*, 2001). RaiA and tetracycline do not interfere in their binding to the A-site (Agafonov *et al.*, 2001). The structure of RaiA from two bacterial species is known. The protein is composed of four β-strands and two α-helices (Parsons *et al.*, 2001; Ye *et al.*, 2002). The arrangement is similar to double-stranded RNA-binding domains, but not close enough so that a binding mode can safely be proposed. The structure of RaiA when bound to the ribosome is not known.

RbfA

Ribosome binding factor A (RbfA) is a small cold-shock adaptation protein in bacteria that is essential for bacterial growth at low temperatures (Jones & Inouye, 1996). It binds to the small subunit and seems essential for the maturation and assembly of the 30S subunit during cold-shock conditions (Bylund *et al.*, 1998). The structure of RbfA is known (Huang *et al.*, 2003).

CsdA

Among the many proteins induced by cold-shock, CsdA is among the larger (70K). It co-purifies with the ribosome. Its most likely function is to unwind double-stranded mRNAs in the absence of ATP (Jones *et al.*, 1996).

10

Inhibitors of Protein Synthesis, Antibiotics, Resistance

Numerous compounds inhibit protein synthesis (Vazquez, 1974). Many of them are known to bind to the ribosome. The inhibitors are to a large extent natural products isolated from different microorganisms. Microorganisms excrete these antibiotics in their fight for living space. Normally, the organism itself is resistant to the antibiotics it produces. Thus, the resistance must have coevolved with the antibiotics with the potential of spreading to other organisms. A number of antibiotics, which target the protein synthesis machinery, are clinically used since they selectively inhibit certain bacteria. The microbial resistance to antibiotics is a serious and growing health problem (Chopra, 2000). The search for new synthetic or semisynthetic inhibitors and antibiotics for which there are no evolved resistance mechanisms, is therefore a major effort (Knowles *et al.*, 2002).

An inhibitor acts like a "spanner in the works." In the ribosome, an antibiotic may inhibit a functional process by binding to an essential binding site or by changing the functional dynamics of the machinery in such a way that the transition to the next state is blocked. Thus, as in the studies of other enzymes, analyses of the translational inhibitors and the resistance against them provide a very good means of understanding the process of translation and to study the interplay between mRNA, tRNA, ribosomal components and translational factors during protein syntheses.

Since the functional sites of the ribosome are primarily composed of rRNA, it is not surprising that antibiotic binding sites are generally located on the rRNA (Cundliffe, 1987; 1990). However, since there are usually multiple copies of the ribosomal RNAs in the genomes, mutations of the ribosomal RNAs can hardly lead to resistance. Thus, antibiotic resistance is frequently due to enzymes that can modify the antibiotic or the rRNA, but resistance can also be caused by mutations in ribosomal proteins, which are usually encoded by only one gene each.

The ribosomal antibiotics inhibit a range of different steps of protein synthesis. We will describe only a limited number of them and focus primarily on those for which the binding sites are identified in detail by crystallography or other means.

It is quite remarkable that antibiotics that cause the ribosome to be inhibited with a filled or blocked A-site induce a cold-shock response, while antibiotics that inhibit the ribosome when the A-site is empty induce heat shock (VanBogelen & Neidhardt, 1990). Evidently, the ribosome acts as a sensor of the state of the cell. Some cold shock proteins that bind to the ribosome are discussed in Section 9.8.

Table 10.1 Ribosomal Inhibitors Mapped by Structural Methods

Inhibitor	Binding Site	References
Inhibitors of initiation		
Edeine	30S, E-site, h24, h28, h44, h45	Pioletti *et al.*, 2001
Decoding site inhibitors		
Streptomycin	30S, h18, h27, h44 and S12	Carter *et al.*, 2000
Paromomycin	30S, h44	Fourmy *et al.*, 1996; 1998; Carter *et al.*, 2000
Gentamycin		Yoshizawa *et al.*, 1998
Inhibitors of aminoacyl-tRNA binding		
Tetracycline	30S	Brodersen *et al.*, 2000 Pioletti *et al.*, 2001
Peptidyl transfer inhibitors		
Anisomycin	50S, C2452, U2504	Hansen *et al.*, 2003
Blasticidin S	50S, 2 sites: G2251 or G2252	Hansen *et al.*, 2003

Table 10.1 (Continued)

Inhibitor	Binding Site	References
Chloramphenicol	1st site: 50S, G2061; C2452, U2504	Schlünzen *et al.*, 2001,
	2nd site: 50S, A2058, A2059	Hansen *et al.*, 2003
Clindamycin	50S	Schlünzen *et al.*, 2001
Puromycin	50S	Nissen *et al.*, 2000; Schmeing *et al.*, 2002; Bashan *et al.*, 2003
Sparsomycin	50S, P-site CCA-end, U2585, A2602	Schlünzen *et al.*, 2001; Hansen *et al.*, 2002b; 2003; Bashan *et al.*, 2003
Virginamycin M	50S, A2062, A2451, C2452	Hansen *et al.*, 2003
Exit tunnel inhibitors		
ABT-773	50S	Schlünzen *et al.*, 2003
Azithromycin	50S	Hansen *et al.*, 2002a; Schlünzen *et al.*, 2003
Carbomycin A	50S	Hansen *et al.*, 2002a
Erythromycin	50S, A2058, A2059,	Schlünzen *et al.*, 2001
Roxithromycin	50S, A2958, A2059	Schlünzen *et al.*, 2001
Spiramycin	50S	Hansen *et al.*, 2002a
Telithromycin	50S, A2059, G2502, A2958	Berisio *et al.*, 2003a
Troleandomycin	50S	Berisio *et al.*, 2003b
Tylosin	50S	Hansen *et al.*, 2002a
Factor related inhibitors		
Translocation		
Hygromycin B	30S, P-site, h44	Brodersen *et al.*, 2000
Spectinomycin	30S, h34, hinge head-body	Carter *et al.*, 2000
Pactamycin	30S, E-site, h23b, h24a,	Brodersen *et al.*, 2000
EF-Tu		
Kirromycin/ Aurodox	Domain interface in EF-Tu, G-III	Vogeley *et al.*, 2001
GE2270A	Domain interface in EF-Tu, G-II	Heffron & Jurnak, 2000
EF-G		
Fusidic acid	EF-G, EF2	
Sordarin	EF2, domain interface	Jørgensen *et al.*, 2003

10.1 INHIBITORS OF INITIATION

Edeine and pactamycin inhibit protein synthesis in all domains of life (Odom *et al.*, 1978). Thus, it is not surprising that they bind to conserved regions of the ribosome. Edeine protects 16S RNA at a site that overlaps with kasugamycin and pactamycin (Woodcock *et al.*, 1991; Mankin, 1997).

Despite very different structures, edeine and pactamycin bind to essentially the same region of the 30S subunit, between the P- and E-sites (Brodersen *et al.*, 2000; Pioletti *et al.*, 2001). However, edeine induces base pairing of G693:C795 (Brodersen *et al.*, 2000), while pactamycin breaks it (Pioletti *et al.*, 2001). They have both been thought to affect initiation, but a more thorough analysis has clarified that only edeine inhibits initiation, whereas pactamycin inhibits translocation (Dinos *et al.*, 2004). They may affect the mobility of the platform during subunit association or displace the mRNA in the E-site region (Carter *et al.*, 2000). This would disturb the SD interaction, which occurs just beyond the E-site codon (Yusupova *et al.*, 2001). Furthermore, both antibiotics interact with bases that are protected by IF3 (Moazed *et al.*, 1995). Edeine and pactamycin seem to act antagonistically (Dinos *et al.*, 2004).

Edeine (Ede)

Edeine is a spermidine-like compound that binds to a region of central importance in the small subunit and interacts with helices h24, h28, h44 and h45 (Pioletti *et al.*, 2001). The aromatic part of Ede makes hydrogen bonds with G926 in h28 like a regular base pair, and A790–A792 in the loop of h24 interact through their sugars with the hydrophobic part of edeine. This leads to a distortion of h24a that induces C795 to base pair with G693 of h23b. Ede inhibits initiation by preventing the binding of fMet-tRNA to the small subunit (Dinos *et al.*, 2004). Addition of Ede after the formation of the initiation complex had no inhibitory effect, since the binding site was already partially occupied. The inhibition is due to a disturbance of recognition of the AUG start codon in the P-site (Pioletti *et al.*, 2001; Dinos *et al.*, 2004). Ede also increases the level of misincorporation (Dinos *et al.*, 2004).

10.2 INTERFERENCE WITH DECODING. DISTORTION OF FIDELITY

A number of antibiotics disturb the decoding mechanism by binding to the small subunit. These are classical observations (Davies *et al.*, 1964). It is not strictly correct to call them inhibitors since they rather interfere with the correct reading of the message and increase the frequency of translation errors. They permit binding not only of cognate, but also near-cognate tRNAs and probably also non-cognate tRNAs. They induce incorporation of the corresponding wrong amino acids. Since the decoding is done in two major steps — the initial recognition and the proofreading steps (see Section 11.4) — one or both steps could be affected.

The antibiotics affecting decoding primarily bind to the decoding center in the neck region of the small subunit close to h44. As discussed in Section 8.6, they affect the balance of the small subunit between a restrictive state and the so-called ribosome ambiguity (*ram*) state, or in structural terms between an open and a closed conformation of the small subunit (see Section 8.6). The main group of antibiotics interfering with the decoding site is the aminoglycosides.

Streptomycin (Str)

One of the most thoroughly characterized antibiotics affecting decoding is streptomycin (Kurland, 1992). The binding of streptomycin to the ribosome leads to error-prone decoding of the message. This is due to the initial selection step as well as the proofreading step (Karimi & Ehrenberg, 1994; 1996). Streptomycin interacts with the phosphate backbone of four different regions of the 16S RNA, primarily nucleotides U13, G526, A915 and C1490 and Lys45 of ribosomal protein S12 (Carter *et al.*, 2000). These residues have previously been identified by analysis of streptomycin protection (Moazed & Noller, 1987), crosslinking (Gravel *et al.*, 1987) and mutagenesis (Montandon *et al.*, 1986; Melancon *et al.*, 1988; Pinard *et al.*, 1993).

The 30S subunit oscillates between the restrictive and *ram* states (see Section 8.6). The fidelity of translation is affected if one state gets stabilized over the other (Allen & Noller, 1989; Lodmell & Dahlberg, 1997). Streptomycin stabilizes the *ram* state, which is the closed conformation of the small subunit and which has increased affinity for non-cognate aminoacyl-tRNA (see Table 8.2). Resistance to streptomycin is primarily due to mutations affecting ribosomal protein S12. Such mutations make the ribosomes restrictive and sometimes hyper-accurate (Kurland *et al.*, 1996). S12 stabilizes the same region as streptomycin. Thus, mutations affecting the loops of the protein that contact nucleotides 908–915, 524–527 and helix h44 can lead to streptomycin resistance (Carter *et al.*, 2000). In these mutants, the *ram* state is destabilized sufficiently that the balance between the two conformational states can be maintained even in the presence of streptomycin. Some of these mutants are streptomycin dependent. In these, the balance toward the restrictive state is so great that streptomycin is needed for efficient translation with sufficiently high fidelity. Revertants from streptomycin dependence have mutations affecting ribosomal proteins S4 and S5 (Kurland *et al.*, 1996; Ogle *et al.*, 2001; 2002; 2003).

For the hyperaccurate mutants, not only is the initial recognition affected but the proofreading as well. This means that even though a cognate tRNA is bound to the A-site, when the GTP hydrolysis is induced and EF-Tu has dissociated, the aminoacyl-tRNA has an increased likelihood to fall off before peptidyl transfer. Thus, the S12 mutation can also disturb the approach of the acceptor part of the tRNA with the aminoacyl moiety to the peptidyl transfer site.

Paramomycin

Paramomycin is another member of the aminoglycoside family of antibiotics. It binds to the decoding site of the small subunit and facilitates a closure of the shoulder domain. More specifically, it binds to the major groove of h44 and induces a conformational change of A1492 and A1493, which get flipped out from the helix h44 (Carter *et al.*, 2000; Ogle *et al.*, 2001; 2002; 2003). Paramomycin then induces misreading of the mRNA due to an increased initial binding affinity for tRNA (Pape *et al.*, 2000). Paromomycin drives the small subunit to adopt the *ram* or closed conformation of the small subunit (see Table 8.7).

The structural studies of paramomycin bound to the 30S subunit have given a leap insight into decoding (Fourmy *et al.*, 1996; Carter *et al.*, 2000; Ogle *et al.*, 2001; 2002; 2003). The ribosome participates in the identification of the cognate mRNA-tRNA interaction by a conformational change and specific hydrogen bonds of the conserved nucleotides G530, A1492 and A1493 (see Section 11.5; Moazed & Noller, 1986; Ogle *et al.*, 2001; 2002). The binding of paramomycin induces this discriminating conformational change regardless of whether the tRNA is cognate or not (Ogle *et al.*, 2001; 2002).

10.3 INHIBITORS OF AMINOACYL-tRNA BINDING

Tetracycline

Tetracycline has been used extensively since the 1940s as a "broad-spectrum" antibiotic, both in human and veterinary medicine (Chopra *et al.*, 1992). This has led to widespread resistance (Salyers *et al.*, 1990; Taylor & Chau, 1996). Tetracycline has one strong binding site on the small ribosomal subunit in addition to several weaker ones (Epe *et al.*, 1987; Kolesnikov *et al.*, 1996).

The strong binding site is at the A-site between the head and shoulder of the small subunit (Brodersen *et al.*, 2000; Pioletti *et al.*, 2001). Tetracycline binds near the position of the anticodon-stem-loop (ASL) of the A-site between the body and the head. It binds to helix h34 (residues 1054–1056 and 1196–1200) and helix h31 (residues 964–967). Tetracycline prevents aminoacyl-tRNA from binding to the A-site (Maxwell, 1967; Geigenmuller & Nierhaus, 1986). However, it does not inhibit the binding of the ternary complex and the dissociation of EF-Tu after GTP hydrolysis (Gordon, 1969). Consequently, the effect of tetracycline binding to the ribosome is that the GTP pool of the cell gets depleted (Brodersen *et al.*, 2000). The tRNA of the ternary complex binds differently from the aminoacyl-tRNA at the A-site after GTP hydrolysis and dissociation of EF-Tu:GDP (see a detailed discussion in Sections 9.3 and 11.4).

It is not clear whether any of the secondary sites has an inhibitory effect, but it was noticed that one of them is close to h27 and near h44 and h11 (Brodersen *et al.*, 2000; Pioletti *et al.*, 2001). This location is known to have effects on accuracy (Lodmell & Dahlberg, 1997).

Resistance against tetracycline is gained by a mutation G1058C in the immediate proximity of the primary binding site for tetracycline on the 16S RNA (Ross *et al.*, 1998). Another type of resistance is due to ribosomal protection proteins, Tet(M) and Tet(O), which are homologous to elongation factors, particularly EF-G (Burdett, 1996; Trieber *et al.*, 1998; Dantley *et al.*, 1998). These proteins eliminate the bound tetracycline by binding to the A-site (see Section 9.8).

10.4 INHIBITORS OF PEPTIDYL TRANSFER

The peptidyl transfer center includes the binding sites for the acceptor ends of the tRNAs in the A- and P-sites. Parts of these sites are the A- and P-loops and the site where peptidyl transfer occurs. There are two hydrophobic crevices, one at the PTC and the other at the entrance to the peptide exit tunnel. Inhibitors of peptidyl transfer can thus be bound to either of these sites.

Puromycin (pur)

A classical inhibitor that acts in the peptidyl transfer site is puromycin (Yarmolinsky & de la Haba, 1959). Puromycin is a mimic of the acceptor end of an aminoacyl-tRNA. It is composed of the terminal adenosine of the tRNA and the amino acid tyrosine linked via an amide bridge instead of an ester bond. The adenine is in the form of dimethyl-adenine. When bound to the A-site, puromycin can act as acceptor of the nascent peptide from the peptidyl tRNA in the P-site. Since it is a minimal mimic of the aminoacyl-tRNA, it does not participate in the continued process but falls off the ribosome together with the peptide. Puromycin is used to analyze whether translocation has occurred and the state at the PTC. If puromycin reacts with the peptide of a tRNA, it does not necessarily mean that the peptidyl-tRNA is fully translocated. Puromycin can react with the peptide of a tRNA in the A/P-state (Zavialov & Ehrenberg, 2003; Sharma *et al.*, 2004).

A number of crystallographic studies related to puromycin have been performed. Puromycin can be incorporated into larger structures. One of these that have been characterized structurally is CCdA-p-Puro, the so-called Yarus tetrahedral intermediate analogue (Welch *et al.*, 1995). Another puromycin derivative is a 13-base-pair minihelix terminated with puromycin (Nissen *et al.*, 2000). In both cases, puromycin is bound to the A-site. In fact, the adenines corresponding to position 76 of the A- and P-site tRNAs bind to the PTC in nearly identical ways, making A-minor interactions (Nissen *et al.*, 2000; Nissen *et al.*, 2001) and are part of the two-fold symmetry of the PTC (Bashan *et al.*, 2003). The dimethyl-adenine of puromycin is hydrogen bonded to G2583 and the 2′-hydroxyl of the ribose is hydrogen bonded to U2585 (Nissen *et al.*, 2000).

The reaction with puromycin and the removal of the nascent peptide does not induce any structural changes in the ribosome that can be observed at 13 Å resolution (Valle *et al.*, 2003b). The reaction leads to a deacylated tRNA in the P-site, which in many respects is indistinguishable from a termination complex, where RF1/2 has hydrolyzed the peptide from the peptidyl-tRNA in the P-site and RF3 has removed RF1/2.

Anisomycin

Anisomycin has a chemical similarity to puromycin and binds to a site that overlaps with the one of puromycin (Hansen *et al.*, 2003). However, the details of the binding differ quite distinctly. The p-methoxyphenyl groups bind at the same crevice in the PTC, but their connections to the sugar moieties are from different directions. Anisomycin also interferes with the binding of P-site substrates.

Blasticidin S

For blastocidin S, two binding sites are observed. The inhibitor has a cytosine base, which can interact with the two guanines of the P-loop (Hansen *et al.*, 2003). The stronger binding site is with G2251 and the weaker with G2252. These sites thus overlap with the CCA-end of the P-site tRNA, which must be the mode of inhibition of blasticidin.

Chloramphenicol

Two different binding sites have been observed for chloramphenicol. In *D. radiodurans* it binds to the PTC crevice (Schlünzen *et al.*, 2001), while in *H. marismortui* it binds in the crevice at the entrance to the exit tunnel (Hansen *et al.*, 2003). Both binding sites are compatible with biochemical studies and the crystallographic findings explain a number of apparently conflicting conclusions (see Hansen *et al.*, 2003, for a discussion).

Sparsomycin

An antibiotic that has received considerable attention is sparsomycin (Goldberg & Mitsugi, 1966; Vazquez, 1979; Cundliffe, 1981). Sparsomycin binds with unusually high affinity and only in the presence of a P-site substrate. It binds to the PTC and is found to stack with A2602 (Porse *et al.*, 1999; Hansen *et al.*, 2002b; 2003; Bashan *et al.*, 2003). It interacts with the CCA-end of the P-site tRNA and extends into the A-site. No major conformational change was observed. Obviously the binding of sparsomycin is incompatible with peptidyl transfer (Hansen *et al.*, 2002b; Bashan *et al.*, 2003). Unexpectedly, sparsomycin was found to induce translocation (Fredrick & Noller, 2003; see Section 11.4).

Virginamycin M

Virginamycin M has a 20-membered ring and is also called streptogramin A. It binds cooperatively with virginamycin S. It binds in the

PTC crevice but extends into the P-site and causes conformational changes in the PTC (Hansen *et al.*, 2003).

10.5 INHIBITORS OF THE EXIT TUNNEL — THE MACROLIDES

The macrolides are based on 12–22-membered lactone rings with different sugar substituents. They inhibit the elongation of the nascent polypeptide and block the exit tunnel (Fig. 10.1; Arevalo *et al.*, 1988). They do not inhibit peptide synthesis or bind to the PTC (Pestka, 1972; Contreras & Vazquez, 1977; Andersson & Kurland, 1987). Ribosomes that are already engaged in the elongation of nascent peptides do not bind or become inhibited by macrolides (Odom *et al.*, 1991). Ribosomes that are inhibited before they have a nascent peptide can only synthesize a few peptide bonds; the length depends on the nature of the inhibitor.

The binding sites of a number of macrolides have been studied by crystallography (Schlünzen *et al.*, 2001; 2003; Hansen *et al.*, 2002a; Berisio

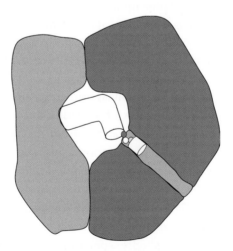

Fig. 10.1 A simplified cross-section of the ribosome showing the peptidyl-tRNA and the exit tunnel. The macrolides inhibit protein synthesis by binding as a plug in the exit tunnel. Only a limited number of amino acids can be incorporated into the nascent peptide. Different macrolides bind somewhat differently in the same general area. Macrolide resistance affects the binding in the narrow part of the tunnel.

et al., 2003b). As expected, the binding area is in the exit tunnel and more particularly in a constriction of the tunnel near proteins L4 and L22. The binding sites partly superimpose on one of the sites for chloramphenicol (Hansen *et al.*, 2003). The peptide synthesized can sometimes be as long as five amino acid residues, but for the larger macrolide carbomycin A, not a single peptide bond can be formed (Mao & Robishaw, 1971; Poulsen *et al.*, 2000). The length of the peptide synthesized primarily depends on the substituents in the C5 position of the lactone ring (Mao & Robishaw, 1971; Poulsen *et al.*, 2000; Hansen *et al.*, 2002a).

16-membered Lactone Rings

A number of 16-membered lactone rings of macrolide antibiotics have been studied by crystallography. Carbomycin A, spiramycin and tylosin have been analyzed at about 3.0 Å resolution (Hansen *et al.*, 2002a). These inhibitors bind immediately adjacent to the PTC and the lactone rings superimpose almost exactly. The main structural difference in the large subunit of *H. marismortui* upon binding of these antibiotics is a movement of A2103 (*E. coli* A2062). The same reorientation of this residue was also seen when a substrate was bound to the P-site (Hansen *et al.*, 2002a).

Unexpectedly, a covalent bond seemed to form between the C6 position of each 16-membered lactone ring and N6 of A2103 (*E. coli* A2062). The bond is most certainly a carbinolamine and not a Schiff base (McGhee & von Hippel, 1977). This was observed as a continuous electron density between each antibiotic and the nucleotide. Any modification of the aldehyde group at position C6 of the lactone ring reduces the inhibitory strength very considerably (see references in Hansen *et al.*, 2002a). No such covalent bond was seen in the case of a 15-membered antibiotic (Hansen *et al.*, 2002a).

The binding of the hydrophobic side of the lactone ring is against an essentially hydrophobic surface of the tunnel wall. A hydrophilic group, N2 of G2099 in *H. marismortui* (2058 in *E. coli*), interrupts the hydrophobic surface, replacing the frequently occurring adenine. The resistance to macrolides is much higher in organisms that have a G in this position than those that have an A. Another mechanism of resistance to macrolides is by enzymes that methylate N6 of A2058 (Vester & Douthwaite, 2001; Hansen *et al.*, 2002a).

15-membered Lactone Rings

One 15-membered lactone-ring antibiotic, the azalide azithromycin, has been studied by crystallography (Hansen *et al.*, 2002a; Schlünzen *et al.*, 2003). Azithromycin is a semisynthetic derivative of erythromycin (Bright *et al.*, 1988). One major difference from the 16-memberered ring macrolides is the absence of the covalent bond to A2103 (*H. marismortui*, Hansen *et al.*, 2000a). The binding site in *D. radiodurans* is somewhat different, but primarily two molecules are bound close to each other. The second binding site is further down the tunnel in close proximity to protein L4, and also near L22 (Schlünzen *et al.*, 2003).

14-membered Lactone Rings

Several macrolide antibiotics with 14-membered lactone rings have been studied using crystallography at around 3.5 Å resolution (Schlünzen *et al.*, 2001; 2003; Berisio *et al.*, 2003a,b). They are clarithromycin, erythromycin, telithromycin, troleandomycin and the ketolide ABT-773. From biochemical analyses, one would expect a similar binding as for the 16- and 15-membered rings, but this is not the case. The conformation and position of the lactone ring and the sugar substituents differ to a surprising extent (Hansen *et al.*, 2002a). Continued studies will reveal the cause.

The 14-membered lactone-ring inhibitor troleandomycin has larger than normal substituents. Thus, it cannot bind like the others but binds rather further into the tunnel. It interacts with helices H35 and H35a as well as with proteins L22 and L32. In the case of L22, the inhibitor occupies part of the binding site for the tip of the β-ribbon. This leads to a conformational change of the extended tip of the protein, which closes the tunnel (Berisio *et al.*, 2003b).

Macrolide Resistance

Resistance to the macrolides is associated with the components of the constriction of the tunnel. Methylation of A2058 or an A to G substitution in this position leads to resistance (Retsema & Fu, 2001). Furthermore, mutations of the extended parts of L4 and L22 in the constriction part of the tunnel can give resistance to macrolides in somewhat different manners. Mutations of L4 prevent the binding of the antibiotics, while mutations of L22 can open the narrow part of the tunnel (Jenni & Ban, 2003). In both cases, the passage of the nascent peptide through the exit tunnel is enabled.

10.6 INHIBITORS OF TRANSLOCATION

Hygromycin B (HygB)

HygB is an aminoglycoside that has little effect on fidelity (Eustice & Wilhelm, 1984a,b). It rather inhibits translocation in bacteria as well as eukarya (Cabanas *et al.*, 1978; Gonzales *et al.*, 1978; Eustice & Wilhelm, 1984b). The affinity of aminoacyl-tRNA for the A-site is increased in the presence of HygB (Eustice & Wilhelm, 1984a). HygB binds close to the A-site on the small subunit (Moazed & Noller, 1987) and close to where other aminoglycosides bind at the top of h44 (Brodersen *et al.*, 2000). HygB binds to the major groove of the helix in a sequence-dependent manner and does not induce any structural alterations of the helix. The binding site of HygB is close to the neck of the small subunit and at a site on h44 close to the A-, P- and E-sites, where there are conformational changes during translocation (Frank & Agrawal, 2000). The antibiotic may inhibit these conformational changes and thereby translocation (Brodersen *et al.*, 2000).

Pactamycin (Ptc)

Pactamycin has been considered as inhibiting the release of initiation factors from the initiation complex (Cohen *et al.*, 1969; Kappen & Goldberg, 1976). Recent findings suggest that the main effect on the inhibition is on the translocation (Dinos *et al.*, 2004). It mimics two nucleotides that are stacked on each other on top of G693 of h23b. The central ring of Ptc mimics the RNA backbone and interacts with C795 and C796 in h24a (Brodersen *et al.*, 2000). It prevents the formation of the G693:C795 base pair (Dinos *et al.*, 2004). The antibiotic binds at a site that is identified as the site for the codon in the E-site. The antibiotic must also prevent the ratchet-like motions, where the head of the small subunit moves with regard to the platform (Valle *et al.*, 2003b). Resistance to Ptc is caused by the mutations A694G, C795U and C796U (Mankin, 1997). This is in good agreement with the binding site of Ptc (Brodersen *et al.*, 2000).

Spectinomycin

Spectinomycin inhibits the translocation of peptidyl-tRNA from the A- to the P-site, catalyzed by EF-G (Bilgin *et al.*, 1990). A crystallographic analysis has identified the binding site of the rigid spectinomycin molecule at the tip of h34 of the small subunit, where it primarily interacts with nucleotides G1064 and C1192 (Carter *et al.*, 2000). Protein S5, in which mutations causing resistance to spectinomycin have been mapped (Wittmann-Liebold & Greuer, 1978), is in the vicinity but not immediately interacting. Translocation involves movements of the head region of the small subunit (Frank & Agrawal, 2000; Valle *et al.*, 2003b; Ogle *et al.*, 2003). It seems likely that h34 and the spectinomycin binding site are close to the pivot point of such movements. Protein S5 is at the domain interface of the small subunit that has a significant role in controlling the open and closed conformations of the decoding site (Carter *et al.*, 2000; Ogle *et al.*, 2003). However, a detailed mechanism for the spectinomycin inhibition is not yet available.

10.7 INHIBITORS OF TRANSLATION FACTORS

Thiostrepton

Thiostrepton is known to inhibit the binding of EF-G to the ribosome (Bodley *et al.*, 1970c). This conclusion has been challenged (Rodnina *et al.*, 1999), but recent investigations suggest that the early conclusions were correct (Cameron *et al.*, 2002). It was found in these studies that thiostrepton inhibits the proper interaction of EF-G with the ribosome in such a way that no GTP hydrolysis occurs. Indeed, it is concluded that EF-G may not bind to the ribosome at all. In contrast to the inhibition of EF-G, thiostrepton stimulates uncoupled GTP hydrolysis by IF2. Thus, the binding sites of EF-G and IF2 on the 50S subunit may not be completely overlapping.

So far, there are no crystallographic studies of the thiostrepton binding site. It is at the base of the L12 stalk of the large subunit and involves residues in the region 1050–1110 in domain II of the 23S RNA (Gale *et al.*, 1981), probably close to nucleotides 1067 and 1095 (Wimberley *et al.*, 1999). This part of the 23S RNA binds the proteins L11 and the pentameric complex L10:(L12)$_4$ (Dijk *et al.*, 1979; Beauclerk *et al.*, 1984; Wimberley *et al.*, 1999). Ribosomes, where L11 is absent, are resistant to thiostrepton (Highland *et al.*, 1975). A methylase that results in thiostrepton resistance uniquely modifies A1067 and prevents thiostrepton from binding (Thompson *et al.*, 1982). A crosslink has also been identified between EF-G and nucleotide A1067 of the 23S RNA (Sköld, 1983). Both thiostrepton and micrococcin prevent the methylation of A1067 (Lenzen *et al.*, 2003). The structure of a complex of thiostrepton with the RNA fragment 1051–1108 has been determined (Lenzen *et al.*, 2003).

Inhibitors Binding to EF-Tu

A number of inhibitors bind to the translation factors. One class comprises inhibitors that lock elongation factors on the ribosome. Thus, for elongation factor Tu, kirromycin and related compounds bind to EF-Tu and the factor is not released, even though the GTP molecule is hydrolyzed (Wolf *et al.*, 1977; Parmeggiani & Swart, 1985). The binding site for aurodox is between domains G and III (Plate 9.9B; see also Section 9.3; Fig. 9.11E; Vogeley *et al.*, 2001). The antibiotic acts like a glue between these domains and does not permit the conformational change that leads to the dissociation of EF-Tu from the ribosome.

The antibiotic GE2270A inhibits the formation of the ternary complex (see Section 9.3; Fig. 9.11D). The structure of the complex with EF-Tu is known (Heffron & Jurnak, 2000). The inhibitor is bound at the interface between domains I and II and the pocket, where the terminal A of the acceptor end of the tRNA binds is occupied by the inhibitor. This explains the mode of inhibition of the antibiotic.

Inhibitors Binding to EF-G and EF2

For EF-G, there is an inhibitor called fusidic acid (FA; Bodley *et al.*, 1969). Its action is related to the way kirromycin inhibits EF-Tu. FA can bind to EF-G only when it is bound with GTP to the ribosome (Baca *et al.*, 1976). EF-G remains firmly attached to the ribosome after GTP hydrolysis if FA is present. A large number of FA resistant mutations are known (Johanson & Hughes, 1994). They are primarily found in three locations of the EF-G molecule, the G-domain, and the interfaces between domains G and III and between domains G and V (Johanson *et al.*, 1996, Laurberg *et al.*, 2000; Nagaev *et al.*, 2001). The binding site for FA has not been experimentally determined. A possible site between the G-domain and domain III has been proposed (Laurberg *et al.*, 2000).

For the fungal factor corresponding to EF-G, EF2, there is an inhibitor that may be compared to FA (Capa *et al.*, 1998; Justice *et al.*, 1998; Dominguez *et al.*, 1999). This inhibitor is called sordarin. By binding to yeast EF2, it locks the factor to the ribosome. The structure of EF2 with a bound sordarin molecule has been determined. The conformation of the complex is quite different from the unliganded molecule or the conformations of EF-G (Jørgensen *et al.*, 2003; Spahn *et al.*, 2004). Presumably, this is a ribosome-binding conformation of EF2 and probably also of EF-G. The sordarin molecule binds to the interface between domains III, IV and V (Jørgensen *et al.*, 2003) at a site where a number of mutations causing sordarin resistance are located. No mutations causing FA resistance are found at this site, suggesting that the binding sites of sordarin and FA are different but somehow related.

By electrospray MS, the proteins that are observed to dissociate from the *E. coli* ribosome are S1, L10, L11 and L12 (Benjamin *et al.*, 1998). Complexes between these proteins and tRNA are also seen. A study of ribosomes with EF-G inhibited by FA or thiostrepton gave surprising results (Hanson *et al.*, 2003). In the ribosomal complex with EF-G and FA, the spectra of the complex of the stalk proteins $L12_4L10$ were enhanced. The inhibition by thiostrepton (see Section 10.7) led to the virtual absence of the stalk proteins from the mass spectra. This inhibitor evidently caused a stronger anchoring of the stalk proteins to the ribosome. Two proteins that became less firmly bound to the thiostrepton-inhibited ribosome are L5 and L18 that primarily interact with the 5S RNA (Horne & Erdmann, 1972). Two different translation inhibitors are associated with the two states of the ribosome. With FA, EF-G gets locked to the ribosome after translocation (Bodley *et al.*, 1970a,b), while thiostrepton prevents EF-G from binding to the ribosome. With thiostrepton, the ribosome must remain in the pre-translocation state (Cameron *et al.*, 2002). Thus, not only does MS show one way of identifying ribosomal states, but it also gives a novel and complementary insight into details of component organization in the ribosome.

11

The Process — Translation

Translation is a process whereby the inherited genomic information is translated into functional proteins, a process that is essential for all cells and organisms. A genome may contain from less than 1000 to more than 30 000 translated genes, depending on the organism. In most genomes, the hereditary material is in the form of DNA. However, certain viruses deviate by containing genomic RNA. The process of translation is usually preceded by transcription and the product of translation, a protein, is frequently transported from where it was synthesized to some other compartment of the cell.

The process of translation occurs on the ribosome in the cytoplasm or in the cellular organelles, mitochondria and chloroplasts. Bacterial translation is the most explored system and the process is the focus of this chapter.

11.1 THE DYNAMICS OF TRANSLATION AND THE RIBOSOME

Translation involves numerous RNA and protein molecules that bind to and dissociate from the ribosome. In addition to these highly dynamic interactions, the ribosome itself is a dynamic enzyme. It is built from two subunits that can move with regard to each other. The 30S subunit has been observed to rotate by about $10°$ with regard to the 50S subunit upon binding of EF-G-GDPNP (Fig. 8.4; Agrawal *et al.*, 1999; Frank & Agrawal, 2000; Frank, 2003; Valle *et al.*, 2003b).

The subunits themselves are also dynamic. The small subunit is built from four domains that are structural units that move with regard to each other. The large subunit may be less flexible since its six rRNA domains are thoroughly woven together. However, the two stalks on each side of the subunit, which are related to the entry and exit of the tRNA, are both highly flexible. The ribosome is a dynamic multi-component ribozyme that can respond to the binding of different ligands by entering different functional states (Spirin, 1969). The translocation step, in particular, needs movements of the mRNA by one codon and the tRNAs from the A- to the P- and from the P- to the E- site by 20 Å or more (Plate 8.3; Yusupov *et al.*, 2001).

The subunit bridges that hold the ribosome together are evidently designed to allow the relative rotation of the subunits. They are either flexible or have alternate binding interactions. Such flexibility was observed for the subunit bridges (B1a, B1b and B2a) that are disordered when studied in isolated subunits (see Section 7.2; Yusupov *et al.*, 2001; Valle *et al.*, 2003b).

11.2 CENTRAL ASSAYS

In the exploration of the ribosomal functions, numerous assays *in vitro* have been developed. Even though many of them are far from the physiological situation, they aim at examining different steps of the ribosomal activity.

One of the classical assays is to use poly-U as messenger to synthesize poly-Phe. Ribosomes with bound mRNA mixed or purified tRNA, F-RS, and translation factors EF-Tu and EF-G are combined. The rate and fidelity of poly-Phe production under varying conditions is measured. One problem with this assay is that EF-Tu hydrolyzes two GTP molecules per functional cycle (Ehrenberg *et al.*, 1990). This deviation is avoided by using more natural mRNAs with a mixture of codons representing different amino acids (Rodnina *et al.*, 1995). Such mRNAs can be a few codons in length, with or without the SD region, or they may correspond to complete genes.

Depending on the ratios and nature of the components in *in vitro* experiments, one can measure a single-round of a certain step, e.g.

translocation, or multiple-turnover (Rodnina *et al.*, 1997). When one component is present in much smaller amounts than the other components, one can either analyze a single-round of reactions or if the limiting component can be recycled, multiple-turnover events. Thus, if EF-G:GTP is present in larger amounts than pretranslocation ribosomes, one can study single-round GTP-hydrolysis or translocation. On the other hand, if EF-G is limiting, but not GTP or pretranslocation ribosomes, multiple-turnover of EF-G activity can be followed.

Studies of peptidyl transfer can be done not only with full tRNA molecules, but also with tRNA fragments. Different fragments of the acceptor CCA-end of the tRNAs can be used in what is called "the fragment reaction" (Traut & Monro, 1964; Monro, 1967). Puromycin is a minimal fragment with only the A-part of the acceptor end. It binds only at the A-site part of the PTC and functions as an acceptor of the nascent peptide in the P-site (see Section 10.4). This leads to termination of protein synthesis and dissociation of puromycin linked to the nascent peptide. This assay is used to explore the occupancy of the A- and P-sites and can be used to establish whether translocation has occurred. Puromycin does not bind to ribosomes if the A-site is occupied. Translocation will allow puromycin to bind and react with the nascent peptide. After peptidyl transfer, the acceptor end of the tRNA in the A-site can slowly and spontaneously move to the hybrid A/P-site, where the CCA-end will base-pair with the P-loop. This will also allow puromycin to react with the nascent peptide (Sharma *et al.*, 2004). The fragment reaction is biologically relevant since it occurs at the same site as the normal peptidyl transfer reaction and is inhibited by the same inhibitors (Moore & Steitz, 2003b).

11.3 INITIATION

Bacterial mRNAs are frequently polycistronic and can contain information for several proteins. Translation is initiated by the binding of an mRNA to the free ribosomal small subunits. It seems necessary to separate the subunits before the mRNA binding, since the mRNA is bound to the interface side and around the neck of the small subunit at a grove that is quite narrow (Yusupova *et al.*, 2001). Poly(U) and mRNAs without a SD sequence can also initiate easily. The ribosome seems to have an affinity for both artificial and natural mRNAs to the proper binding site.

Subsequently, the pre-initiation complex forms. It consists of the small subunit, the mRNA, the initiator tRNA (fMet-tRNA^{fMet}), and the initiation factors IF1, IF2:GTP and IF3 (Gualerzi & Pon, 1990). The binding of IF3 is probably the initial step of initiation since this factor prevents the formation of the 70S ribosome (Grunberg-Manago *et al.*, 1975) and may assist in the removal of the deacylated tRNA from the small subunit (Karimi *et al.*, 1999). Once the pre-initiation complex is properly formed, the 50S subunit will associate, GTP will be hydrolyzed and the initiation factors will be released. The elongation phase of translation can begin (Fig. 11.1).

For correct initiation, the initiation codon, a methionine AUG (or GUG) codon, needs to be selected and bound at the ribosomal P-site. In bacteria, the selection of the initiation methionine codon over elongation

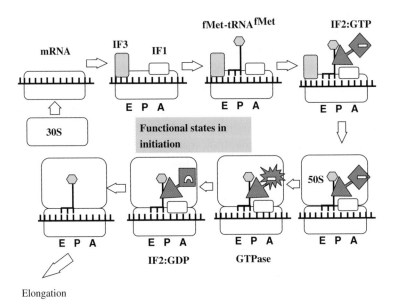

Fig. 11.1 Initiation of bacterial translation. The small subunit binds a messenger RNA and subsequently initiation factors IF1 and IF3. IF1 binds to the A-site and the N-terminal domain of IF3 binds to the E-site. This assists the binding of fMet-tRNA^{fMet} to the P-site. Subsequently, IF2:GTP binds to the pre-initiation complex and catalyzes the joining of the large subunit. In this process, IF2 interacts with IF1 and the initiator tRNA (Roll-Mecak *et al.*, 2001). IF3 has to be ejected to allow the joining of the subunits. The GTPase of IF2 is activated and this factor and IF1 leave the ribosome. The initiator tRNA is left in the P-site of the 70S ribosome and can engage in elongation.

methionine codons or out of frame AUGs is done through a specific interaction of a part of the mRNA preceding the AUG initiation codon with the small subunit RNA. A segment of nucleotides rich in A and G of the mRNA forms base pairs with a complementary part of the 3'-end of the 16S RNA (Fig. 8.1A; Shine & Dalgarno, 1974). This interacting region is variable in length and location in relation to the initiation codon (Fig. 8.1; see Section 8.1; Schulzaberger *et al.*, 2001).

The two domains of IF3 bind near the initiator tRNA (Plate 9.8; Dallas & Noller, 2001). The N-terminal domain binds to the 30S part of the E-site (Dallas & Noller, 2001) and the C-terminal domain is located where the A-site finger (bridge B2b or H69) contacts the small subunit (McCutcheon *et al.*, 1999; Dallas & Noller, 2001). With this binding, one can rationalize from the observations that IF3 prevents the premature joining of the ribosomal subunits (Grunberg-Manago *et al.*, 1975). It also prevents initiation at other AUG codons and it stimulates the binding of fMet-tRNA^{fMet} to the small subunit (La Teana *et al.*, 1993). The initiation factors are needed to guide and catalyze the steps of translation initiation.

IF1 binds to the decoding part of the A-site and prevents the initiator tRNA from binding there (Plate 9.7; Moazed *et al.*, 1995; Carter *et al.*, 2001). Furthermore, it induces a conformational change of the small subunit that may represent a transition state between subunit association and dissociation (Carter *et al.*, 2001; Ramakrishnan, 2002).

Subsequently, the initiator-tRNA binds to the small subunit guided by IF1 in the A-site and the N-terminal domain of IF3 in the E-site (Carter *et al.*, 2001; Dallas & Noller, 2001). Initiator tRNA binds to the P-site, contrary to all other incoming tRNAs, which bind to the A-site. Previous models for initiation have suggested that IF2 carries fMet-tRNA^{fMet} to the ribosome. This corresponds to the situation in archaea and eukaryotes, where eIF2 carries the initiator tRNA to the ribosome. However, eIF2 does not correspond to bacterial IF2. In bacteria, no factor is known to transport the initiator tRNA to the ribosome. The task of IF2 in bacteria and eIF5B in archaea and eukaryotes is to stabilize the bound initiator tRNA and catalyze the joining of the small with the large subunit (Pestova *et al.*, 2000; Lee *et al.*, 2002).

Tomsic *et al.* (2000) studied the binding of the 50S subunit to the initiation complex. Their surprising finding was that whether the nucleotide was GTP, GDPNP or GDP, the association of the large subunit proceeded quickly, as did also the release of IF2, allowing elongation to start.

However, Antoun *et al.* (2003) identified problems in these experiments. They observed that GDP prevented fast association of the large subunit with the initiation complex as well as the start of elongation. GDPNP, on the other hand, allowed fast association of the large subunit, but the start of elongation was very much slowed down due to the lack of dissociation of IF2. As for the other tGTPases, GTP was needed for IF2 to permit rapid association of subunits and release of the factor to allow the start of elongation.

A reasonable model for initiation can be proposed. IF1 and IF3 have important roles to guide the initiator tRNA into the P-site by occupying the A- and E-sites, respectively. The detailed structural understanding of the role of IF2/eIF5B for subunit joining remains to be explored. The binding site for the G-domain of the tGTPases is normally the large subunit, but it is absent during this phase of initiation. Domain II, on the other hand, interacts with the small subunit (see Section 9.1). In addition, IF2 interacts with IF1 in the A-site (Fig. 11.1; Choi *et al.*, 2000) and due to the long helix, domain IV stretches across the IF1 in the A-site to interact with the initiator tRNA in the P-site of the small subunit (Fig. 9.8; Roll-Mecak *et al.*, 2001). When these interactions are proper, IF2/eIF5B will act like a trap, where the G domain has affinity for the GTPase associated region (GAR) at the base of the L12 stalk of the large subunit. The factor will bridge the ribosomal subunits and bring them together. By this procedure, only properly initiated small subunits with an initiator tRNA bound to the P-site will be joined with a large subunit. The initiation complex reacts properly with the amino acid of a ternary complex or puromycin. If IF2 remains bound to the 70S ribosome in complex with GDPNP, fMet cannot form a peptide bond with puromycin (Antoun *et al.*, 2003). Since there is no reason to believe that IF2 interacts with the PTC, the alternative is that fMet is not properly located in the P-site part of PTC, until IF2 has dissociated. The joining of the large subunit to the small subunit and the dissociation of the initiation factors conclude the initiation part of translation (Antoun *et al.*, 2003).

One of the roles of IF3 is to prevent association of the subunits (Grunberg-Manago *et al.*, 1975). Thus, it has to be ejected before the large subunit can associate with the initiation complex. The dissociation of IF1 may be associated with GTP hydrolysis and dissociation of IF2 from the ribosome. Limited structural insights on the steps of initiation are available.

11.4 ELONGATION

In each cycle of elongation, one amino acid is incorporated into the nascent peptide. The *in vivo* rate of translation is about 20 amino acids incorporated per second and ribosome (Kjeldgaard & Raussing, 1974). This means that each complete cycle of elongation is performed in 50 msek. Figure 11.2 shows a summary of the steps involved in elongation. Further details are discussed below.

At the start of the elongation cycle, the ribosome is in the post-translocation state with fMet-tRNA or a peptidyl tRNA in the P-site. The tRNA is bound with its anticodon to the codon at the region of the neck of the 30S subunit, while the acceptor end is at the PTC of the large subunit. The nascent peptide is situated in the peptidyl exit tunnel. One

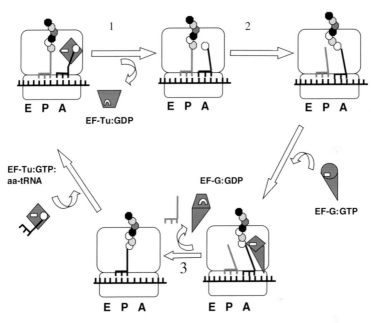

Fig. 11.2 A summary of elongation in protein synthesis. The cycle starts at the bottom left. A peptidyl-tRNA is bound to the P-site. EF-Tu introduces a new aminoacyl-tRNA in the A/T-state and dissociates after GTP hydrolysis (transition state 1), allowing the tRNA to bind to the A-site. In transition state 2, the acceptor end of the A-site tRNA has moved into the PTC close to the P-site tRNA and participates in peptidyl transfer. The translocation of the peptidyl-tRNA from the A-site to P-site is catalyzed by EF-G (transition state 3). The elongation cycle needs two catalysts (EF-Tu and EF-G). The peptidyl transfer step is spontaneous.

line of observation says that there is no tRNA in the E-site of fully active ribosomes (Semenkov *et al.*, 1996; Stark *et al.*, 2002). Another view is that there is an allosteric relationship between the E- and A-sites. In the presence of a deacylated tRNA bound to the E-site, only cognate ternary complexes have enough affinity for the A-site to induce the necessary conformational changes that lead to the release of the E-site tRNA (Nierhaus, 1990; Ramakrishnan, 2002). The crystallographic structure of the 70S ribosome shows a tRNA remaining bound to the E-site through several steps of purification (Yusupov *et al.*, 2001). Apparently, the affinity of the E-site for deacylated tRNA is significant and the interplay between the sites needs to be reconsidered.

Binding of Aminoacyl-tRNA — Decoding

Fidelity related states of the small subunit

A central theme in translation is fidelity. Incorrect translation of the messages may lead to catastrophical consequences for the cell. The fidelity includes: identification of the correct initiation codon, maintaining the correct reading frame (see review by Stahl *et al.*, 2002 and references therein), the correct translation of each codon and termination of protein synthesis at a stop codon.

When a codon of the mRNA binds a cognate tRNA in the A-site, the small subunit undergoes a conformational change from an open to a closed form (Plate 8.9; see also Section 8.6; Ogle *et al.*, 2002; 2003). Some mutations that affect fidelity prevent or induce this conformational transition. Furthermore, antibiotics that stabilize the closed form will lead to error-prone reading. Thus, streptomycin (see Section 10.2) stabilizes the closed form (Carter *et al.*, 2000; Ogle *et al.*, 2002; 2003). Mutations in S12 that cause streptomycin resistance or streptomycin dependence may lead to hyperaccurate ribosomes (Kurland *et al.*, 1996). These mutations destabilize the closed form (Ogle *et al.*, 2002). Revertant mutations that lead to a fidelity close to normal are found in proteins S4 and S5 (Andersson *et al.*, 1986). The closed form of the small subunit leads to breakage of the interactions between S4 and S5 (Plate 8.9; Clemons *et al.*, 1999; Carter *et al.*, 2000; Ogle *et al.*, 2002; 2003). Mutations that break the interaction between these proteins stabilize the closed form and are thus of the *ram* (ribosome ambiguity) phenotype (Ogle *et al.*, 2002).

One region that has been identified as important for the error frequency is helix h27. The base-pairing of this switch helix can change between two alternatives. Lodmell & Dahlberg (1997) found that mutations that stabilize the base-pairing of nucleotides 910–912 with nucleotides 885–887 are restrictive or "hyper-accurate," and that mutations where the alternative base-pairing is with nucleotides 888–890 have the *ram* phenotype. Cryo-EM work comparing these mutations found large-scale structural differences in the ribosome (Gabashvili *et al.*, 1999). Ogle *et al.* (2002) compared crystallographically cognate with near-cognate codon-anticodon interactions on the small subunit without detecting conformational alterations of the switch helix. Thus, in normal translation, the switch helix does not change conformation between cognate and non-cognate interaction of the codon and anticodon.

Wild type and a hyperaccurate mutant (a streptomycin dependent mutation in protein S12, G92D) of 70S ribosomes from *E. coli* were compared with 70S ribosomes from *T. thermophilus* 70S (Yusupov *et al.*, 2001) in a low-resolution crystallographic study (about 9 Å) and structural differences were analyzed (Vila-Sanjurjo *et al.*, 2003). The ribosomes, that did not have a cognate tRNA bound in the A-site, had the open conformation of the 30S subunit, confirming the results from studies of small subunits alone (Ogle *et al.*, 2002). The head of the small subunit in the three available crystal structures of 70S particles was always bent toward the central protuberance compared with isolated small subunits. h27 in all these structures had the *ram* type of base-pairing (Vila-Sanjurjo *et al.*, 2003). This contradicts the hypothesis that the base-pairing of the switch helix would be in the hyper-accurate form for hyperaccurate mutants (Lodmell & Dahlberg, 1997).

The mechanisms for decoding

In the cell, most of the aminoacyl-tRNA is bound to EF-Tu:GTP. The elongation cycle begins with the binding of a ternary complex, aminoacyl-tRNA:EF-Tu:GTP, into the T-site of the ribosome (see Chapter 8.2; Moazed & Noller, 1989). The recognition of the codon by the anticodon of the tRNA, or the decoding, is a central process in protein synthesis. It is essential that high fidelity is maintained. The error rate in translation is in the order of 10^3–10^4 (Kurland *et al.*, 1996). The difference in affinity between the cognate and near-cognate codon-anticodon pairs is not enough to explain the low error frequency (Grosjean *et al.*, 1978; Kurland, 1992). This can be illustrated by the fact that the base-pairing of an incorrect codon-anticodon pair could be more stable than a correct one. Thus, the cognate interaction between the UUU codon and the phenylalanine anticodon GAA is less stable than the non-cognate interaction between the cysteine codon UGC and the arginine anticodon GCG (Ramakrishnan, 2002). Nevertheless, on the ribosome, the cognate interaction is favored.

Two different types of mechanisms that can explain the high fidelity of translation have been discussed. The current understanding of the system suggests that the ribosome uses both. The two mechanisms are:

(1) **Geometrical recognition**. The properties of the decoding site are such that the base-pairing is geometrically screened by the ribosome. This is possible if the correct codon-anticodon pairing leads to conformational changes of the decoding site as described in Section 8.6 (Ogle *et al.*, 2001; 2003; Ramakrishnan, 2002).

(2) **Kinetic proofreading**. The rates of binding and dissociation favor cognate tRNAs over non-cognate or near-cognate tRNAs. In the kinetic proofreading mechanism, the tRNA recognition is done in two steps: the initial recognition step and the proofreading step. They are separated by the irreversible hydrolysis of GTP by EF-Tu (Hopfield, 1974; Ninio, 1974; Ogle *et al.*, 2003). The accuracy of the process is the product of the accuracy of the individual steps.

Initial recognition

In the initial selection step, EF-Tu in a ternary complex binds regardless of whether there is a codon or not exposed in the A-site or whether it is cognate or non-cognate. This has been clarified by fluorescence spectroscopy (Rodnina *et al.*, 1993; 1995a,b; 1996). In this initial binding, the anticodon is not directed toward the codon (Fig. 11.3A). Subsequently, the base-pairing of the codon and anticodon is matched by kinking of the tRNA between the anticodon stem and the D-stem. The bent conformation has not previously been seen in tRNAs and may thus be strained and unstable. Indeed, it leads to a short phosphate distance between residues 10 and 47 (Valle *et al.*, 2003a).

In case the codon does not match the anticodon, the binding affinity for the tRNA remains low and the ternary complex is likely to fall off. The number of ternary complexes tested in this way in each cycle of elongation can be quite large.

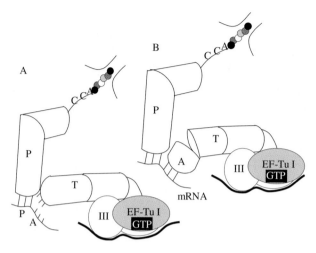

Fig. 11.3 A simplified representation of the binding of an aminoacyl-tRNA in complex with EF-Tu:GTP to the decoding site of the ribosome, according to Valle *et al.* 2003b. **(A)** The initial binding to the T-site, where the main interaction between EF-Tu and the ribosome takes place. The binding occurs regardless of the presence or absence of mRNA or whether the codon is cognate or non-cognate. **(B)** The anticodon stem makes a kink with regard to the D-stem and bends to interact with the codon of the mRNA. The cognate codon-anticodon interaction induces GTP hydrolysis.

At the core of the initial recognition and the proofreading are the codon-anticodon interactions in the decoding site. A number of universally conserved bases of the 16S RNA have been identified by footprinting experiments to be involved in the decoding site (see Table 8.1). These are G530, A1492 and A1493 (Moazed & Noller, 1986; 1990; Powers & Noller, 1990). NMR experiments on a fragment of helix h44 have given structural insights (Fourmy *et al.*, 1996; Yoshizawa *et al.*, 1999). Crystallographic experiments of binding cognate and near-cognate ASLs to the small subunit have provided a very interesting picture of how the ribosome participates in the decoding. Thus, correct Watson-Crick base-pairing between the mRNA and the cognate tRNA induces conformational changes of A1492 and A1493 of h44 as well as of G530 of the 16S RNA (Plate 11.1; Carter *et al.*, 2000; Schlünzen *et al.*, 2000; Yusupov *et al.*, 2001; Ogle *et al.*, 2001; 2002; 2003). The two adenines flip out from their normal configuration and insert into the minor groove of the codon-anticodon helix. Such insertion of adenines into the minor groove has been observed in many places of the ribosomal RNAs (Nissen *et al.*, 2001).

The inserted bases hydrogen bond to the 2'-OH groups of both riboses of the base-pair (Plate 11.2). This is possible only for Watson-Crick base-pairing. In this case, A1493 hydrogen bonds to the first base-pair of the codon-anticodon helix in the A-site. In a similar manner, the second base-pair interacts with A1492 and G530. The wobble base-pair of the A-site is not checked in the same way and does not have the same strict requirements for base-pairing, but here the ribose of the codon nucleotide hydrogen bonds to G530 and contacts a magnesium ion as well (Ogle *et al.*, 2001; 2002). These interactions in the initial recognition complex remain when EF-Tu has dissociated (Valle *et al.*, 2003a).

A naïve belief could be that the initial binding requires the interaction between codon and anticodon. However, it turns out that this multi-step process described by Rodnina *et al.* (1996) is not the rate-limiting step even with a physiological concentration of non-cognate tRNAs.

If the antibiotic paramomycin is bound at the decoding site, the fidelity is significantly reduced (Pape *et al.*, 2000). The antibiotic induces A1492 and A1493 to be oriented as if the codon-anticodon match was perfect (Plate 11.3; Ogle *et al.*, 2001; 2002; 2003). The conformational changes of these three bases are associated with a closure of the 30S subunit (Ogle *et al.*, 2002; 2003).

GTP hydrolysis

A cognate interaction between the codon on the small subunit and the anticodon of an intact tRNA, contrary to tRNA fragments, is required for efficient GTP hydrolysis (Rodnina *et al.*, 1994; Piepenburg *et al.*, 2000). Measurements of the rates of dissociation of tRNAs from the ribosome as well as the rate of GTPase activation have shown that cognate tRNAs have a slower dissociation and a faster GTPase activation than non-cognate tRNAs (Pape *et al.*, 1999; Gromadski & Rodnina, 2004).

In case of a good match, EF-Tu is induced to hydrolyze its bound GTP to GDP and inorganic phosphate (Fig. 11.4). The signal must pass from the decoding site on the small subunit to the GTPase-associated region (GAR) on the large subunit. The signal to hydrolyze GTP could be transmitted through the tRNA. Alternatively, the signal could pass from the decoding site through the ribosome to the GAR. H69 of the 23S RNA is situated in such a way that it could mediate this signal (Plates 8.4C, 9.12C and 9.13F; Bashan *et al.*, 2003; Valle *et al.*, 2003a). Evidently, GAR and EF-Tu both need to go through conformational changes to hydrolyze

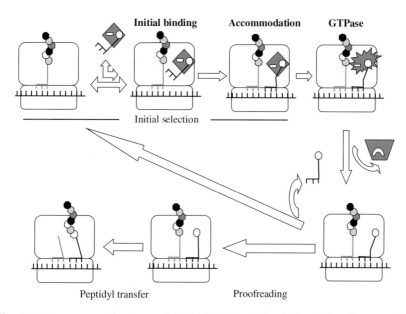

Fig. 11.4 A summary of aminoacyl-tRNA binding to the A-site of the ribosome. The initial selection is composed of two steps, initial binding and accommodation. In the proofreading step, the tRNA can fall off the ribosome or proceed to peptidyl transfer.

GTP, since ribosome binding *per se* is not enough to induce GTP hydrolysis. The details of the interaction between GAR and the factor are not fully understood. How L12 is involved is not known (see Sections 7.4 and 9.1). However, the induction that results in the activation of the water molecule at the γ-phosphate and stabilization of the transition state are essential ingredients (see Sections 9.1 and 9.3).

The best view of this state comes from cryo-EM studies of the ternary complex of EF-Tu with aminoacyl tRNA and GTP bound to the ribosome (Plates 9.12, 9.13; Stark *et al.*, 1997; 2002; Valle *et al.*, 2002; 2003a). The studies were done with the help of the inhibitor kirromycin that locks EF-Tu on the ribosome even though the GTP is hydrolyzed (Wolf *et al.*, 1977). This structure represents a state when the GTP hydrolysis has just occurred. Thus, EF-Tu is in the state just preceding the transition to its nonbinding GDP conformation.

The cognate interaction induces a closed conformation of the small subunit (Ogle *et al.*, 2002; 2003). This leads to a movement of the GAR of the large subunit by about 7 Å, while the SRL maintains its position (Valle *et al.*, 2003a). The GAR moves toward the body of the large subunit and approaches protein L16. For the large subunit, one could use the terminology of open and closed state as well. The open state is before the binding of EF-Tu and the closed state is with the stalled ternary complex (Valle *et al.*, 2003a).

A small difference in the conformation of the isolated factor with and without kirromycin has been identified (Vogeley *et al.*, 2001; Nielsen *et al.*, 2004). Furthermore, the cryo-EM structure of the complex, when it is bound to the ribosome, does not fully agree with the crystallographic structure of the ternary complex (Nissen *et al.*, 1995; 1998; Stark *et al.*, 2002; Valle *et al.*, 2002; 2003a; Nielsen *et al.*, 2004). A relative movement of the tRNA with regard to EF-Tu is observed (Valle *et al.*, 2003a). However, the conformation of the factor, when it is bound to the ribosome, is not altered from the crystallographic study of the ternary complex except for the switch I region that is outside electron density (Valle *et al.*, 2003a; Nielsen *et al.*, 2004). This probably means that switch I, or the effector loop, is flexible in this state. What this means in relation to GTP hydrolysis is not known.

Proofreading

When the GTP molecule is hydrolyzed, the EF-Tu:GDP complex has a conformation with low affinity for the aminoacyl-tRNA and the ribosome. Thus, it dissociates from the tRNA and the ribosome (Yokosawa *et al.*, 1975). The acceptor stem and the aminoacyl moiety of the tRNA that were bound to EF-Tu are located far from the PTC (Stark *et al.*, 1997; Valle *et al.*, 2003a) and can now be reoriented into the A-site of the PTC on the large subunit, while retaining the interaction with the codon (Fig. 11.5). It appears as if the tRNA acts like a spring where the strain in the kink between the anticodon stem and the D-stem is relieved, and the tRNA regains its normal L-shape as has been seen in most tRNA structures (Valle *et al.*, 2003a). In this movement, the ASL maintains the position it has already adopted (Valle *et al.*, 2003a).

This process coincides with the second step of binding of aminoacyl-tRNA to the A-site, the proofreading step (Fig. 11.4; Ruusala *et al.*, 1982). An incorrect (non-cognate) match of the anticodon to the codon increases the likelihood that the aminoacyl-tRNA will dissociate before its amino acid is properly located in the PTC of the ribosome. The affinity for the cognate tRNA is considerably higher than near- or non-cognate tRNAs (Rodnina *et al.*, 2000). The peptidyl moiety in the P-site can react only if the aminoacyl moiety in the A-site is properly located. If these steric requirements are not

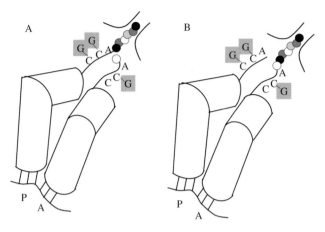

Fig. 11.5 (A) The state after GTP hydrolysis and dissociation of EF-Tu:GDP. The kink between the ASL and the D-stem has closed and allowed the tRNA to swing into the A-site next to the peptidyl-tRNA in the P-site. Even though the tRNAs are primarily related by a sideways movement, the CCA-ends are related by a 180° rotation and interact with the A- and P-loops, respectively. The nascent peptide is in the exit tunnel. **(B)** The state immediately after peptidyl transfer. At a subsequent step, the 3′-ends of the tRNAs are relocated such that the tRNAs become bound to the P/E- and A/P-states.

satisfied, maybe due to improper base-pairing in the decoding site (75 Å away), the reaction will be slowed down to the extent that the aminoacyl tRNA may fall off long before peptidyl transfer can occur (Liljas, 1990).

Peptidyl Transfer

The crystallographic studies have shown that the PTC is located in a pocket on the interface side of the large subunit at the mouth of the peptide exit tunnel that goes through the subunit and emerges on the external side. It is entirely clear that peptidyl transfer is catalyzed without the direct assistance of the ribosomal proteins (see Section 8.3; Noller *et al.*, 1992; Ban *et al.*, 2000; Nissen *et al.*, 2000; Harms *et al.*, 2001). The PTC is solely composed of RNA and thus the ribosome must be considered as a ribozyme.

For enzymatic activities, proximity and correct orientation of the reactants are major factors. The binding at an active site must lead to an optimal positioning of the reactants. For many enzymes, there are groups of the enzymes, coenzymes or metal ions that actively participate in the process. These aspects are obviously of interest with regard to the ribosomal PTC.

After the initial recognition of a cognate ternary complex, EF-Tu dissociates. Then the kink between the anticodon stem and the D-stem reanneals and the aminoacyl moiety of the A-site tRNA swings into the PTC (Figs. 11.4 and 11.5; Valle *et al.*, 2003a). If the aminoacyl-tRNA does not fall off the ribosome in the proofreading process, the peptide on the peptidyl-tRNA in the P-site can be transferred to the aminoacyl group. This leads to a peptidyl-tRNA in the A-site and a deacylated tRNA in the P-site. The peptidyl transfer reaction is quite rapid for a cognate tRNA (>100 sec^{-1}; Rodnina *et al.*, 2000).

This reaction is essentially the reverse of breaking a peptide bond by a protease. The α-amino group on the aminoacyl-tRNA in the A-site needs to be deprotonated and in its neutral state to be able to attack the carbonyl carbon at the ester linkage between the P-site tRNA and the nascent peptide (Fig. 11.5). To what extent does the ribosome depend on acid base catalysis?

In the transition from being bound to a tRNA to making a new peptide bond, the carbonyl carbon goes through an intermediate tetrahedral state during which the carbonyl oxygen develops into an oxyanion that could be stabilized in something corresponding to the oxyanion hole in proteases. The ribosome is likely to provide groups that stabilize this transition state. However, there is no evidence that a covalent intermediate with the ribosome is formed in this process.

Attempts to characterize the catalytic site and the reaction have been done with the aid of different inhibitors and tRNA fragments bound to the large subunit (Table 11.1). The approach has been both chemical and crystallographic. The 5.5 Å resolution structure of the *T. thermophilus* ribosome is a very important reference point, since it provides the positions for the complete tRNAs in the A-, P- as well as the E-sites (Yusupov *et al.*, 2001). Peptidyl transfer catalyzed by the 70S ribosome is orders of magnitude faster that the reaction on the large subunit (see Moore & Steitz, 2003b for a summary). Evidently, there are conformational differences that need to be explored.

An attractive ligand for crystallographic studies of PTC was the so-called Yarus inhibitor, CCdApPuromycin, which aims at mimicking the transition state for peptidyl transfer (Welch *et al.*, 1995). This inhibitor clearly identified the PTC and many of the interactions in the active center (Nissen *et al.*, 2000; Bashan *et al.*, 2003). However, it was later realized that the interactions of the phosphate of the inhibitor that replaces the tetrahedral carbon of the peptide might be misleading compared with studies of substrates or products (Hansen *et al.*, 2002b; Agmon *et al.*,

Table 11.1 Crystallographic Studies of tRNA Binding to the PTC

Species	Resolution (Å)	A-site	P-site	References
H. marismortui 50S	3.1	CCdA-phospate-puromycin		Nissen et al., 2000
	3.1	CC-puromycin-peptide	CCA	Schmeing et al., 2002
	3.0	CCA-pcb	CCA-pcb	Hansen et al. 2002b
	2.8	sparsomycin	CCA-pcb	
	3.0	aa-Minihelix		
D. radiodurans 50S	3.7	Acceptor stem mimic (ASM, 35 nucleotides)		Bashan et al., 2003
	3.7	Sparsomycin		
	3.6	ASM+ Sparsomycin		
	3.7	CCA-phosphate-puromycin		
T. thermophilus 70S	5.5		tRNA^{MetF}	Yusupov et al., 2001
	6.5	tRNA^{Phe}	tRNA^{MetF}	
	7.5		ASL	

2003; Bashan et al., 2003; Jenni & Ban, 2003). One of the phosphate oxygens was assumed to correspond to the carbonyl oxygen at the reaction site and located in an oxyanion hole, where A2451 (E. coli numbering is used if nothing other is said) seemed to play an important role (Nissen et al., 2000). However, subsequent complexes show that the carbonyl oxygen is oriented in the opposite direction. Thus, A2451 could not be involved in stabilizing the oxyanion intermediate as initially proposed (Nissen et al., 2000; Moore & Steitz, 2003a). More likely, a stabilizing interaction is formed with U2585 (Hansen et al., 2002b). Unfortunately, there is no structure with substrate tRNAs or tRNA fragments in both the A- and P-sites. Attempts were made but the tRNA fragments react when bound to the active large subunits yielding products (Schmeing et al., 2002). A superposition of the structures available has led to an insight into the mechanism of peptidyl transfer.

The α-amino group of the aminoacyl-tRNA in the A-site is suitably oriented close to the carbonyl carbon of the ester linking the peptide to the P-site tRNA (Fig. 11.6; Hansen *et al.*, 2002b). No metal ion has been identified to participate in the process. The groups that can form hydrogen bonds to, and stabilize, the uncharged NH_2 form of the α-amino group are N3 of A2451, the 2'-OH of A2451 or the 2'-OH of the terminal ribose of the P-site tRNA (Hansen *et al.*, 2002b; Bashan *et al.*, 2003). If any of these groups has an elevated pK_a, this would assist in the deprotonation of the α-amino group (Moore & Steitz, 2003a). The 2'-OH of A76 in the P-site is known to be important, since tRNAs that have a terminal 2'-deoxyadenosine can be charged with an amino acid. They are active as A-site substrates, but they cannot donate their nascent polypeptide to an A-site tRNA (Quiggle *et al.*, 1981). It is not clear whether the role of the 2'-OH of A76 is to position the attacking α-amino group or whether it performs some other yet unidentified task (Moore & Steitz, 2003a,b).

It is clear that the peptidyl transfer activity is pH dependent with a pK_a of 7.5 (Katunin *et al.*, 2002; Rodnina & Wintermeyer, 2003). A2451 of the 23S rRNA is completely conserved and was initially suggested to serve as a general acid-base during peptide bond formation (Nissen *et al.*, 2000; Muth *et al.*, 2000). This would require that N3 has an elevated pK_a-value. The suggested role of A2451 has caused considerable debate (see Moore & Steitz, 2003a,b; Jenni & Ban, 2003 for summaries). A quantum chemical study of the effects of an adenine anion assisting peptidyl transfer, indicates that the reaction rate would be very significantly increased (Suarez & Merz, 2001). The chemical reactivity of N3 of A2451 as well as mutational analysis, could be interpreted as an activation of the α-amino group as a general base but possibly also with a conformational change of the PTC with the same pK_a (Bayfield *et al.*, 2001; Muth *et al.*, 2001; Polacek *et al.*, 2001; Thompson *et al.*, 2001; Xiong *et al.*, 2001). Sievers *et al.* (2004) find that the rate of peptidyl transfer is enhanced by 2×10^7-fold on the ribosome due to lowering of the entropy of activation. This could be due to the positioning of the substrates in the PTC or the removal of the reaction from bulk water.

The Acceptor Ends and Hybrid States

Footprinting experiments of whole tRNA molecules interacting with complete ribosomes, suggest that the CCA-end of the P-site tRNA is moved to the E-site after peptidyl tRNA transfer. Thus, the deacylated

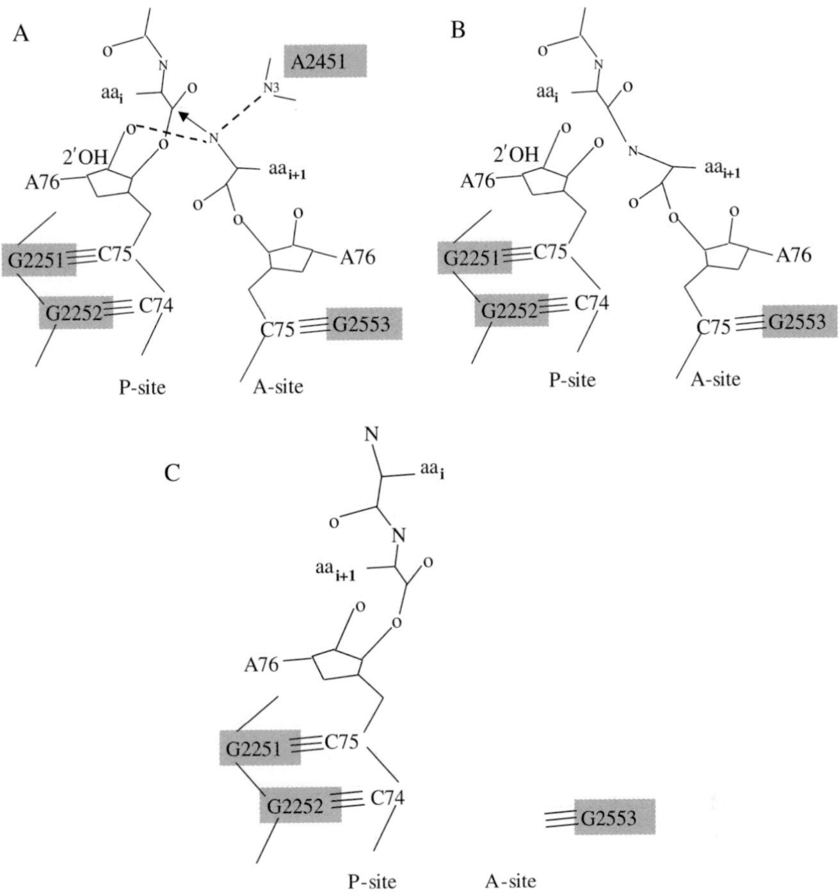

Fig. 11.6 Peptidyl transfer. The acceptor ends of the tRNAs in the P-site (*left*) and A-site (*right*). The base-pairing with the ribosomal RNA to the P- and A-loops as well as the approximate rotational symmetry of 180° between the two sites is illustrated. **(A)** Two potential hydrogen bonds to the α-amino group of the aminoacyl residue bound to the tRNA in the A-site are marked with dashed lines. These hydrogen bonds may be involved in abstracting a proton from the amino group to enable it to form a tetrahedral intermediate with the carbonyl carbon between the nascent polypeptide and the P-site tRNA. **(B)** Peptidyl transfer has occurred leaving the peptide bound to the A-site tRNA. **(C)** The acceptor end with the peptide has been translocated to the P-site. In this process, it undergoes an 180° rotation.

tRNA has been observed to move to the hybrid P/E-site (Moazed & Noller, 1989). Likewise, the CCA end of the peptidyl-tRNA in the A-site has been observed to move into the hybrid A/P-state and interact with the P-loop (G2251 and G2252; Moazed & Noller, 1989).

The CCA-end of the P-site tRNA base-pairs with the P-loop (G2251 and G2252) of the large subunit (Fig. 11.6; Samaha *et al.*, 1995; Green *et al.*, 1998; Nissen *et al.*, 2000; Yusupov *et al.*, 2001; Hansen *et al.*, 2002b; Bashan *et al.*, 2003). Likewise, the CCA part of the A-site tRNA base-pairs with G2553 of the A-loop (Green *et al.*, 1998; Nissen *et al.*, 2000; Bashan *et al.*, 2003). While a sideways movement relates the tRNAs in the A- and P-sites, the CCA-ends are related by an approximate 180° rotation with a slight forward movement in the direction of the peptide exit tunnel (Nissen *et al.*, 2000; Yusupov *et al.*, 2001; Schmeing *et al.*, 2002; Hansen *et al.*, 2002b; Bashan *et al.*, 2003; Agmon *et al.*, 2003).

Crystals of *H. marismortui* 50S subunits are active in peptidyl transfer despite an unfavorable pH (Schmeing *et al.*, 2002). Compared with the *in vivo* rate of 70S ribosomes, the puromycin reaction catalyzed by the large subunit is several orders of magnitude slower (Maden *et al.*, 1968; Moore & Steitz, 2003a). In a crystallographic study, where CCA-Phe-caproic acid-biotin (CCA-pcb) was bound to the P-site and CC-puromycin was bound to the A-site, the electron density showed CC-puromycin-pcb in the A-site and the deacylated CCA to the P-site. Thus, the structure reveals the situation after peptidyl transfer. The CCA fragments of the products remain in essentially the same positions as the substrates (Schmeing *et al.*, 2002; Hansen *et al.*, 2002b). Evidently, these crystallographic results demonstrate that movement into the hybrid states is not simultaneous with peptidyl transfer (Green *et al.*, 1998; Schmeing *et al.*, 2002). Valle *et al.* (2003b) have made similar observations. Their cryo-EM studies suggest that the peptidyl-tRNA remains in the A-site after peptidyl transfer. Furthermore, if the CCA-end had moved from the A- to P-site after peptidyl transfer, one would expect that puromycin could react with the peptidyl-tRNA, but this requires translocation by EF-G (Borowski *et al.*, 1996). Furthermore, Wower *et al.* (2000) have shown that the CCA-end of the P-site tRNA does not move spontaneously to the E-site. In conclusion, the movement of the acceptor ends of A- and P-site tRANAs to hybrid states is at best slow.

The Tyr and Phe residues of the product CC-puromycin-pcb are covalently linked but also remain close to their substrate positions. The side chains are oriented in opposite directions and the peptide is in a β-type of configuration (Schmeing *et al.*, 2002). The peptide produced then also agrees with the two-fold symmetry of the PTC. There is no interaction between these amino acids and the ribosome, but further into the exit tunnel, the nascent peptide will interact with the ribosome.

The Main Steps during Aminoacyl-tRNA Binding and Peptidyl Transfer

The main steps of aminoacyl-tRNA binding and peptidyl transfer may be summarized as follows:

(1) Initial recognition. The ternary complex of EF-Tu:GTP:aa-tRNA binds to the T-site in which EF-Tu interacts with the factor-binding region of the ribosome.

(2) The binding induces a kink between the anticodon and D-stems of the tRNA. This allows the anticodon to interact with the codon in the A-site. This binding for the tRNA is called the A/T-state. In case of a cognate interaction, nucleotides of the 16S RNA stabilize the binding and stimulate the subsequent steps.

(3) A non-cognate ternary complex falls off and a new complex is tried.

(4) A cognate interaction at the decoding site of the small subunit is stabilized by the ribosome and a conformational signal is induced that leads to GTP hydrolysis by EF-Tu at the GAR of the large subunit.

(5) EF-Tu:GDP dissociates from the tRNA and the ribosome.

(6) The kink of the strained aminoacyl-tRNA closes. The tRNA swings into the A-site and places its aminoacyl moiety in the PTC.

(7) The peptide is transferred from the tRNA in the P-site to the aminoacyl moiety on the tRNA in the A-site. The acceptor ends of the two tRNA molecules remain in their sites.

Translocation

Is translocation a spontaneous process inherent in the ribosome?

The final step of elongation is translocation of the peptidyl-tRNA from the A-site to the P-site and the movement of mRNA by three nucleotides to expose the next codon in the A-site. Inoue-Yokosawa *et al.* (1974) presented a classical model. Normally EF-G:GTP catalyzes this process by hydrolysis of GTP to GDP and inorganic phosphate (Ishitsuka *et al.*, 1970; Kaziro, 1978; Rodnina *et al.*, 1997). However, a slow rate of translocation has been observed without factors (Pestka, 1969; Gavrilova & Spirin, 1971; Gavrilova *et al.*, 1976; Southworth *et al.*, 2002). These results have been questioned on the ground that energy is normally required for similar activities. Small amounts of EF-G could possibly lead to the results observed. However, even in the presence of fusidic acid (see Section 10.7), a slow rate of translocation was observed (Spirin, 1978).

In a discussion of translocation, it is necessary to define what is meant by translocation. The process has several steps. The key step that defines translocation is the movement of a peptidyl-tRNA from the A- to the P-site. In this step, a new codon is exposed in the A-site. We will use these movements to define translocation.

A dramatic new insight comes from the observation that the antibiotic sparsomycin induces translocation in the absence of EF-G and GTP (Fredrick & Noller, 2003). Sparsomycin binds at the PTC of the large subunit and inhibits peptidyl transfer (See Section 10.4; Hansen *et al.*, 2002b; Bashan *et al.*, 2003). Several antibiotics that inhibit EF-G-dependent translocation also inhibit translocation by sparsomycin (Fredrick & Noller, 2003). These include viomycin, paramomycin, neomycin, streptomycin, and spectinomycin (see Sections 10.2 and 10.6). These inhibitors target the subunit interface. The fact that the same inhibitors inhibit both these inducers of translocation suggests that the mechanisms must be similar. Inhibitors like thiostrepton and fusidic acid specifically inhibit EF-G, but are not related to the binding or action of sparsomycin. Other antibiotics with no effect on EF-G-dependent translocation inhibit translocation catalyzed by sparsomycin. These are lincomycin, spectinomycin, carbomycin A and chloramphenicol (Fredrick & Noller, 2003). These inhibitors prevent the interaction of sparsomycin with the peptidyl-tRNA in the PTC (Schlünzen *et al.*, 2001; Hansen *et al.*, 2002b).

In crystal structure analyses of the large subunits, sparsomycin stabilizes the binding of peptidyl-tRNA but blocks the access of aminoacyl-tRNA to the peptidyl-tRNA by binding to part of the A-site (Schlünzen *et al.*, 2001; Hansen *et al.*, 2002b). The sparsomycin molecule interacts with the ribose-phosphate backbone of C75 and A76 in the P-site as well as with the nascent peptide.

Small acceptor end fragments of tRNA can bind either to the A- or the P-site. In case the fragment contains an amino acid with a free amino group, the A-site is preferred, perhaps because of the lack of suitable interactions in the P-site (Hansen *et al.*, 2002b). Acceptor end fragments of peptidyl tRNA bind to either site, but in the presence of sparsomycin, only the P-site can be occupied (Moazed & Noller, 1991; Hansen *et al.*, 2002b).

After peptidyl transfer between tRNA molecules, the main difference to the preceding state is that the nascent peptide is connected to the A-site tRNA (Schmeing *et al.*, 2002). In a subsequent step, the acceptor end (CCA-peptide) rotates by 180° to be placed in the P-site (Hansen *et al.*, 2002b; Bashan *et al.*, 2003), yielding a peptidyl tRNA in the A/P-hybrid site base-pairing to the P-loop (Moazed & Noller, 1989). Since translocation is spontaneous, the movements of the acceptor or CCA-ends of the tRNAs must also be spontaneous. Most likely, the process is slow and may need to be stimulated or catalyzed to proceed at a physiological rate. Sparsomycin stabilizes the binding of the acceptor end in the P-site (Hansen *et al.*, 2002b). This apparently leads to a complete translocation by sparsomycin of the peptidyl tRNA into the P-site (Fredrick & Noller, 2003).

Some conformational changes of the ribosome upon binding of sparsomycin are known. A2602 undergoes a movement to stack with the aromatic ring of sparsomycin (Hansen *et al.*, 2002b). This movement could possibly be transmitted through the ribosome with the effect that the tRNA is translocated. However, it is also possible that the tRNA itself has a major role. Once the acceptor end is stabilized in the P-site, the stacking interaction between the acceptor end and the rest of the tRNA may lower the activation energy to translocate the rest of the tRNA and mRNA to the P-site (Hansen *et al.*, 2002b; Fredrick & Noller, 2003). A structural analysis of 70S ribosomes with a peptidyl-tRNA and sparsomycin remains an interesting task.

Gavrilova and Spirin (1971; 1972) found that p-chloromercuribenzoate stimulated spontaneous translocation. Since this reagent modifies thiols, it could hardly have a role in the PTC, which has no contribution of protein molecules. Protein S12, which has several thiols, was early on identified as the main site (Gavrilova *et al.*, 1974). From its location in the decoding site of the small subunit and its possibility of affecting the conformational changes of the small subunit, one could imagine that it could stimulate translocation.

It is remarkable that the binding of the small sparsomycin molecule to the large subunit can be transmitted over the large distance to the small subunit, leading to translocation of the mRNA (Southworth & Green, 2003). Likewise, it is remarkable that S12 or EF-G can catalyze the same process by acting on the small subunit far from the PTC where sparsomycin is bound. Fredrick and Noller (2003) suggest that one of the universally conserved uridines, U2584 or U2585, could play the same role as sparsomycin in the case of the EF-G catalyzed translocation. The binding of sparsomycin changes the balance between the restrictive and *ram* states of the ribosome and leads to translocation (see Section 8.6). This may suggest that the fine functional balance of the ribosome may have evolved from a slow and spontaneous translocation, stimulated by the binding of specific small molecules to the situation, where a translation factor efficiently catalyzes translocation once peptidyl transfer is performed.

One major difference is that, once sparsomycin is bound, it remains bound and becomes an inhibitor, while EF-G:GDP is released permitting further rounds of peptidyl transfer and translocation. However, it is evident that translocation is intrinsic to the ribosome as was suggested long ago (Pestka, 1969; Gavrilova & Spirin, 1971; Gavrilova *et al.*, 1976). Since translocation is spontaneous, the pre-translocation state must have higher energy than the post-translocation state with a moderate energy barrier.

EF-G Catalyzed Translocation

EF-G binds to the ribosome in complex with GTP (Nishizuka & Lipmann, 1966; Kaziro, 1978). EF-G identifies ribosomes in the pre-translocation state (Zavialov & Ehrenberg, 2003), not to waste GTP molecules unnecessarily, since as soon as EF-G binds to the ribosome, it hydrolyzes its bound GTP molecule (Rodnina *et al.*, 1997). Accurate translocation of mRNA requires a peptidyl-tRNA bound to the A-site. However, if aminoacyl-tRNA or deacylated tRNA is translocated from the A- to P-site, the movement of the mRNA becomes uncoupled (Zavialov & Ehrenberg, 2003). It is therefore inaccurate (Fredrick & Noller, 2002).

The classical view of translocation is that EF-G acts as a molecular switch like other GTPases (Vetter & Wittinghofer, 2001). According to this mechanism, it binds to the ribosome with GTP, translocates the peptidyl-tRNA, deacylated tRNA and mRNA, hydrolyzes its GTP and subsequently falls off the ribosome in complex with GDP. Spirin (2002) has summarized the arguments for this view. According to an alternative view, EF-G, like several ATPases, may act like a mechano-mechanical protein or a motor protein (Rodnina *et al.*, 1997; Cross, 1997). The main observation supporting this mechanism is that EF-G hydrolyzes its GTP shortly after it has bound to the ribosome (Rodnina *et al.*, 1997). Translocation was found to be slower than GTP hydrolysis.

The crystallographic work has clarified the structure of EF-G and EF2 in several different conformations off the ribosome (Czworkowski *et al.*, 1994; Ævarsson *et al.*, 1994; Al-Karadaghi *et al.*, 1996; Laurberg *et al.*, 2000, 2004; Jørgensen *et al.*, 2003). The structural details are discussed in Section 9.3. The three N-terminal domains, G, G' and II form a block whose interactions remain (Laurberg *et al.*, 2000; Jørgensen *et al.*, 2003). The tRNA mimicking domains (IV and V) go through dramatic movements. Domain III, the most mobile domain, moves independently (Jørgensen *et al.*, 2003). The cryo-EM studies of EF-G bound to the ribosome confirm and extend these observations. Two types of complexes have been studied; EF-G has been studied bound to the ribosome in complex with GDPNP or in complex with GDP and fusidic acid (FA; Agrawal *et al.*, 1998, 1999; Frank & Agrawal, 2000; 2001; Stark *et al.*, 2000; Frank, 2003; Gao *et al.*, 2003; Valle *et al.*, 2003b). In addition, yeast ribosomes have been studied in complex with yeast EF2 and sordarin (Gomez-Lorenzo *et al.*, 2000).

Ribosomes in complex with EF-G or EF-2 studied by cryo-EM show the factor bound to essentially the same site as the ternary complex between EF-Tu, GTP and aminoacyl-tRNA (Stark *et al.*, 2002; Valle *et al.*, 2002; 2003b). This agrees with the similar shape of the ternary complex and EF-G (Nissen *et al.*, 1995; see Sections 9.3 and 9.6). However, it is clear that the orientation of the tRNA mimicry block of EF-G, domains IV and V differs from the crystal structures of EF-G (Agrawal *et al.*, 1998; Wriggers *et al.*, 2000; Gao *et al.*, 2003; Valle *et al.*, 2003b). The crystal structure of EF2 with sordarin, however, seems to fit better with the cryo-EM observations (Gomez-Lorenzo *et al.*, 2000; Jørgensen *et al.*, 2003).

Peptidyl transfer occurs without large conformational changes and the products remain very close to the positions of the substrates (Schmeing *et al.*, 2002; Valle *et al.*, 2003b). The subsequent translocation leads to the sideways movement of the peptidyl tRNA and deacylated tRNA from the A- and P-sites to the P-site and E-sites, respectively (Fig. 11.7). This is preceded by the rotational movement of the CCA-ends of the tRNAs to the hybrid A/P and P/E-states (Moazed & Noller, 1989). The P/E-state has also been observed during the ratchet-like rotation with a deacylated tRNA in the P-site (Valle *et al.*, 2003b).

The base-pairing to the A- and P-loops guides the orientations of the CCA-ends. When the CCA-end of the peptidyl-tRNA is rotated the nascent peptide is also rotated, for the part nearest to the peptidyl tRNA by 180°, for the rest maybe less. It is also moved a slight distance into the exit tunnel (Bashan *et al.*, 2003). The rotation can be performed without

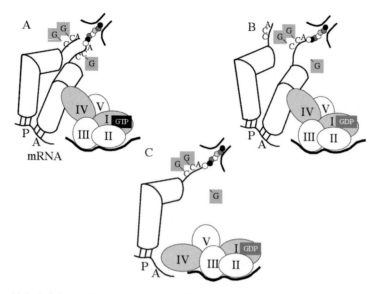

Fig. 11.7 A Schematic representation of translocation. **(A)** EF-G in complex with GTP binds to the ribosome with a peptidyl tRNA in the A-site. GTP hydrolysis occurs rapidly. **(B)** The first part of the tRNAs that move are the acceptor ends. Thus, the peptidyl tRNA adopts the A/P-state. **(C)** Subsequently, EF-G undergoes a conformational change such that the peptidyl-tRNA is translocated. At translocation, domain IV occupies the decoding part of the A-site.

sterical clashes. Two nucleotides close to the two-fold rotation axis come into close contact with the rotating moiety. These are A2602 and U2585. The large conformational freedom of A2602 may be related to the rotational movement of the rotating moiety (Bashan *et al.*, 2003).

Domains IV of EF-G and EF2 binds to the decoding part of the A-site in all structures studied, with GDPNP, GDP and FA or sordarin. This means that with GDPNP, the ribosome is in the post-translocation state even without GTP hydrolysis. In complex with FA, EF-G is not able to dissociate from the ribosome despite GTP hydrolysis. These interesting observations cannot alone discriminate between the two models for translocation. The classical model, where the translocation precedes GTP hydrolysis (Spirin, 2002), agrees with the cryo-EM observations. However, in the alternative model, where GTP hydrolysis precedes translocation, even GDPNP is observed to slowly lead to translocation (Rodnina *et al.*, 1997; Zavialov & Ehrenberg, 2003).

Conformational changes in subunit orientation and subunit structure during translocation have been characterized by cryo-EM. They are caused by the binding of EF-G and are essential parts of translocation. The conformational changes are similar to a ratchet motion (Agrawal *et al.*, 1999; Frank & Agrawal, 2000; Gao *et al.*, 2003). In this motion, the relative orientation of the two subunit changes by about 10° (Valle *et al.*, 2003b). Simultaneously, the structure of the small subunit is changed. The movements of the head domain and shoulder may be the most significant. The tip of domain IV of EF-G reaches the top part of h44 that undergoes a lateral movement of about 8 Å parallel to the path of the mRNA (VanLoock *et al.*, 2000; Valle *et al.*, 2003b). This leads to an opening and closing of the path for the mRNA in the structure of the 70S ribosome in a way that would simplify translocation of the mRNA.

A full insight into the interplay between EF-G and the ribosome that leads to translocation is still lacking. However, a likely scenario is the following (Fig. 11.8):

(1) EF-G in complex with GTP (an unknown conformation) binds to pre-translocation ribosomes (the conformation at binding is not known).

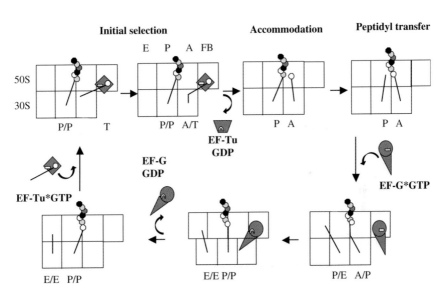

Fig. 11.8 A summary of the elongation illustrating also the ratchet movement caused by the action of EF-G.

(2) Activation of the GTPase. Interactions and conformational changes of the ribosome and EF-G. L12 CTD stabilizes the transition state in some unknown way.

(3) GTP hydrolysis (Rodnina *et al.*, 1997). The detailed mechanism is unknown.

(4) Ratchet-like rotation of subunits into a strained state. Whether the tRNAs move into the hybrid states at this stage is not known. The peptide is accessible to react with puromycin when the acceptor end of the A-site tRNA has moved into the P-site.

(5) Pi is released from EF-G:GDP. This step occurs simultaneously but independently of translocation. For certain mutants of L12, the Pi release is slow.

(6) Translocation. During translocation, domain IV of EF-G will swing into the decoding part of the A-site.

(7) EF-G:GDP dissociates from the ribosome.

(8) In the second step of the ratchet-like movement, the subunits return to normal orientation.

(9) A new ternary complex (EF-Tu:GTP:aminoacyl-tRNA) can bind to the post-translocation ribosomes and start a new cycle of elongation.

In case of EF-G:GTP with FA, the process will halt, at step 7. EF-G:GDPNP will also proceed to this step even though no GTP can be hydrolyzed. In both cases EF-G will remain bound to the ribosome and unable to dissociate.

The Relationship between EF-Tu and EF-G

All tGTPases are expected to bind to the same site, the GTPase site, of the ribosome. This is mirrored by the fact that they all have two domains in common, the G-domain and domain II interacting with the 50S and 30S subunits, respectively. The mimicry between the ternary complex and EF-G is an extension of this similarity and they can be regarded as twin molecules that cause the ribosome to oscillate between two different states, the pre- and the post-translocation states, by binding to the same site (Fig. 11.9). A number of observations have been accumulated to support this hypothesis (Table 11.2).

In the large family of GTPases, there is a considerable variation in the size and sequence of the effector loop. This is most certainly due to its

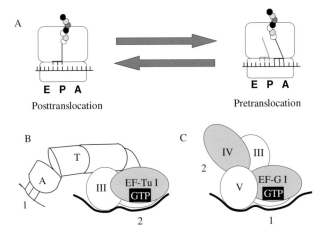

Fig. 11.9 (A) The ribosome oscillates between the pre-and post-translocation states. EF-Tu and EF-G have reciprocal roles in binding to overlapping sites on the ribosome. **(B)** EF-Tu binds to the post-translocated ribosome and hydrolyzes its GTP molecule (2) only when a cognate interaction (1) has occurred between the codon and anticodon in the decoding site. **(C)** EF-G hydrolyzes its bound GTP (1) shortly after it has bound to the pre-translocated ribosome. Subsequently translocation occurs and domain IV of EF-G moves into the decoding region (2).

Table 11.2 The Reciprocal Mode of Action of EF-Tu and EF-G

- The factors bind to overlapping sites of the ribosome.
- EF-Tu and EF-G make the ribosome alternate between two different conformational states. EF-Tu binds to post-translocation ribosomes, whereas EF-G binds to pre-translocation ribosomes.
- A correct match of codon-anticodon in the decoding site induces GTP hydrolysis for EF-Tu. For EF-G:GTP, hydrolysis occurs directly upon binding. After translocation, domain IV binds to the decoding site. The order of events for the two factors is reversed (Fig. 11.9).
- EF-G:GDP dissociates from a site that its mimic EF-Tu:aa-tRNA:GTP binds to.
- If the effector loop of EF-Tu replaces that of EF-G, the factor can still bind to the ribosome and hydrolyze GTP. However, the factor cannot translocate (Kolesnikov & Gudkov, 2002).
- The trypsin susceptibility of L12 is different when EF-Tu is bound to the ribosome with GDPNP or kirromycin. This trypsin sensitivity is opposite to what is found when EF-G is bound with GDPNP or FA (Gudkov & Gongadze, 1984; Gudkov & Bubunenko, 1989).

interaction with very different receptors. However, the effector loops of EF-Tu and EF-G have a high degree of identity and similarity. Thus, one could expect that this loop could be exchanged between the two proteins.

This is, however, not possible. A hybrid EF-G with the effector loop of EF-Tu does not support protein synthesis despite the fact that it binds to the ribosome and can hydrolyze GTP (Kolesnikov & Gudkov, 2002). Presumably, despite the similarities, there are important differences between the effector loops, which enable EF-Tu to interact with post-translational ribosomes and EF-G with pre-translocation ribosomes.

11.5 TERMINATION

The termination of protein synthesis depends on the exposure of one of the three stop codons UAA, UAG and UGA in the decoding part of the A-site. In bacteria, the Class I release factor RF1 responds to the stop codons UAA and UAG, while RF2 responds to UAA and UGA. The release factors decode the stop codons and hydrolyze the completed peptide from the P-site tRNA. The hydrolysis may be induced directly or indirectly (Nakamura & Ito, 2003). Eukarya have a single decoding factor, eRF1, without any structural relation to the bacterial release factors (see Section 9.4). The termination factor, RF3, in bacteria and eukarya catalyzes the dissociation of the decoding release factors from the ribosome. This factor is lacking in some species (see Table 9.2).

In the crystal structures of the Class 1 release factors, one could identify a resemblance to tRNA (Plate 9.14). In eRF1, the location of GGQ (see Section 9.4) was at one extreme of the Y-shaped structure and the potential decoding region at the extreme end of another domain (Song *et al.*, 2000). The structure of bacterial RF2 made it clear that the eukaryotic and bacterial release factors have no structural similarity. Secondly, the two supposed functional sites, corresponding to the extreme ends of the tRNA, were no more than 25–30 Å apart (Vestergaard *et al.*, 2001).

Contrary to expectation, RF2 has no structural resemblance to tRNA molecules when bound to the ribosome, neither in shape nor in the manner it is bound as observed by cryo-EM (Klaholz *et al.*, 2003; Rawat *et al.*, 2003). A large conformational change of the protein from the crystal structure to the bound state was identified. The domains of RF2 could be oriented in such a way that they both satisfy the cryo-EM maps and place the functional regions in their expected locations. Thus, a functional relationship to tRNA remains (Nakamura & Ito, 2003). RF2 can operate in the PTC and activate a water molecule to hydrolyze the ester bond between the tRNA and the completed peptide. The common sequence GGQ is placed at the PTC. The "anticodon mimicry" region (Ito *et al.*, 2000) is in the decoding site of the small subunit.

As for other functions of translation, one can expect that the ribosome participates actively. Thus, mutations of A2602 in the PTC reduce the peptide release activity significantly, whereas mutations affecting the other residues in this site (A2451 and U2585) have little effect (Polacek *et al.*, 2003; Nakamura & Ito, 2003).

The role of RF3 is to release RF1 and RF2 from the ribosome (Fig. 11.10). RF3 is a tGTPase with the classical G domain and a domain II (Ævarsson, 1995). Its mode of action is somewhat unorthodox. The ribosome first acts like a GEF to replace GDP with GTP and subsequently as a GAP to induce GTP hydrolysis. RF3:GDP binds to termination complexes

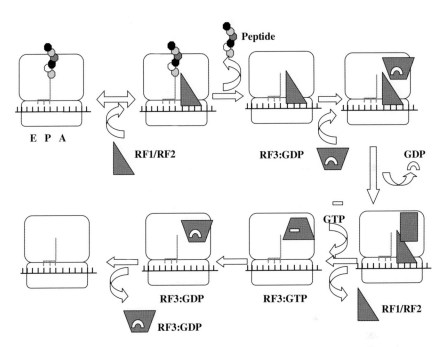

Fig. 11.10 The steps involved in termination of protein synthesis. The ribosome on the top left has a stop codon in the A-site. RF1 or RF2, depending on the stop codon, binds to the A-site and releases the peptide. To remove RF1/RF2, a tGTPase, RF3, is normally needed. It binds in complex with GDP to the ribosome that functions as a GEF and replaces GDP with GTP. The associated conformational change of the factor and the ribosome will cause RF1/RF2 to dissociate. Subsequently, RF3 hydrolyzes its GTP to GDP and dissociates from the ribosome.

with ribosome-bound RF1/2 and a deacylated tRNA in the P-site (Zavialov *et al.*, 2001; 2002; Zavialov & Ehrenberg, 2003). If the peptide is not hydrolyzed, RF3 cannot exchange its GDP for GTP, but only replace its GDP with GDP (Zavialov & Ehrenberg, 2003). When the peptide is hydrolyzed, RF3 can exchange its GDP for GTP, release RF1/2 from the ribosome and subsequently hydrolyze its GTP to GDP and be released from the terminated ribosome. Evidently, when the peptide is released from the P-site tRNA, the ribosome undergoes a conformational change allowing the nucleotide exchange. Whether or not an RF of Class 1 or the ribosome acts as nucleotide exchange factors has not yet been established. No structural insights are available for RF3, neither alone nor when bound to the ribosome.

11.6 RIBOSOME RECYCLING

When the ribosome has reached a stop codon and the peptide is released, the extensive machinery of the ribosome has to be reused. For this, the mRNA and the deacylated tRNA have to be released. Since the mRNA is threaded through the tunnel between the subunits and in intimate contact with the neck region of the small subunit and the deacylated tRNA has tight interactions in the P-site, the subunits would probably have to open up or possibly even dissociate. For a new initiation, they have to be separated since the initiation is done with the small subunit alone. The ribosome-recycling factor (RRF) participates in this process together with EF-G. The footprinting of the RRF on the ribosome using the Fe-EDTA method and cryo-EM suggests that RRF binds in a very unexpected manner across both the A- and P-sites (Lancaster *et al.*, 2002; Agrawal *et al.* 2004). The deacylated tRNA would in that case not be able to bind to the P-site, but may have moved to the P/E-site.

T. thermophilus RRF (ttRRF) does not function in *E. coli* (Toyoda *et al.*, 2000). Mutants that would make ttRRF functional in *E. coli* were concentrated around the hinge between the two domains (Toyoda *et al.*, 2000; Ito *et al.*, 2002). In particular, a deletion of five C-terminal residues was part of this group of mutants (Fujiwara *et al.*, 1999). The mutations around the hinge region suggest that the flexibility of the molecule is functionally important, but no sufficiently detailed model is available at present.

The interaction with EF-G has been analyzed by various means. RRF from *Mycobacterium tuberculosum* will not function in *E. coli*. However, if this factor is complemented by EF-G from the same species function is regained (Rao & Varshney, 2001). This is evidence for a direct contact during the functional cycle. The same pair-wise dependence also exists for the *T. thermophilus* factors (Ito *et al.*, 2002). Mutants in *E. coli* EF-G enabled a functional interaction with ttRRF. These mutations are localized to one side of domain IV, the domain that also is essential for translocation. However, these mutations are distinct from mutations in domain IV that lead to a low rate of translocation (see Table 9.7). Ribosomes pretreated with EF-G in complex with GTP or GDP will not prevent RRF from binding since EF-G will dissociate. However, in ribosomes with EF-G:GDPNP or EF-G:GTP and fusidic acid, EF-G would compete with the binding of RRF (Kiel *et al.*, 2003). In these states, EF-G binds to the decoding site with the ribosomes in a post-translocational state.

In conclusion, RRF and EF-G are in contact on the ribosome during some stage. After translocation, EF-G can prevent the binding of RRF to the ribosome. It is unlikely that EF-G undergoes other movements than the translocational movement. Even though different point mutations on domain IV show that different surfaces are important for translocation of tRNAs and the mRNA and interaction with RRF, the same movement of EF-G is responsible both for translocation and RRF release.

12

Protein Folding and Targeting

12.1 FOLDING OF THE NASCENT CHAIN

A newly synthesized protein is exposed to a crowded environment in the cell. The protein concentration is about 300 mg/ml and the environment is hostile due to a large number of proteolytic enzymes in the cell. If the protein is not properly folded it may aggregate or be degraded in the cell. Thus, proper folding of the nascent chain is an important aspect of translation. The exit tunnel of the ribosome is quite narrow and whether it can permit folding of the nascent chain is under discussion (Gilbert *et al.*, 2004). The length of this tunnel is about 100 Å and its width, 10–20 Å (Nissen *et al.*, 2000; Harms *et al.*, 2001; Gabaswili *et al.*, 2001). Between 44 and 72 amino acid residues of the nascent chain can be contained in the tunnel, depending on the secondary structure of the completed protein (Kramer *et al.*, 2001).

The folding of many proteins may commence as they pass through the peptide exit tunnel of the ribosome, and may be completed spontaneously when they emerge from the exit tunnel (Hardesty & Kramer, 2001). However, in many cases, chaperones are needed for the proper folding of the emerging polypeptide (Hartl & Hayer-Hartl, 2002). Several types of chaperones are induced by thermal stress. This has given them the name heat shock proteins (Hsp). Some chaperones interact directly with the ribosome and the nascent chain both in prokaryotic and eukaryotic systems (for a review, see Frydman, 2001). The sites identified by the chaperones are short but exposed hydrophobic regions. In bacteria, the nascent chain interacts primarily with small or "holding" chaperones that bind to ribosomes. Trigger factor (TF) and DnaK belong to this class (Deuerling *et al.*, 1999; Teter *et al.*, 1999). These are monomeric proteins that primarily prevent the aggregation of the growing polypeptide.

There are interesting parallels between the chaperones of bacteria and eukarya (Frydman, 2001). There are potential TF homologues and several well known Hsp70 and Hsp40 versions in eukarya (Frydman, 2001). In many cases, the interactions with these chaperones are sufficient for the proper folding of the protein. In other cases, interactions with the more complex oligomeric ring-shaped chaperonins like GroES/Hsp60 are needed for the proper folding of the protein. These are not known to interact with the ribosome. For a detailed discussion of the functions of chaperones, see Frydman (2001).

Trigger Factor

Trigger factor is a bifunctional protein. It belongs to the class Hsp70-like chaperones (Frydman, 2001) and is a peptidyl-prolyl-*cis/trans* isomerase (PPIase) that participates in the binding and correct folding of nascent peptides (Bukau *et al.*, 2000). TF can be crosslinked to nascent polypeptides (Schaffitzel *et al.*, 2001).

The structure of the ribosome binding region of TF (TFrb) is available (Kristensen & Gajhede, 2003). The protein is composed of three domains. The N-terminal domain (residues 1–144) is the TFrb, the middle domain (residues 145–247) is the PPIase, with homology to FK506-binding proteins, while the function of the C-terminal domain (247–432) is unknown (Stoller *et al.*, 1995; Hesterkamp *et al.*, 1996; 1997).

TF can be crosslinked to L23 and L29 at the opening of the exit tunnel on the external surface of the large subunit (Kramer *et al.*, 2002b). A mutation in L23 can destroy the interaction with TF (Kramer *et al.*, 2002b). TFrb is an elongated structure with a positively charged end, which contains the TF-signature sequence motif (FRK; Kristensen & Gajhede, 2003).

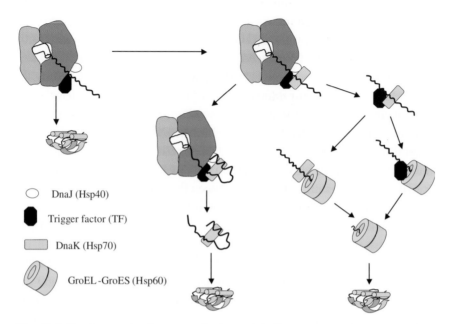

Fig. 12.1 The interaction between the bacterial ribosome and the nascent chain with chaperones (after Frydman *et al.,* 2001).

Observations by neutron scattering indicate that TF binds to the ribosome as a homo-dimer (Blaha *et al.,* 2003). From the crystal structure, there are several possibilities that this could be realized (Kristensen & Gajhede, 2003). When bound to the ribosome, TF is probably the first chaperone to interact with hydrophobic segments of the nascent peptide (Fig. 12.1; Kramer *et al.,* 2002b).

The affinity of TF for different peptides was investigated (Patzelt *et al.,* 2001). It was found that TF binds to stretches of eight amino acid residues, with a preference for peptides with aromatic and positively charged amino acids. Binding is quite weak. The binding site is the PPIase domain of TF, but there is no special preference for peptides containing prolyl residues. TF has a central role in identifying nascent peptides that need assistance to be folded as well as peptides that are addressed for export (see Section 12.2).

DnaK

DnaK belongs to the Hsp70 family of chaperones and is a very abundant ATPase (Herendeen *et al.,* 1979). Its role overlaps with that of TF

(Teter *et al.*, 1999; Deuerling *et al.*, 1999). In bacteria, DnaJ (a member of the Hsp40 family of chaperones) assists TF. DnaJ induces the ATPase activity, and GrpE catalyzes the nucleotide exchange of DnaK (Suh *et al.*, 1999). The J-domain of DnaJ induces ATP hydrolysis in DnaK that leads to a high affinity for the polypeptide. The nucleotide exchange assisted by GrpE leads to polypeptide release (Harrison *et al.*, 1997). Both TF and DnaK can be crosslinked to nascent polypeptides (Schaffitzel *et al.*, 2001). Deletions of the genes for both TF and DnaK are lethal (Teter *et al.*, 1999; Deuerling *et al.*, 1999).

A nascent peptide that interacts with TF may also interact with DnaK (Albanese & Frydman, 2002). DnaK is not recruited to ribosomes that lack TF (Kramer *et al.*, 2002a).

12.2 TRANSPORT OF THE PRODUCT

The translated proteins are frequently aimed for other compartments of the cell than where they were synthesized. Proteins are synthesized in the cytoplasm and then transported to their final destination (Walter & Blobel, 1980; 1982). This is the case for all exported proteins regardless of the kingdom of living things.

The transport of proteins involves complex transport systems. This includes the soluble signal recognition particle (SRP) in the cytoplasm, the signal recognition particle receptor (SR), which is a membrane-associated protein and the translocon, which is a membrane channel or protein-conducting channel (PCC) that permits export of the protein to its final destination (Keenan *et al.*, 2001). The ribosome interacts with several components of this transport machinery. The processes involved are only partly characterized.

SRP is composed of an RNA molecule and differing number of proteins (see Keenan *et al.*, 2001 for a review). In bacteria, the RNA component is called 4.5S RNA (120 nucleotides), whereas in archaea and eukarya, it is 7S RNA (Table 12.1). The 7S RNA (about 300 nucleotides) is composed of two domains, the Alu domain (absent in 4.5S RNA) and the S domain, part of which corresponds to the 4.5S RNA (Poritz *et al.*, 1988). The number of proteins of SRP varies from one to six, depending on the species (Table 12.1). The structures of some of these proteins and fragments of the SRP RNA are known (Table 12.1). One protein is completely conserved in all species. It is called SRP54 or Ffh in bacteria. The M-domain of

Table 12.1 Components of SRP, SR and Translocon in the Three Domains of Life

	Bacteria	Archaea	Eukarya	Comments
SRP				
RNA	4.5S	7S	7S	7S RNA has Alu- and S-domains. The 4.5S RNA from eubacterial corresponds to part of the S-domain from eukarya.
Proteins			SRP9	Heterodimers of SRP9:SRP14
			SRP14	bind to the Alu-domain.
		SRP19	SRP19	Binds to S-domain.
	Ffh	SRP54	SRP54	GTPase. Binds to S-domain. Depends on the binding of SRP19 in eukarya and archaea.
			SRP68	Heterodimers of SRP68/SRP72
			SRP72	bind to the S-domain.
SR				
	FtsY		SRα	GTPase
			SRβ	GTPase, membrane bound
Translocon				
	SecYEG		Sec61αβγ	A heterotrimeric pore-forming protein complex.

Bacterial and archaeal components of SRP and SR for which the structures are known

Component	Species	References
Ffh	*T. aquaticus*	Keenan *et al.*, 1998
4.5S (domain IV): Ffh (M-domain)	*E. coli*	Batey *et al.*, 2000
FtsY, N- and G-domains	*E. coli*	Montoya *et al.*, 1997
Ffh-FtsY complex	*T. aquaticus*	Focia *et al.*, 2004
		Egea *et al.*, 2004
SecYEG	*E. coli*	Breyton *et al.*, 2002
SecYEβ	*M. jannashii*	van den Berg *et al.*, 2004

Ffh/SRP54 binds to helix 8 of the S-domain of 7S RNA, which in 4.5S RNA is called domain IV (Schmitz *et al.*, 1999). The protein is a GTPase and is composed of three domains, where the central one is the classical G-domain (Freymann *et al.*, 1997). The hydrophobic signal peptide associates with the third or M-domain of the GTPase that also binds to domain IV of 4.5S RNA (Keenan *et al.*, 1998; Batey *et al.*, 2000).

Ffh/SRP54 binds to protein L23 just outside the exit tunnel of the ribosome (Pool *et al.*, 2002). This binding is then very similar to the binding of trigger factor and the two proteins compete for binding to the ribosome (Beck *et al.*, 2000; Albanese & Frydman, 2002; Gu *et al.*, 2003; Ullers *et al.*, 2003).

In bacteria, SR is a single soluble protein, FtsY, which can loosely attach to the membrane (Lurink *et al.*, 1994), while in eukarya there are two proteins, SRa and SRb (Keenan *et al.*, 2001; Table 12.1). SRa corresponds to FtsY and SRb has one membrane-spanning helix. All SR proteins are GTPases and FtsY of SR is a close homologue of Ffh of SRP, despite their different functional roles (Keenan *et al.*, 2001).

The protein to be transported has an N-terminal sequence, of 20–30 generally hydrophobic amino acids, (or "tag" called the signal peptide; von Heine, 1985). When the signal peptide of the nascent polypeptide emerges from the exit tunnel of the bacterial ribosomes, it binds to the M-domain of Ffh in its GTP conformation (Keenan *et al.*, 1998; Batey *et al.*, 2000). Further translation is then temporarily arrested in a process that includes the binding site for tGTPases on the ribosome. The distance between the polypeptide exit tunnel and the GTPase binding site is more than 150 Å. Evidently, SRP has to be quite extended. Ffh in SRP and the ribosome then form a complex with FtsY in their GTP conformations (Montoya *et al.*, 2000). A eukaryotic complex of 80S ribosome with the SRP has been investigated, and the result shows how the SRP complex stretches from the polypeptide exit channel to the binding site for the tGTPases (Plate 12.1; Halic *et al.*, 2004). This large complex associates with the translocon (Fig. 12.2). The interaction between Ffh and FtsY is due to the N- and G-domains of both molecules (Montoya *et al.*, 2000). The two homologous proteins form a symmetric heterodimer with the two GTP molecules located in a composite active site (Focia *et al.*, 2004; Egea *et al.*, 2004). Significant conformational changes of the proteins occur upon interaction. This interaction leads to the interaction with PCC or translocon (see Mori & Ito, 2001 for a review). The ribosome and the translocon jointly induce GTP hydrolysis by the two or three GTPases, which leads to the dissociation of the SRP and SR parts of the complex. The translational arrest is then relieved and the co-translational transport can proceed.

The translocon or protein-conducting channel (PCC) is composed of a number of proteins. PCC is a membrane channel with a narrow pore ring (van den Berg *et al.*, 2004). In eukarya, it is called Sec61 and composed of

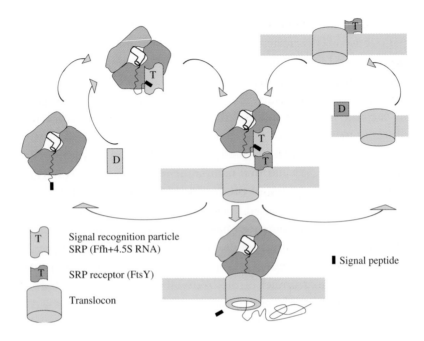

Fig. 12.2 SRP and SR participate to transport a protein out of the cytoplasm (after Bates *et al.*, 2000; Keenan *et al.*, 2001). The letters T and D on the symbols for the SRP and the SRP receptor represent GTP and GDP, respectively.

trimers of Sec61α-, β- and γ-subunits (for a review, see Rapoport *et al.*, 1996). In bacteria and archaea, the protein corresponding to the α-subunit is called SecY and the γ-subunit is called SecE. The β-subunit of eukarya and archaea shows no homology to the SecG subunit in bacteria. A cryo-EM structure has been determined of the *E. coli* pore complex has been determined (Breyton *et al.*, 2002) and a crystallographic structure is available of the complex from the archeon *M. jannashii* (van den Berg *et al.*, 2004). SecY has 10 TM helices displaying a pseudo symmetry parallel with the membrane and with five helices in each symmetric part. The other two subunits have one TM helix each in the archaea, but SecG has two TM helices. The pore goes right through the SecY subunit and the cytoplasmic side has a 20–25 Å wide tapering funnel down to a constriction. Six residues, primarily isoleucines, form the constriction.

Obviously, the ion permeability barrier of the membrane must be maintained despite this channel. This seems to be due to a short helix called the plug located in the constriction and that may be movable (van den Berg *et al.*, 2004). The hydrophobic part of the signal peptide of the peptide to be transported may form a helix and a likely binding site has been identified in the funnel. The binding of the signal sequence triggers the opening of the channel.

Cryo-EM studies of eukaryotic ribosomes in complex with PCC show a disk of density associated with the exit tunnel of the ribosomes (Beckmann *et al.*, 1997; 2001). The disk has a diameter of about 100 Å. This is a strong support for the ribosomal tunnel being the path for the nascent peptide. A gap of about 15 Å between the ribosome and the tunnel is observed. The protrusions on the cytoplasmic side from the C-terminal half of SecY would explain this (van den Berg *et al.*, 2004). The size of the disk suggests that there is more than one trimer of Sec61α-, β- and γ-subunits (Beckmann *et al.*, 2001). The joint results from the studies of the *E. coli* and *M. jannashii* complexes suggest that it may be dimers of the trimers that are seen at the ribosomal tunnel and that only one is engaged in translocating a polypeptide (Breyton *et al.*, 2002; van den Berg *et al.*, 2004).

13

Evolution of the Translation Apparatus

When we study the evolution of translation, the currently living organisms give us a fragmented representation of what has passed. The genetic code is universal with marginal variations, the latter primarily in the mitochondria of higher eukaryotes. This suggests that the translation apparatus must have been one of the first molecular systems that "crystallized" at a very early stage before the evolution of species (Woese & Fox, 1977; Woese, 2002). This primarily includes the canonical genetic code that must have emerged before the last universal ancestor (Knight *et al.*, 2001). This also includes the tRNAs, the tRNA synthetases, the ribosomal proteins and some of the translation factors (Harris *et al.*, 2003).

The key molecules of the translation apparatus are RNA molecules, the mRNA, tRNA and rRNA. The main functional sites of the ribosome are composed almost entirely of rRNA. This certainly strengthens the arguments for an early world where RNA was the dominant biological polymer. As Crick (1968) suggested, the original ribosome might have been composed of RNA alone. Numerous chemical reactions needed to provide the building blocks for such a world can be catalyzed by ribozymes (Joyce, 2002). tRNAs, rRNAs and early ribosomal proteins may have co-evolved in the primitive cells (Woese, 1998), but so also small RNA molecules and antibiotic progenitors that could regulate translation (Davies, 1990; Dinos *et al.*, 2004).

13.1 EVOLUTION OF CODONS, tRNAS AND tRNA SYNTHETASES

The aminoacyl-tRNA synthetases (aaRS) must have evolved with the genetic code. The two classes of aaRS form two groups of enzymes with evolutionary relationships within each group. One might imagine that the two families have emerged from two early single domain enzymes with less specific charging of tRNAs (Schimmel *et al.*, 1993). With one exception, the class of aaRS to which a certain amino acid belongs is constant throughout evolution (see Table 5.2; Ribas de Pouplana & Schimmel, 2001). Each subclass is likely to have had a common ancestor. If the current set of enzymes with their subclass is arranged as in Table 13.1, one can see that pairs of relatively similar amino acid residues can be grouped together.

X-ray structures of aaRS complexes show that pairs of Classes I and II aaRS and of the same subclass can bind to opposite sides of the acceptor stem of the tRNA without steric clash for related amino acids (Ribas de Pouplana & Schimmel, 2001). As discussed (Chapter 5), enzymes of Class I attach the amino acid on the 2'-hydroxyl, whereas enzymes of Class II charge the tRNA on the 3'-hydroxyl. The synthetases might have evolved

Table 13.1 The Symmetry of the aaRS Subclasses (from Ribas de Pouplana & Schimmel, 2001)

Classes/Subclass I		II	
a	M	P	Hydrophobic
	V	T	
	L	A	
	I	G	
	C	S	
	R	H	
b	E	D	Charged and amidated
	Q	N	
	K	K	
c	Y	F	Aromatic
	W		

as pairs to protect the charged tRNA in hostile environments (Ribas de Pouplana & Schimmel, 2001). This might suggest an early genetic code with fewer amino acids and a higher ambiguity between related amino acids. The expansion to the present canonical genetic code probably occurred through duplication and mutations of genes for both tRNAs and synthetases. Such an evolution should lead to related codons for related amino acids. Ribas de Pouplana and Schimmel (2001) find support for this. Thus, the codons for the aromatic residues (Subclasses Ic and IIc) all begin with U (Fig. 4.1). The codons for Subclasses Ib and IIb (charged and amidated residues) all share A for the second base. Asn and Gln may have evolved at a late stage from their acidic origins, Asp and Glu, respectively (Skoulobris *et al.*, 2003). Four of the six Class Ia and four of the six Class IIa synthetases have a common middle base in their codons, U and C, respectively. They are differentiated primarily by their first codon. An early primitive genetic code could have had a smaller set of amino acids, with paired tRNA synthetases and a low charging specificity that partly depended on the availability of amino acids.

13.2 EVOLUTIONARY RELATIONSHIP BETWEEN tRNAS AND RIBOSOMAL RNAS

The ribosomal RNAs provide excellent material to study the evolutionary relationships between species. This led Woese *et al.* (1990) to the conclusion that there are three main domains among living organisms: Bacteria, Archaea and Eukarya. Even though there are large variations in the size and sequence of ribosomal RNAs from different organisms, there are universally conserved regions, now known to be associated with the functional centers. It may seem remarkable that the most conserved regions of the rRNA are single-stranded (Noller & Woese, 1981). However, with the understanding that these regions interact primarily with other RNA molecules, the need for their conservation is evident. The number of helices and loops varies greatly between the rRNAs from different species. The size range of the rRNAs is also highly variable. The smallest rRNAs found so far are the 9S and 12S RNA molecules of trypanosomal mitochondria. The corresponding ribosomes have not been isolated and investigated. The range of proteins belonging to these ribosomes is not known. However, a number of potential ribosomal genes are identified (Tittawella *et al.*, 2003).

Some examples of the universally conserved interplay between the tRNAs and rRNA can illustrate the features of translation that were frozen at an early stage. The tRNA molecules are not only highly conserved in their general organization. The long L-shaped molecule is universal. In addition, the acceptor end, the CCA, is universal. In the ribosome, this is paralleled by the conserved A- and P-loops that basepair with the CCA-ends of the tRNAs in the A- and P-sites. In the decoding site, the residues G530, A1492 and A1493 are conserved due to their essential function in maintaining high fidelity.

13.3 EVOLUTIONARY RELATIONSHIP OF RIBOSOMAL PROTEINS

The database of ribosomal protein sequences is rapidly growing, primarily due to the number of genomes that have been completely sequenced. From sequence comparisons, it is evident that more than 50% of the bacterial ribosomal proteins have homologues in chloroplasts, archaea as well as eukarya (Table 6.4). Complete investigations of ribosomal proteins from mitochondria have not yet been performed. The list of conserved proteins may increase when the structures of archaeal or eukaryotic small subunits become known, since the protein sequences have often diverged to such an extent that a homology can no longer be predicted from sequence information. The recognition of sequence motifs or structural and functional correspondence may extend the fraction of proteins conserved between the different kingdoms.

Bairoch *et al.* provide a database of ribosomal proteins that is continuously updated (http://www.expasy.org/cgi-bin/lists?ribosomp.txt). Table 6.4 is primarily taken from this database. Lecompte *et al.* (2002) make a thorough comparison of the sequences of the ribosomal proteins found in all completed genome sequences (Fig. 13.1). A large fraction of the proteins are found in all three domains. Furthermore, all proteins in archaeal ribosomes except one (Lxa) are also found in eukarya. However, there are no proteins that are common only to bacteria and archaea or bacteria and eukarya. Thus, with regard to protein composition, archaea and eukarya are closely related, whereas the rRNAs of bacteria and archaea are comparable in size.

As stated above, the amino acid sequences give limited possibilities to determine whether proteins are related. Thus, the identification of

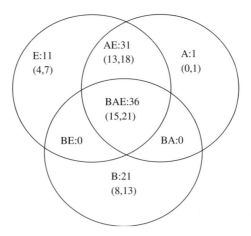

Fig. 13.1 Diagram modified from Lecompte *et al.* (2002) showing the distribution of r-proteins between Bacteria (B), Archaea (A) and Eukarya (E). The number of common proteins is given after the letters. The numbers related to the small and large subunits are given within parenthesis. The only differences compared with the diagram by Lecompte *et al.* (2002) are two additional protein pairs between H50S and D50S that were found to have the same structure and the same location (Harms *et al.*, 2001; Agmon *et al.*, 2003).

proteins related by fold and location in the structures of the 50S subunits from *H. marismortui* and *D. radiodurans* (Ban *et al.*, 2000; Harms *et al.*, 2001) adds to the number of homologous proteins. The clearest case is L16b, which corresponds to L10e (Harms *et al.*, 2002; see also Chapter 6).

Harris *et al.* (2003) have investigated the universally conserved genes in all the available genomes. No more than 80 such genes were found. Only 50 show the same phylogenetic relationship as the rRNAs. Of these, 37 encode proteins or factors that are physically associated with the ribosome in modern organisms: ribosomal proteins, IF1, IF2, EF-Tu and EF-G. The ribosomal proteins found in this investigation (14 in the small subunit and 15 in the large subunit) are among the proteins found in all the domains (see Table 6.4; http://www.expasy.org/cgi-bin/lists?ribosomp.txt; Lecompte *et al.*, 2002). However, some proteins do not share the phylogenetic relationship with the rRNAs and may thus have been subjected to lateral gene transfer. Among the additional universally conserved proteins are methionine peptidase, which removes the initiator methionine from a newly synthesized polypeptide, and XaaPro amino peptidase (PepP, in *E. coli*). In addition, three proteins (SecY, Ffh and FtsY) involved in protein insertion into membranes or protein export are universally conserved and show phylogenetic relationship with the rRNA (Harris *et al.*, 2003). They also interact with the ribosome.

Furthermore, structural studies of ribosomal proteins show that several different ribosomal proteins have the same or closely related folds and may have a common origin. The relationship of ribosomal proteins to proteins having other functions is also revealed by structural studies. More sequence and structural data may clarify whether this is due to a common evolutionary origin.

13.4 THE RNA WORLD

The hypothesis of an early RNA world (see Gesteland *et al.*, 1999) is a tempting thought with the current knowledge of the translation machinery. The central components are all RNA molecules that have evolved to produce proteins: the mRNA, tRNAs, and rRNAs. Together they contain the functional components needed for translation. Evidently, the common ancestor must have contained the small number of universally conserved proteins, with a major fraction of them participating in translation. A number of additional proteins could have existed in this common ancestor, but after the divergence have been deleted in some domains of life.

Components of the translation system have probably had a long evolution in the primordial soup: the genetic code, tRNAs, some early charging molecules (ribozymes?) and rRNA. With the charges of the phosphates in the RNA, there must have been a need of charge neutralization and stabilization of the structures. The positively charged tails of the ribosomal proteins may be remnants of the early ribosomes. Indeed many of the universally conserved proteins have such tails or loops penetrating into the RNA. These may later have gained structural domains which all are located on the surface of the ribosome. The common factors, IF1, IF2, EF-Tu and EF-G have evolved to add to the speed and fidelity of translation.

Structural Characteristics of Ribosomal Proteins. The domain arrangement, fold, tails, loops, hinges, location and interactions. The table originates from the work on 30S subunits from *T. thermophilus* by Carter *et al.* (2000); Brodersen *et al.* (2002) and Schlüntzen *et al.* (2000); on 50S subunits from *H. marismortui* by Ban *et al.* (2000); Klein *et al.* (2004) and 50S subunits from *D. radiodurans* by Harms *et al.* (2001)

Protein	Domain Arrangement	Domain Fold[1]	Tails, Loops and Hinges	Location, Primary Interactions with Proteins and RNA Helices, Function
S1	1–185	OB-fold		Binding to 30S subunit
	6 repeats of about 70 amino acids			
S2	(1–97, 160–240) 2 domains	α-helical	(98–159) Coiled-coil α-loop (241–256) disordered	Back of 30S. At hinge head-body, h35–37, h26
S3	NTD (7–106)	α + β	(1–6) N-terminal tail	Back of head; 3' major domain. S10, S14 h16, h34–35, h38
	CTD (107–207)	α + β; β-meander	(208–239) disordered	
S4	I (1–98, 196–209)	α-helical	(1–40) binds a Zn through 4 Cys	Back of shoulder. S5, five-way junction of h3 h4, h16, h17, h18
	II(99–195)	α + β		
S5	NTD(1–74)	ββɓα-meander	(15–27) Loop extension of β-strands	Back of body: h1, h2, h26, h28, h34–h36
	CTD (75–147)	α + β		
S6	(1–92)	Double split-βαβ	(93–101) C-terminal tail	Platform. S18, h22, h23
S7	(15–155)	α + β-ribbon	(1–14) N-terminal tail	Head. h23b, h28, h43; Junction of h29, h30 h41, h42
S8	NTD(1–63)	Split βαβ		Back of body. (S2), S5, h20–h22, h25
	CTD (74–130)	Double split βαβ	(63–81) Connecting loop	

(Continued)

Appendix I (Cont'd)

Protein	Domain Arrangement	Domain Fold¹	Tails, Loops and Hinges	Location, Primary Interactions with Proteins and RNA Helices, Function
S9	(1–105)	α + β sandwich	(106–128) C-terminal tail	Top of head. S7, h39, h41, h29–h31, h38, h43
S10	(1–42, 68–105)	Double split-βαβ βββα-meander	(43–67) β-ribbon	Top of head. h39, h31, h34, h41, h43
S11	(1–111)	Like L18	(112–129) C-terminal tail	Top of platform. S18, h23, h23b h23, h24, h45
S12	(25–135)	OB-fold	(1–24) N-terminal tail	Body. Interface to 50S. S8, S17, h3, h5, h11 h19, h27, hl8, h44
S13	(1–64)	α-helical	(67–126) Long helical C-terminal tail; (1–61) Complete protein is α-helical tail	Head. S19, h30–h31, h42, h41–h42
S14				Head. h31–h34, h38, h42, h43
S15	(1–88)	Four helix bundle		Base of platform. Bridge B4. Junction of h20–h22; h23a, h24
S16	(1–79)	β + α	(80–88) C-terminal tail	Lower part of body. h4, h7, h12, h15, h17
S17	(1–85)	OB-fold	(23–41) extended β-ribbon; (85–105) C-terminal tail	Junction of 5' and central domain below platform. h7, h11, h20–h21, h27
S18	(21–82)	α-helical	(1–20) N-terminal tail	Platform. S6, h22, h23, h26 S11
S19	(8–80)	α + β	(83–88) C-terminal tail; (1–10) N-terminal tail	Head. S13, h30–h33, h42. Alternate in bridge?
S20	(11–76)	Three helix bundle	(77–93) C-terminal tail; (1–10) N-terminal tail; 93–106 C-terminal tail	Bottom of 30S. h6–h?), h11, h13

S21				
THX				
L1	I (1–66, 159–228)	Split β–α–β Rossmann fold	(1–27) Complete protein	Top of head. h30, h41–h43
	II (71–158)			L1 protuberance. H76–H78
L2	(62–130)	OB-fold	(1–61) N-terminal tail	Interface, on L1 side. Strong contact with L37ae. H66, Bridge B7b
	(131–194)	β-barrel, SH3-like		
L3	NTD	β-barrel, domain II of tGTPases	(195–237) C-terminal tail (1–22)* N-terminal tail	Edge of interface. Below L6 on L12 side. L13 L14, H94, H96, H100–101
	CTD	Antiparallel α + β	(206–260)* Internal loop	
L4		Mixed α + β	(41–105)*	External side on L1 side. L18e, L37e, H19–20 H28, H46
L5		α + β	(170–190)*	Central protuberance. L18. H3 of 5S RNA and H84. Bridge B1b
L6	NTD (1–79)	Split β–α–β		Below L12 stalk. L13, H97
	CTD (80–177)	Split β–α–β		
L9	NTD (1–40)	Split β–α–β	The two domains are separated by a connecting helix (41–73)	Below L1 stalk
	CTD (75–149)	Split β–α–β		
L10	NTD (1–72)			Base of L12 stalk. Binds both dimers of L12
L11	CTD (75–141)	Homeodomain		Base of L12 stalk
L12	NTD (1–37)	Helical hairpin	Flexible hinge (37–52)	L12 stalk protuberance. Two dimers bind to L10
	CTD (53–120)	Split β–α–β		

(Continued)

Appendix I (Cont'd)

Protein	Domain Arrangement	Domain Fold[1]	Tails, Loops and Hinges	Location, Primary Interactions with Proteins and RNA Helices, Function
L13		Mixed α + β		External side. L3, L6, H41
L14	NTD (1–88)	β-barrel		Interface side. L3. Interprotein β-sheet with L19 or L24e. Bridges B5 and B8
	CTD (103–122)			
L15		Like L18e	(1–60)* N-terminal tail	External side below L1 stalk. L18e, L32e H29, H31
L16	(1–171)*	Split β–α–β		In crevice between central protuberance and L12 stalk. H38, H89
L17				
L18		Mixed α + β	(1–20)* N-terminal tail	Central protuberance. L5, L21e. H1 of 5S RNA
L19	NTD		Long hinge	Interface side. Interprotein β-sheet with L14
	CTD			
L20				
L21				
L22	(1–113)	Split β–α–β	(120–139)* Long β-ribbon	External side at peptide exit tunnel. Central loop, H2
L23		Split β–α–β	Tail in *D* radiodurans	External side at peptide exit tunnel. L29 L39e, H50–51
L24	(1–91)	β-barrel SH3-like	(1–16)* N-terminal tail	External side. H7, H19–20
L25	(92–176)	β-barrel?		Behind central protuberance
	(213–253)			Between central protuberance and L11 A-site tRNA
L27				
L28				
L29		α-helical		External side. L23, H5–7

L30	(1–60)	Split β–α–β	External side. H40–41
L31			
L32		Zinc finger motif	
L33			
L34			
L35			
L36		Zinc finger motif	No corresponding protein in *H. marismortui*
L7Ae	(1–119)*	Mixed α + β	Below L1 stalk. Strong contact with L15e
		domain III of eRF1	Sequence and structural homology to
			15.5 kDa spliceosomal protein and L30e
L10e			Corresponds to L16
L15e	(1–67, 98–169)*	Split β–α–β	Edge of interface on L1 side. Strong contact
	(68–97)*		with L7Ae. L44e, H10–11, H15, H21
	(170–194)* C-terminal tail		
L18e	(1–115)*	Like L15	External surface. L4, L15, H27–31. No
			corresponding protein in bacteria. The long
			H30 in *H. marismortui* is absent in *D.*
			radiodurans
L19e	(1–51)*	α-helical	Bottom of interface surface. H47–48, H63
	(91–143)*		Replaced in bacteria by a longer H59
	(52–90)*		
L21e	(1–23)* N-terminal tail		External surface below central protuberance
			L18, H86
	(24–95)*	β-barrel SH3-like	Interface side. Interprotein β-sheet with L14
L24e	(4–56)*	Zinc finger motif	Bottom of external surface
L31e	(7–88)*	Split β–α–β	External surface. L15, H46
L32e	(95–236)*	Mixed α + β	
	(109–179)*		Interface side. Strong contact with L2
L37Ae	(1–72)*	Zinc finger motif	Internal. L4, L39e
L37e	(1–56)*	Zinc finger motif	External surface. L23, L37e
L39e	(1–49)*	α-helical	
L44e	(1–24, 62–92)*	Zinc finger motif	L15e, H88

*Residue numbers refer to *H. marismortui*.

Appendix II

Structural Determination of Individual Ribosomal Proteins or Complexes
(adapted from reviews by Al-Karadaghi *et al.*, 2000a,b)

Protein	Isolated Protein	Reference
S1	*E. coli*	Sengupta *et al.*, 2001
S4	*B. stearothermophilus*	Davies *et al.*, 1998; Markus *et al.*, 1998
S5	*B. stearothermophilus*	Ramakrishnan & White, 1992
S6	*T. thermophilus*	Lindahl *et al.*, 1994; Agalarov *et al.*, 2000
S7	*B. stearothermophilus*	Hosaka *et al.*, 1997
	T. thermophilus	Wimberly *et al.*, 1997
S8	*B. stearothermophilus*	Davies *et al.*, 1996a
	T. thermophilus	Nevskaya *et al.*, 1998; Berglund *et al.*, 1997
S15	*T. thermophilus*	Agalarov *et al.*, 2000
	B. stearothermophilus	Clemons *et al.*, 1998
S16	*T. thermophilus*	Allard *et al.*, 2000
S17	*B. stearothermophilus*	Golden *et al.*, 1993a; Jaishree *et al.*, 1996
S18	*T. thermophilus*	Agalarov *et al.*, 2000
S19	*T. thermophilus*	Helgstrand *et al.*, 1999
L1	*T. thermophilus*	Nikonov *et al.*, 1996
	M. jannashii	Nevskaya *et al.*, 2000
	M. thermolithotrophicus	Nevskaya *et al.*, 2002
	S. acidocaldarius	Nikulin *et al.*, 2003

(Continued)

Structural Determination of Individual Ribosomal Proteins or Complexes
(adapted from reviews by Al-Karadaghi *et al.*, 2000a,b)

L2	*B. stearothermophilus*	Nakagawa *et al.*, 1999
L4	*T. maritima*	Worbs *et al.*, 2000
L6	*B. stearothermophilus*	Golden *et al.*, 1993b
L9	*B. stearothermophilus*	Hoffman *et al.*, 1994
L11	*T. maritima*	Markus *et al.*, 1997; Wimberly *et al.*, 1999;
	B. stearothermophilus	Conn *et al.*, 1999
L12	*E. coli*	Leijonmarck *et al.*, 1980; Bocharov *et al.*, 2004
	T. maritima	Wahl *et al.*, 2000
L14	*B. stearothermophilus*	Davies *et al.*, 1996b
L18	*T. thermophilus*	Woestenenk *et al.*, 2002
	B. stearothermophilus	Turner & Moore, 2004
L22	*T. thermophilus*	Unge *et al.*, 1998
L23	*T. thermophilus*	Ohman *et al.*, 2003
L25	*E. coli*	Stoldt *et al.*, 1998; Lu & Steitz, 2000
	T. thermophilus	Fedorov *et al.*, 2001
L30	*B. stearothermophilus*	Wilson *et al.*, 1986
	T. thermophilus	Fedorov *et al.*, 1999

References

Some Useful Addresses on the Internet

A data-base providing information on the genetic code and its variations is provided by Elzanowski & Ostell at:
http://www.ncbi.nlm.nih.gov/Taxonomy/Utils/wprintgc.cgi?mode=c

A data base about ribosomal crosslinks (Baranov *et al.*, 1998; 1999) is found at:
http://www.mpimg-berlin-dahlem.mpg.de/%7Eag_ribo/ag_brimacombe/drc/defaultlow.html

There is also another database that is directly relevant to translation:
http://recode.genetics.utah.edu

A data base of the sequences of ribosomal proteins by Bairoch *et al.* at:
http://www.expasy.org/cgi-bin/lists?ribosomp.txt)

Protein Data Bank, PDB; http://www.rcsb.org/pdb

References

Abdulkarim, F., Liljas, L., and Hughes, D. 1994. Mutations to kirromycin resistance occur in the interface of domains I and III of EF-Tu.GTP. *FEBS Lett.* **352**:118–122.

Abel, K., Yoder, M.D., Hilgenfeld, R., and Jurnak, F. 1996. An α to β conformational switch in EF-Tu from *Escherichia coli*. *Structure* **4**:1153–1159.

Agafonov, D.E., Kolb, V.A., and Spirin, A.S. 1997. Proteins on ribosome surface: measurements of protein exposure by hot tritium bombardment technique. *Proc. Natl. Acad. Sci. USA* **94**:12892–12897.

Agafonov, D.E., Kolb, V.A., Nazimov, I.V., and Spirin, A.S. 1999. A protein residing at the subunit interface of the bacterial ribosome. *Proc. Natl. Acad. Sci. USA* **96**:12345–12349.

Agafonov, D.E., Kolb, V.A., and Spirin, A.S. 2001. Ribosome associated protein that inhibits translation at the aminoacyl-tRNA binding stage. *EMBO Rep.* **2**:399–402.

Agalarov, S.C., Sridhar Prasad, G., Funke, P.M., *et al.* 2000. Structure of the S15, S6, S18-rRNA complex: assembly of the 30S ribosome central domain. *Science* **288**:107–113.

Agmon, I., Auerbach, T., Baram, D., *et al.* 2003. On peptide bond formation, translocation, nascent protein progression and the regulatory properties of ribosomes. *Eur. J. Biochem.* **270**:2543–2556.

Agrawal, V., and Kishan, K.V. 2003. OB-fold: growing bigger with functional consistency. *Curr. Prot. Pept. Sci.* **4**:195–206.

Agrawal, R.K., Penczek, P., Grassucci, R.A., *et al.* 1996. Direct visualization of A-, P-, and E-site transfer RNAs in the *Escherichia coli* ribosome. *Science* **271**:1000–1002.

Agrawal, R.K., Penczek, P., Grassucci, R.A., and Frank, J. 1998. Visualization of elongation factor G on the *Escherichia coli* 70S ribosome: the mechanism of translocation. *Proc. Natl. Acad. Sci. USA* **95**:6134–6138.

Agrawal, R.K., Heagle, A.B., Penczek, P., *et al.* 1999. EF-G-dependent GTP hydrolysis induces translocation accompanied by large conformational changes in the 70S ribosome. *Nat. Struct. Biol.* **6**:643–647.

Agrawal, R.K., Sharma, M.R., Kiel, M.C., *et al.* 2004. Visualization of ribosome-recycling factor on the *Escherichia coli* ribosome: functional implications. *Proc. Natl. Acad. Sci. USA* **101**:8900–8905.

Alexander, R.W., and Schimmel, P. 2001. Domain-domain communication in aminoacyl-tRNA synthetases. *Prog. Nucl. Acid Res. Mol. Biol.* **69**:317–349.

Albanese, V., and Frydman, J. 2002. Where chaperones and nascent polypeptides meet. *Nat. Struct. Biol.* **9**:716–718.

Al-Karadaghi, S., Ævarsson, A., Garber, M., *et al.* 1996. The structure of elongation factor G in complex with GDP: conformational flexibility and nucleotide exchange. *Structure* **4**:555–565.

Al-Karadaghi, S., Davydova, N., Eliseikina, I., *et al.* 2000a. Ribosomal proteins and their structural transitions on and off the ribosome. In *The Ribosome: Structure, Function, Antibiotics and Cellular Interactions*, Eds. R.A. Garrett, S.R. Douthwaite, *et al.*, pp. 65–72. ASM Press, American Society for Mocrobiology, Washington, D.C.

Al-Karadaghi, S., Kristensen, O., and Liljas, A. 2000b. A decade of progress in understanding the structural basis of protein synthesis. *Prog. Biophys. Mol. Biol.* **73**:167–193.

Allard, P., Rak, A.V., Wimberly, B.T., *et al.* 2000. Another piece of the ribosome: solution structure of S16 and its location in the 30S subunit. *Struct. Fold Des.* **8**:875–882.

Allen, P.N., and Noller, H.F. 1989. Mutations in ribosomal proteins S4 and S12 influence the higher order structure of 16S ribosomal RNA. *J. Mol. Biol.* **208**:457–468.

Andersen, G.R., Pedersen, L., Valente, L., *et al.* 2000. Structural basis for nucleotide exchange and competition with tRNA in the yeast elongation factor complex eEF1A:eEF1Balpha. *Mol. Cell* **6**:1261–1266.

Andersen, G.R., Nissen, P., and Nyborg, J. 2003. Elongation factors in protein biosynthesis. *Trends Biochem. Sci.* **28**:434–441.

Andersson, S., and Kurland, C.G. 1987. Elongating ribosomes *in vivo* are refractory to erythromycin. *Biochimie* **69**:901–904.

Andersson, D.I., Andersson, S.G., and Kurland, C.G. 1986. Functional interactions between mutated forms of ribosomal proteins S4, S5 and S12. *Biochimie* **68**:705–713.

Antoun, A., Pavlov, M.Y., Andersson, K., *et al.* 2003. The roles of initiation factor IF2 and guanosine triphosphate in initiation of protein synthesis. *EMBO J.* **22**:5593–5601.

Aoki, H., Dekany, K., Adams, S.L., and Ganoza, M.C. 1997. The gene encoding the elongation factor P protein is essential for viability and is required for protein synthesis. *J. Biol. Chem.* **272**:32254–32259.

Arai, K.I., Kawakita, M., and Kaziro, Y. 1972. Studies on polypeptide elongation factors from *Escherichia coli*. II. Purification of factors Tu-guanosine diphosphate, Ts, and Tu-Ts, and crystallization of Tu-guanosine diphosphate and Tu-Ts. *J. Biol. Chem.* **247**:7029–7037.

Arai, N., and Kaziro, Y. 1975. Mechanism of the ribosome-dependent uncoupled GTPase reaction catalyzed by polypeptide chain elongation factor G. *J. Biochem.* (Tokyo) **77**:439–447.

Aravind, L., and Koonin, E.V. 1998. The HD domain defines a new superfamily of metal-dependent phosphohydrolases. *Trends Biochem. Sci.* **23**:469–472.

Arevalo, M.A., Tejedor, F., Polo, F., and Ballesta, J.P. 1988. Protein components of the erythromycin binding site in bacterial ribosomes. *J. Biol. Chem.* **263**:58–63.

Atkins, J.F., and Gesteland, R.F. 2001. mRNA readout at 40. *Nature* **414**:693.

Augustin, M.A., Reichert, A.S., Betat, H., *et al.* 2003. Crystal structure of the human CCA-adding enzyme: insights into template-independent polymerization. *J. Mol. Biol.* **328**:985–994.

Ævarsson, A. 1995. Structure-based sequence alignment of elongation factors Tu and G with related GTPases involved in translation. *J. Mol. Evol.* **41**:1096–1104.

Ævarsson, A., Brashnikov, E., Garber, M., *et al.* 1994. Three-dimensional structure of the ribosomal translocase: elongation G from *Thermus thermophilus*. *EMBO J.* **13**: 3669–3677.

Baca, O.G., Rohrbach, M.S., and Bodley, J.W. 1976. Equilibrium measurements of the interactions of guanine nucleotides with *Escherichia coli* elongation factor G and the ribosome. *Biochemistry* **15**:4570–4574.

Bailey-Serres, J., Vangala, S., Szick, K., and Lee, C.-H.K. 1997. Acidic phosphoprotein complex of 60S ribosomal subunit of maize seedling roots. *Plant Physiol.* **114**: 1293–1305.

Ballesta, J.P.G., and Remacha, M. 1996. The large ribosomal subunit stalk as a regulatory element of the eukaryotic translational machinery. *Prog. Nucl. Acid Res. Mol. Biol.* **55**:157–193.

Ballesta, J.P.G., Rodriguez-Gabriel, M.A., Bou, G., *et al.* 1999. Phosphorylation of the yeast ribosomal stalk. Functional effects and enzymes involved in the process. *FEMS Microbiol. Rev.* **23**:537–550.

Ballesta, J.P.G., Guarinos, E., Zurdo, J., *et al.* 2000. In *The Ribosome: Structure, Function, Antibiotics and Cellular Interactions*, Eds. R.A. Garrett, S.R. Douthwaite, A. Liljas, *et al.*, pp. 115–125. ASM Press, American Society for Mocrobiology, Washington, D.C.

Ban, N., Freeborn, B., Nissen, P., *et al.* 1998. A 9 Å resolution X-ray crystallographic map of the large ribosomal subunit. *Cell* **93**:1105–1115.

Ban, N., Nissen, P., Hansen, J., *et al.* 1999. Placement of protein and RNA structures into a 5 Å-resolution map of the 50S ribosomal subunit. *Nature* **400**:841–847.

Ban, N., Nissen, P., Hansen, J., *et al.* 2000. The complete atomic structure of the large ribosomal subunit at 2.4 Å resolution. *Science* **289**:905–920.

Baranov, P.V., Sergiev, P.V., Dontsova, O.A., *et al.* 1998. The database of the *E. coli* ribosomal cross-links (DRC). *Nucl. Acids Res.* **26**:187–189.

Baranov, P.V., Kubarenko, A.V., Gurvich, O.L., *et al.* 1999. The database of the *E. coli* ribosomal cross-links: an update. *Nucl. Acids Res.* **27**:184–185.

Baranov, P.V., Gesteland, R.F., and Atkins, J.F. 2002a. Recoding: translational bifurcations in gene expression. *Gene* **286**:187–201.

Baranov, P.V., Gesteland, R.F., and Atkins, J.F. 2002b. Release factor 2 frameshifting sites in different bacteria. *EMBO Rep.* **3**:373–377.

Bargis-Surgey, P., Lavergne, J.-P., Gonzalo, P., *et al.* 1999. Interaction of elongation factor eEF2 with ribosomal P proteins. *Eur. J. Biochem.* **262**:606–611.

Baron, C., Heider, J., and Böck, A. 1993. Interaction of translation factor SELB with the formate dehydrogenase H selenopolypeptide mRNA. *Proc. Natl. Acad. Sci. USA* **90**: 4181–4185.

Barta, A., Steiner, G., Brosius, J., *et al.* 1984. Identification of a site on 23S ribosomal RNA located at the peptidyl transferase center. *Proc. Natl. Acad. Sci. USA* **81**:3607–3611.

Bashan, A., Agmon, I., Zarivach, R., *et al.* 2003. Structural basis of the ribosomal machinery for peptide bond formation, translocation, and nascent chain progression. *Mol. Cell* **11**:91–102.

Batey, R.T., Rambo, R.P., Lucast, L., *et al.* 2000. Crystal structure of the ribonucleoprotein core of the signal recognition particle. *Science* **287**:1232–1239.

Battiste, J.L., Pestova, T.V., Hellen, C.U., and Wagner, G. 2000. The eIF1A solution structure reveals a large RNA-binding surface important for scanning function. *Mol. Cell* **5**:109–119.

Bayfield, M.A., Dahlberg, A.E., Schulmeister, U., *et al.* 2001. A conformational change in the ribosomal peptidyl transferase center upon active/inactive transition. *Proc. Natl. Acad. Sci. USA* **98**:10096–10101.

Beauclerk, A.A.D., Cundliffe, E., and Dijk, J. 1984. The binding site for ribosomal protein complex L8 within 23S RNA of *Escherichia coli*. *J. Biol. Chem.* **259**: 6559–6563.

Beck, K., Wu, L.F., Brunner, J., and Muller, M. 2000. Discrimination between SRP- and SecA/SecB-dependent substrates involves selective recognition of nascent chains by SRP and trigger factor. *EMBO J.* **19**:134–143.

Becker, H.D., Min, B., Jacobi, C., *et al.* 2000. The heterotrimeric *Thermus thermophilus* Asp-tRNA(Asn) amidotransferase can also generate Gln-tRNA(Gln). *FEBS Lett.* **476**:140–144.

Beckmann, R., Bubeck, D., Grassucci, R., *et al.* 1997. Alignment of conduits for the nascent polypeptide chain in the ribosome-Sec61 complex. *Science* **278**:2123–2126.

Beckmann, R., Spahn, C.M., Eswar, N., *et al.* 2001. Architecture of the protein-conducting channel associated with the translating 80S ribosome. *Cell* **107**:361–372.

Belitsina, N.V., Glukhova, M.A., and Spirin, A.S. 1975. Translocation in ribosomes by attachment-detachment of elongation factor G without GTP cleavage: evidence from a column-bound ribosome system. *FEBS Lett.* **54**:35–38.

Benjamin, D.R., Robinson, C.V., Hendrick, J.P., *et al.* 1998. Mass spectrometry of ribosomes and ribosomal subunits. *Proc. Natl. Acad. Sci. USA* **95**:7391–7395.

Benson, T.E., McCroskey, M.C., Cialdella, J.I., *et al.* 2000. Structure of *S. aureus* elongation factor P: another tRNA mimic in protein translation? Oral and Poster Presentation at Structural Aspects of Protein Synthesis 2. Rensselaerville, New York.

Berchtold, H., Reshetnikova, L., Reiser, C.O., *et al.* 1993. Crystal structure of active elongation factor Tu reveals major domain rearrangements. *Nature* **365**:126–132.

Berestowskaya, N.H., Vasiliev, V.D., Volkov, A.A., and Chetverin, A.B. 1988. Electron microscopy study of Q beta replicase. *FEBS Lett.* **228**:263–267.

Berg, P. 1961. Specificity in protein synthesis. *Ann. Rev. Biochem.* **30**:293–324.

Berglund, H., Rak, A., Serganov, A., *et al.* 1997. Solution structure of the ribosomal RNA binding protein S15 from *Thermus thermophilus*. *Nat. Struct. Biol.* **4**:20–23.

Berisio, R., Harms, J., Schluenzen, F., *et al.* 2003a. Structural insight into the antibiotic action of telithromycin against resistant mutants. *J. Bacteriol.* **185**:4276–4279.

Berisio, R., Schluenzen, F., Harms, J., *et al.* 2003b. Structural insight into the role of the ribosomal tunnel in cellular regulation. *Nat. Struct. Biol.* **10**:366–370.

Bernabeau, C., and Lake, J.A. 1982. Nascent polypeptide chains emerge from the exit domain of the large ribosomal subunit: immune mapping of the nascent chain. *Proc. Natl. Acad. Sci. USA.* **79**:3111–3115.

Bilgin, N., Richter, A.A., Ehrenberg, M., *et al.* 1990. Ribosomal RNA and protein mutants resistant to spectinomycin. *EMBO J.* **9**:735–739.

Bilgin, N., Claesens, F., Pahverk, H., and Ehrenberg, M. 1992. Kinetic properties of *Escherichia coli* ribosomes with altered forms of S12. *J. Mol. Biol.* **224**:1011–1027.

Biou, V., Shu, F., and Ramakrishnan, V. 1995. X-ray crystallography shows that translational initiation factor IF3 consists of two compact alpha/beta domains linked by an alpha-helix. *EMBO J.* **14**:4056–4064.

Birse, D.E., Kapp, U., Strub, K., *et al.* 1997. The crystal structure of the signal recognition particle Alu RNA binding heterodimer, SRP9/14. *EMBO J.* **16**: 3757–3766.

Björk, G.R. 1995. Biosynthesis and function of modified nucleosides. In *tRNA: Structure, Biosynthesis, and Function*, Eds. D. Söll, and U. RajBhandary, pp. 165–205.

Blaha, G., Wilson, D.N., Stroller, G., *et al.* 2003. Localization of the trigger factor binding site on the ribosomal 50S subunit. *J. Mol. Biol.* **326**:887–897.

Blobel, G., and Sabatini, D.D. 1970. Controlled proteolysis of nascent polypeptides in rat liver cell fractions. I. Location of the polypeptides within ribosomes. *J. Cell Biol.* **45**:130–145.

Bocharov, E.V., Gudkov, A.T., and Arseniev, A.S. 1996. Topology of the secondary structure elements of ribosomal protein L7/L12 from *E. coli* in solution. *FEBS Lett.* **379**:291–294.

Bocharov, E.V., Gudkov, A.T., Budovskaya, E.V., and Arseniev, A.S. 1998. Conformational independence of N- and C-domains in ribosomal protein L7/L12 and in the complex with protein L10. *FEBS Lett.* **423**:347–350.

Bocharov, E.V., Sobol, A.G., Pavlov, K.V., *et al.* 2004. From structure and dynamics of protein L7/L12 to molecular switching in ribosome. *J. Biol. Chem.* **279**:17697–17706.

Bodley, J.W., Zieve, F.J., Lin, L., and Zieve, S.T. 1969. Formation of the ribosome-G factor-GDP complex in the presence of fusidic acid. *Biochem. Biophys. Res. Commun.* **37**:437–443.

Bodley, J.W., Zieve, F.J., and Lin, L. 1970a. Studies on translocation. IV. The hydrolysis of a single round of guanosine triphosphate in the presence of fusidic acid. *J. Biol. Chem.* **245**:5662–5567.

Bodley, J.W., Zieve, F.J., Lin, L., and Zieve, S.T. 1970b. Studies on translocation. 3. Conditions necessary for the formation and detection of a stable ribosome-G factor guanosine diphosphate complex in the presence of fusidic acid. *J. Biol. Chem.* **245**:5656–5661.

Bodley, J.W., Lin, L., and Highland, J.H. 1970c. Studies on translocation. VI. Thiostrepton prevents the formation of a ribosome-G factor-guanine nucleotide complex. *Biochem. Biophys. Res. Commun.* **41**:1406–1411.

Borowski, C., Rodnina, M.V., and Wintermeyer, W. 1996. Truncated elongation factor G lacking the G domain promotes translocation of the 3′-end but not of the anticodon domain of peptidyl-tRNA. *Proc. Natl. Acad. Sci. USA* **93**:4202–4206.

Boublik, M., Hellmann, W., and Roth, H.E. 1976. Localization of ribosomal proteins L7/L12 in the 50S subunit of *Escherichia coli* ribosomes by electron microscopy. *J. Mol. Biol.* **107**:479–490.

Bourne, H.R., Sanders, D.A., and McCormick, F. 1990. The GTPase superfamily: a conserved switch for diverse cell functions. *Nature* **248**:125–132.

Bourne, H.R., Sanders, D.A., and McCormick, F. 1991. The GTPase superfamily: conserved structure and molecular mechanism. *Nature* **349**:117–127.

Brenner, S., Jacob, F., and Meselson, M. 1961. An unstable intermediate carrying information from genes to ribosomes for protein synthesis. *Nature* **190**:576–581.

Bretscher, M. 1968. Translocation in protein synthesis: a hybrid structure model. *Nature* **218**:675–677.

Breyton, C., Haase, W., Rapoport, T.A., *et al.* 2002. Three-dimensional structure of the bacterial protein-translocation complex SecYEG. *Nature* **418**:662–665.

Bright, G.M., Nagel, A.A., Bordner, J., *et al.* 1988. Synthesis, *in vitro* and *in vivo* activity of a novel 9-deoxo-9a-AZA-9a-homoerythromycin A derivative — a new class of macrolide antibiotics, the azalides. *J. Antibiot.* **41**:1029–1047.

Brodersen, D.E., Clemons Jr., W.M., Carter, A.P., *et al.* 2000. The structural basis for the action of the antibiotics tetracycline, pactamycin, and hygromycin B on the 30S ribosomal subunit. *Cell* **103**:1143–1154.

Brodersen, D.E., Clemons Jr., W.M., Carter, A.P., *et al.* 2002. Crystal structure of the 30S ribosomal subunit from *Thermus thermophilus*: structure of the proteins and their interactions with 16S RNA. *J. Mol. Biol.* **316**:725–768.

Brodersen, D.E., and Ramakrishnan, V. 2003. Shape can be seductive. *Nat. Struct. Biol.* **10**:78–80.

Brosius, J., Palmer, M.L., Kennedy, P.J., and Noller, H.F. 1978. Complete nucleotide sequence of a 16S ribosomal RNA gene from *Escherichia coli*. *Proc. Natl. Acad. Sci. USA* **75**:4801–4805.

Brosius, J., Dull, T.J., and Noller, H.F. 1980. Complete nucleotide sequence of a 23S ribosomal RNA gene from *Escherichia coli*. *Proc. Natl. Acad. Sci. USA* **77**:201–204.

Brown, C.M., and Tate, W.P. 1994. Direct recognition of mRNA stop signals by *Escherichia coli* polypeptide chain release factor two. *J. Biol. Chem.* **269**: 33164–33170.

Brune, M., Hunter, J.L., Corrie, J.E., and Webb, M.R. 1994. Direct, real-time measurement of rapid inorganic phosphate release using a novel fluorescent probe and its application to actinomycin subfragment 1 ATPase. *Biochemistry* **33**:8262–8271.

Bubunenko, M.G., and Gudkov, A.T. 1990. Elongation factors Tu and G change their conformation on interaction with ribosomes. *Biomed. Sci.* **1**:127–132.

Buckingham, R.H., Grentzmann, G., and Kisselev, L. 1997. Polypeptide chain release factors. *Mol. Microbiol.* **24**:449–456.

Buglino, J., Shen, V., Hakimian, P., and Lima, C.D. 2002. Structural and biochemical analysis of the Obg GTP binding protein. *Structure* **10**:1581–1592.

Bukau, B., Deuerling, E., Pfund, C., and Craig, E.A. 2000. Getting newly synthesized proteins into shape. *Cell* **101**:119–122.

Bult, C.J., White, O., Olsen, G.J., *et al.* 1996. Complete genome sequence of the methanogenic archaeon, *Methanococcus jannaschii. Science* **273**:1058–1073.

Burdett, V. 1991. Purification and characterization of Tet(M), a protein that renders ribosomes resistant to tetracycline. *J. Biol. Chem.* **266**:2872–2877.

Burdett, V. 1996. Tet(M)-promoted release of tetracycline from ribosomes is GTP dependent. *J. Bacteriol.* **178**:3246–3251.

Burma, D.P., Srivastava, A.K., Srivastava, S., and Dash, D. 1985. Interconversion of tight and loose couple 50S ribosomes and translocation in protein synthesis. *J. Biol. Chem.* **260**:10517–10525.

Bushuev, V.N., Gudkov, A.T., Liljas, A., and Sepetov, N.F. 1989. The flexible region of protein L12 from bacterial ribosomes studied by proton nuclear magnetic resonance. *J. Biol. Chem.* **264**:4498–4505.

Bycroft, M., Hubbard, T.J., Proctor, M., *et al.* 1997. The solution structure of the S1 RNA binding domain: a member of an ancient nucleic acid-binding fold. *Cell* **88**:235–242.

Bylund, G.O., Wipemo, L.C., Lundberg, L.A., and Wikström, P.M. 1998. RimM and RbfA are essential for efficient processing of 16S rRNA in *Escherichia coli. J. Bacteriol.* **180**:73–82.

Cabanas, M.J., Vazquez, D., and Modolell, J. 1978. Inhibition of ribosomal translocation by aminoglycoside antibiotics. *Biochim. Biophys. Res. Commun.* **83**:991–997.

Cabedo, H., Macain, F., Villarroya, M., *et al.* 1999. The *Escherichia coli* trmE (mnmE) gene, involved in tRNA modification, codes for an evolutionary conserved GTPase with unusual biochemical properties. *EMBO J.* **18**:7063–7076.

Caldon, C.E., Yoong, P., and March, P.E. 2001. Evolution of a molecular switch: universal bacterial GTPases regulate ribosome function. *Mol. Microbiol.* **41**:289–297.

Cameron, D.M., Thompson, J., March, P.E., and Dahlberg, A.E. 2002. Initiation factor IF2, thiostrepton and micrococcin prevent the binding of elongation factor G to the *Escherichia coli* ribosome. *J. Mol. Biol.* **319**:27–35.

Capa, L., Mendoza, A., Lavandera, J.L., *et al.* 1998. Translation elongation factor 2 is part of the target for a new family of antifungals. *Antimicrob. Agents Chemother.* **42**:2694–2699.

Capecchi, M.R., and Klein, H.A. 1969. Characterization of three proteins involved in polypeptide chain termination. *Cold Spring Harb. Symp. Quant. Biol.* **34**:469–477.

Capel, M.S., Engelman, D.M., Freeborn, B.R., *et al.* 1987. A complete mapping of the proteins in the small ribosomal subunit of *Escherichia coli. Science* **238**:1403–1406.

Carter, A.P., Clemons, W.M., Brodersen, D.E., *et al.* 2000. Functional insights from the structure of the 30S ribosomal subunit and its interactions with antibiotics. *Nature* **407**:340–348.

Carter, A.P., Clemons Jr., W.M., Brodersen, D.E., *et al.* 2001. Crystal structure of an initiation factor bound to the 30S ribosomal subunit. *Science* **291**:498–501.

Cashel, M., Gentry, D.R., Hernandez, V.J., and Vinella, D. 1996. In *Escherichia coli and Salmonella thyphimurium: Cellular and Molecular Biology*, Eds. F.C. Neidhardt, *et al.*, pp. 1458–1496, American Society for Microbiology, Washington, D.C.

Caskey, T., Scolnick, E., Tompkins, R., *et al.* 1969. Peptide chain termination, codon, protein factor, and ribosomal requirements. *Cold Spring Harb. Symp. Quant. Biol.* **34**:479–488.

Casparson, T. 1941. Studien über den Eiweissunsatz der Zelle. *Naturwissenschaften* **29**:33–43.

Cate, J.H., Yusupov, M.M., Yusupova, G.Z., *et al.* 1999. X-ray crystal structures of 70S ribosome functional complexes. *Science* **285**:2095–2104.

Cavdar Koc, E., Burkhart, W., Blackburn, K., *et al.* 2001a. The small subunit of the mammalian mitochondrial ribosome. Identification of the full complement of ribosomal proteins present. *J. Biol. Chem.* **276**:19363–19374.

Cavdar Koc, E., Burkhart, W., Blackburn, K., *et al.* 2001b. The large subunit of the mammalian mitochondrial ribosome. Analysis of the complement of ribosomal proteins present. *J. Biol. Chem.* **276**:43958–43969.

Cech, T.R., Zaug, A.J., and Grabowski, P.J. 1981. *In vitro* splicing of the ribosomal RNA precursor of tetrahymena: involvement of the guanosine nucleotide in the excision of the intervening sequence. *Cell* **27**:487–496.

Chamberlin, M., and Berg, P. 1962. Deoxyribonucleic acid-directed synthesis of ribonucleic acid by an enzyme from *Escherichia coli*. *Proc. Natl. Acad. Sci. USA* **48**:81–94.

Chen, X., Court, D.L., and Ji, X. 1999. Crystal structure of Era: a GTPase-dependent cell cycle regulator containing an RNA binding motif. *Proc. Natl. Acad. Sci. USA* **96**:8396–8401.

Chirgadze, YuN., Nikonov, S.V., Brazhnikov, E.V., *et al.* 1983. Crystallographic study of elongation factor G from *Thermus thermophilus* HB8. *J. Mol. Biol.* **168**:449–450.

Choi, S.K., Olsen, D.S., Roll-Mecak, A., *et al.* 2000. Physical and functional interaction between eukaryotic orthologs of prokaryotic translation initiation factors IF1 and IF2. *Mol. Cell. Biol.* **20**:7183–7191.

Choli, T., Franceschi, F., Yonath, A., and Wittmann-Liebold, B. 1993. Isolation of a new ribosomal protein from the *Thermophilic eubacteria, Thermus thermophilus, T. aquaticus* and *T. flavus*. *Biol. Chem. Hoppe Seyler.* **374**:377–383.

Chopra, I. 2000. New drugs for superbugs. *Microbiol. Today* **27**:4–6.

Chopra, I., Hawkey, P.M., and Hinton, M. 1992. Tetracyclines, molecular and clinical aspects. *J. Antimicrob. Chemother.* **29**:245–277.

Christensen, S.K., Mikkelsen, M., Pedersen, K., and Gerdes, K. 2001. RelE, a global inhibitor of translation, is activated during nutritional stress. *Proc. Natl. Acad. Sci. USA* **98**:14328–14333.

Clemons, W.M. Jr., Davies, C., White, S.W., and Ramakrishnan, V. 1998. Conformational variability of the N-terminal helix in the structure of ribosomal protein S15. *Structure* **6**:429–438.

Clemons, W.M. Jr., May, J.L.C., Wimberly, B.T., *et al.* 1999. Structure of a bacterial 30S ribosomal subunit at 5.5 Å resolution. *Nature* **400**:833–840.

Clemons, W.M. Jr., Brodersen, D.E., McCutcheon, J.P., *et al.* 2001. Crystal structure of the 30S ribosomal subunit from *Thermus thermophilus*: purification, crystallisation and structure determination. *J. Mol. Biol.* **310**:829–845.

Cohen, L.B., Goldberg, I.H., and Herner, A.E. 1969. Inhibition by pactamycin of the intiation of protein synthesis. Effect on the 30S ribosomal subunit. *Biochemistry* 8:1327–1335.

Conn, G.L., Draper, D.E., Lattman, E.E., and Gittis, A.G. 1999. Crystal structure of a conserved ribosomal protein-RNA complex. *Science* 284:1171–1174.

Connell, S.R., Trieber, C.A., Dinos, G.P., *et al.* 2003. Mechanism of Tet(O)-mediated tetracycline resistance. *EMBO J.* 22:945–953.

Contreras, A., and Vazquez, D. 1977. Cooperative and antagonistic interactions of peptidyl-tRNA and antibiotics with bacterial ribosomes. *Eur. J. Biochem.* 74:539–547.

Cool, R.H., and Parmeggiani, A. 1991. Substitution of histidine-84 and the GTPase mechanism of elongation factor Tu. *Biochemistry* 30:362–366.

Cowgill, C.A., Nichols, B.G., Kenny, J.W., *et al.* 1984. Mobile domains in ribosomes revealed by proton nuclear magnetic resonance. *J. Biol. Chem.* 259:15257–15263.

Craigen, W.J., Cook, R.G., Tate, W.P., and Caskey, C.T. 1985. Bacterial peptide chain release factors: conserved primary structure and possible frameshift regulation of release factor 2. *Proc. Natl. Acad. Sci. USA* 82:3616–3620.

Crick, F.H.C. 1958. On protein synthesis. *Symp. Soc. Exp. Biol.* 12:138–163.

Crick, F.H.C. 1962. The genetic code. *Sci. Am.* 207:66–74.

Crick, F.H.C. 1963. On the genetic code. *Science* 139:461–464.

Crick, F.H.C. 1966a. The genetic code: Yesterday, today and tomorrow. *Cold Spring Harb. Symp. Quant. Biol.* 31:1–9.

Crick, F.H.C. 1966b. Codon-anticodon pairing: the wobble hypothesis. *J. Mol. Biol.* 19:548–555.

Crick, F.H.C. 1968. The origin of the genetic code. *J. Mol. Biol.* 38:367–379.

Crick, F.H.C., Barnett, L., Brenner, S., and Watts-Tobin, R. 1961. General nature of the genetic code for proteins. *Nauchni Tr. Vissh. Med. Inst. Sofia Nat.* 192:1227–1232.

Cross, R.A. 1997. A protein-making motor protein. *Nature* 385:18–19.

Culver, G.M., and Noller, H.F. 2000. Directed hydroxyl radical probing of RNA from iron(II) tethered to proteins in ribonucleoprotein complexes. *Methods Enzymol.* 318:461–475.

Culver, G.M., Cate, J.H., Yusupova, G.Z., *et al.* 1999. Identification of an RNA-protein bridge spanning the ribosomal subunit interface. *Science* 285:2133–2136.

Cundliffe, E. 1980. Antibiotics and prokaryotic ribosomes: action, interaction and resistance. In *Ribosomes: Structure Function and Genetics*, Eds. G. Chambliss, G.R. Craven, J. Davies, *et al.*, pp. 555–581. University Park Press, Baltimore.

Cundliffe, E. 1981. In *Antibiotic Inhibitors of Ribosome Function*, Eds. E.F. Gale, E. Cundliffe, P.E. Reynolds, *et al.*, Wiley, London, New York, Sydney, Toronto.

Cundliffe, E. 1987. On the nature of antibiotic binding sites in ribosomes. *Biochimie* 69:863–869.

Cundliffe, E. 1990. Recognition sites for antibiotics within rRNA. In *The Ribosome: Structure, Function and Evolution*, Eds. W.E. Hill, A. Dahlberg, R.A. Garrett, *et al.*, pp. 479–490. ASM Press, Washington, DC.

Curnow, A.W., Tumbula, D.L., Pelaschier, J.T., *et al.* 1998. Glutamyl-tRNA(Gln) amido-transferase in *Deinococcus radiodurans* may be confined to asparagine biosynthesis. *Proc. Natl. Acad. Sci. USA.* 95:12838–12843.

Cusack, S. 1995. Eleven down and nine to go. *Nat. Struct. Biol.* 2:824–831.

Cusack, S., Berthet-Colominas, C., Hartlein, M., *et al.* 1990. A second class of synthetase structure revealed by X-ray analysis of *Escherichia coli* seryl-tRNA synthetase at 2.5 Å *Nature* **347**:347–255.

Czworkowski, J., Wang, J., Steitz, T.A., and Moore, P.B. 1994. The crystal structure of elongation factor G complexed with GDP, at 2.7 Å resolution. *EMBO J.* **13**:3661–3668.

Dabbs, E. 1986. Mutant studies on the prokaryotic ribosome. In *Structure, Function, and Genetics of Ribosomes*, Eds. B. Hardesty and G. Kramer, pp. 733–748. Springer-Verlag, New York.

Dabbs, E.R., Ehrlich, R., Hasenbank, R., *et al.* 1981. Mutants of *Escherichia coli* lacking ribosomal protein L1. *J. Mol. Biol.* **149**:553–578.

Dabbs, E.R., Hasenbank, R., Kastner, B., *et al.* 1983. Immunological studies of *Escherichia coli* mutants lacking one or two ribosomal proteins. *Mol. Gen. Genet.* **192**:301–308.

Dahlquist, K.D., and Puglisi, J.D. 2000. Interaction of translation initiation factor IF1 with the *E. coli* ribosomal A-site. *J. Mol. Biol.* **299**:1–15.

Dallas, A., and Noller, H.F. 2001. Interaction of translation initiation factor 3 with the 30S ribosomal subunit. *Mol. Cell* **4**:855–864.

Dantley, K.A., Dannelly, H.K., and Burdett, V. 1998. Binding interaction between Tet(M) and the ribosome: requirements for binding. *J. Bacteriol.* **180**:4089–4092.

Das, S., and Maitra, U. 2000. Mutational analysis of mammalian translation initiation factor 5 (eIF5): role of interaction between β subunit of 2IF2 and eIF5 in eIF5 function *in vitro* and *in vivo*. *Mol. Cell. Biol.* **20**:3942–3950.

Davies, J. 1990. What are antibiotics? Archaic functions for modern activities. *Mol. Microbiol.* **4**:1227–1232.

Davies, J., Gilbert, W., and Gorini, L. 1964. Streptomycin, suppression, and the code. *Proc. Natl. Acad. Sci. USA* **51**:883–890.

Davies, C., Gerstner, R.B., Draper, D.E., *et al.* 1998. The crystal structure of ribosomal protein S4 reveals a two-domain molecule with an extensive RNA-binding surface: one domain shows structural homology to the ETS DNA-binding motif. *EMBO J.* **17**:4545–4558.

Davies, C., Ramakrishnan, V., and White, S.W. 1996a. Structural evidence for specific S8-RNA and S8-protein interactions within the 30S ribosomal subunit: ribosomal protein S8 from *Bacillus stearothermophilus* at 1.9 A resolution. *Structure* **4**:1093–1104.

Davies, C., White, S.W., and Ramakrishnan, V. 1996b. The crystal structure of ribosomal protein L14 reveals an important organizational component of the translational apparatus. *Structure* **4**:55–66.

Daviter, T., Wieden, H.J., and Rodnina, M. 2003. Essential role of histidine 84 in elongation factor Tu for the chemical step of GTP hydrolysis on the ribosome. *J. Mol. Biol.* **332**:689–699.

Davydova, N., Streltsov, V., Wilce, M., *et al.* 2002. L22 ribosomal protein and effect of its mutation on ribosome resistance to erythromycin. *J. Mol. Biol.* **322**:635–644.

Decoster, E., Vassal, A., and Faye, G. 1993. MSS1, a nuclear-encoded mitochondrial GTPase involved in the expression of COX1 subunit of cytochrome c oxidase. *J. Mol. Biol.* **232**:79–88.

Dennis, P., and Nomura, M. 1974. Stringent control of ribosomal gene expression in *Escherichia coli*. *Proc. Natl. Acad. Sci. USA* **71**:3819–3823.

Deuerling, E., Schulze-Specking, A., Tomoyasu, T., *et al.* 1999. Trigger factor and DnaK cooperate in folding of newly synthesized proteins. *Nature* **400**:693–696.

Dey, D., Bochkariov, D.E., Jokhadze, G.G., and Traut, R.R. 1998. Cross-linking of selected residues in the N- and C-terminal *Escherichia coli* protein L7/L12 to other ribosomal proteins and the effect of elongation factor Tu. *J. Biol. Chem.* **273**: 1670–1676.

Dijk, J., and Littlechild, J. 1979. Purification of ribosomal proteins from *Escherichia coli* under nondenaturing conditions. *Methods Enzymol.* **59**:481–502.

Dijk, J., Garrett, R.A., and Müller, R. 1979. Studies of the binding of the ribosomal protein complex L7/12-L10 and protein L11 to the 5'-one third of the 23S RNA: a functional center of the 50S subunit. *Nucl. Acids Res.* **6**:2717–2730.

Dinos, G., Wilson, D.N., Teraoka, Y., *et al.* 2004. Dissecting the ribosomal inhibition mechanisms of edeine and pactamycin: the universally conserved residues G693 and C795 regulate P-site RNA binding. *Mol. Cell* **13**:113–124.

Dock-Bregeon, A.C., Sankaranarayanan, R., Romby, P., *et al.* 2001. Transfer RNA mediated editing in threonyl-tRNA synthetase: the class II solution to the double discrimination problem. *Cell* **103**:877–884.

Dominguez, J.M., Gomez-Lorenzo, M.G., and Martin, J.J. 1999. Sordarin inhibits fungal protein synthesis by blocking translocation differently to fusidic acid. *J. Biol. Chem.* **274**:22423–22427.

Donner, D., Villems, R., Liljas, A., and Kurland, C.G. 1977. Guanosinetriphosphatase activity dependent on elongation factor Tu and ribosomal protein L7/L12. *Proc. Natl. Acad. Sci. USA* **75**:3192–3195.

Doudna, J.A., and Cech, T.R. 2002. The chemical repertoire of natural ribozymes. *Nature* **418**:222–228.

Draper, D.E., and von Hippel, P.H. 1978. Nucleic acid binding properties of *Escherichia coli* ribosomal protein S1. I. Structure and interactions of binding site I. *J. Mol. Biol.* **122**:321–338.

Draper, D.E., and Reynaldo, L.P. 1999. RNA binding strategies of ribosomal proteins. *Nucl. Acids Res.* **27**:381–388.

Dubochet, J., Adrian, M., Chang, J.J., *et al.* 1988. Cryo-electron microscopy of vitrified specimens. *Quant. Rev. Biophys.* **21**:129–228.

Eckerman, D.J., and Symons, R.H. 1978. Sequence at the site of attachment of an affinity-label derivative of puromycin on 23S ribosomal RNA of *E. coli* ribosomes. *Eur. J. Biochem.* **82**:225–234.

Edebjerg, J., Douthwaite, S.R., Liljas, A., and Garrett, R.A. 1990. Characterization of the binding sites of protein L11 and the L10.(L12)$_4$ pentameric complex in the GTPase domain of 23S ribosomal RNA from *Escherichia coli*. *J. Mol. Biol.* **213**:275–288.

Egea, P.F., Shan, S.O., Napetschnig, J., *et al.* 2004. Substrate twinning activates the signal recognition particle and its receptor. *Nature* **427**:215–221.

Ehrenberg, M., Rojas, A.-M., Weiser, J., and Kurland, C.G. 1990. How many EF-Tu molecules participate in aminoacyl-tRNA binding and peptide bond formation in *Escherichia coli* translation? *J. Mol. Biol.* **211**:739–749.

Elf, J., Nilsson, D., Tenson, T., and Ehrenberg, M. 2003. Selective charging of tRNA isoacceptors explains patterns of codon usage. *Science* **300**:1718–1722.

Elkon, K.B., Parnassa, A.P., and Foster, C.L. 1985. Lupus autoantibodies target ribosomal P proteins. *J. Exp. Med.* **162**:459–471.

Endo, Y., Mitsui, K., Motizuki, M., and Tsurugi, K. 1987. The mechanism of action of ricin and related toxic lectins on eukaryotic ribosomes. The site and the characteristics of the modification in 28S ribosomal RNA caused by the toxins. *J. Biol. Chem.* **262**:5908–5912.

Epe, B., Woolley, P., and Hornig, H. 1987. Competition between tetracycline and tRNA at both P- and A-sites of the ribosome of *Escherichia coli*. *FEBS Lett.* **213**: 443–447.

Eriani, G., Delarue, M., Poch, O., Gangloff, J., and Moras, D. 1990. Partition of tRNA synthetases into two classes based on mutually exclusive sets of sequence motifs. *Nature* **347**:203–206.

Estevez, A.M., and Simpson, L. 1999. Uridine insertion/deletion RNA editing in trypanosome mitochondria — a review. *Gene* **240**:247–260.

Eustice, D.C., and Wilhelm, J.M. 1984a. Fidelity on the eukaryotic codon-anticodon interaction: interference by aminoglycoside antibiotics. *Biochemistry* **23**:1462–1467.

Eustice, D.C., and Wilhelm, J.M. 1984b. Mechanism of action of aminoglycoside antibiotics in eukaryotic protein synthesis. *Agents Chemother.* **26**:53–60.

Fakunding, J.L., Traut, R.R., and Hershey, J.W. 1973. Dependence of initiation factor IF-2 activity on proteins L7 and L12 from *Escherichia coli* 50S ribosomes. *J. Biol. Chem.* **248**:8555–8559.

Fedorov, R., Nevskaya, N., Khairullina, A., *et al.* 1999. Structure of ribosomal protein L30 from *Thermus thermophilus* at 1.9 Å resolution: conformational flexibility of the molecule. *Acta Cryst.* **D55**:1827–1833.

Fedorov, R., Meshcheryakov, V., Gongadze, G., *et al.* 2001. Structure of ribosomal protein TL5 complexed with RNA provides new insights into the CTC family of stress proteins. *Acta Cryst.* **D57**:968–976.

Feng, L., Stathopoulos, C., Ahel, I., *et al.* 2002. Aminoacyl-tRNA formation in the extreme thermophile *Thermus thermophilus*. *Extremophiles* **6**:167–174.

Ferré-D'Amaré, A.R. 2003. RNA-modifying enzymes. *Curr. Opin. Struct. Biol.* **13**:49–55.

Focia, P.J., Shepotinovskaya, I.V., Seidler, J.A., and Freymann, D.M. 2004. Heterodimeric GTPase core of the SRP targeting complex. *Science* **303**:373–377.

Forchhammer, K., Leinfelder, W., and Böck, A. 1989. Identification of a novel translation factor necessary for the incorporation of selenocysteine into protein. *Nature* **342**:453–456.

Fourmy, D., Recht, M.I., Blanchard, S.C., and Puglisi, J.D. 1996. Structure of the A-site of *Escherichia coli* 16S ribosomal RNA complexed with an aminoglycoside antibiotic. *Science* **274**:1367–1371.

Fourmy, D., Yoshizawa, S., and Puglisi, J.D. 1998. Paromomycin binding induces a local conformational change in the A-site of 16S rRNA. *J. Mol. Biol.* **277**:333–345.

Frank, J. 2003. Electron microscopy of functional ribosome complexes. *Biopolymers* **68**:223–233.

Frank, J., and Agrawal, R.K. 1998. The movement of tRNA through the ribosome. *Biophys. J.* **74**:589–594.

Frank, J., and Agrawal, R.K. 2000. A ratchet-like inter-subunit reorganization of the ribosome during translocation. *Nature* **406**:318–322.

Frank, J., and Agrawal, R.K. 2001. Ratchet-like movements between the two ribosomal subunits: their implications in elongation factor recognition and tRNA translocation. *Cold Spring Harb. Symp. Quant. Biol.* **66**:67–75.

Frank, J., Verschoor, A., and Boublik, M. 1981. Computer averaging of electron micrographs of 40S ribosomal subunits. *Science* **214**:1353–1355.

Frank, J., Verschoor, A., Li, Y., *et al.* 1995. A model of the translational apparatus based on a three-dimensional reconstruction of the *Escherichia coli* ribosome. *Biochem. Cell Biol.* **73**:757–765.

Frank, J., Radermacher, M., Penczek, P., *et al.* 1996. SPIDER and WEB: processing and visualization of images in 3D electron microscopy and related fields. *J. Struct. Biol.* **116**:190–199.

Fredrick, K., and Noller, H.F. 2002. Accurate translocation of mRNA by the ribosome requires a peptidyl group or its analog on the tRNA moving into the 30S P-site. *Mol. Cell.* **9**:1125–1131.

Fredrick, K., and Noller, H.F. 2003. Catalysis of ribosomal translocation by sparsomycin. *Science* **300**:1159–1162.

Freistroffer, D.V., Pavlov, M.Y., MacDougall, J., *et al.* 1997. Release factor RF3 in *E. coli* accelerates the dissociation of release factors RF1 and RF2 from the ribosome in a GTP-dependent manner. *EMBO J.* **16**:4126–4133.

Freymann, D.M., Keenan, R.J., Stroud, R.M., and Walter, P. 1997. Structure of the conserved GTPase domain of the signal recognition particle. *Nature* **385**:361–364.

Friesen, J.D., Fiil, N.P., Parker, J.M., and Haseltine, W.A. 1974. A new relaxed mutant of *Escherichia coli* with an altered 50S ribosomal subunit. *Proc. Natl. Acad. Sci. USA* **71**:3465–3469.

Frolova, L.Y., Tsivkovskii, R.Y., Sivolobova, G.F., *et al.* 1999. Mutations in the highly conserved GGQ motif of class 1 polypeptide release factors abolish ability of human eRF1 to trigger peptidyl-tRNA hydrolysis. *RNA* **5**:1014–1020.

Frolova, L., Seit-Nebi, A., and Kisselev, L. 2002. Highly conserved NIKS tetrapeptide is functionally essential in eukaryotic translation termination factor eRF1. *RNA* **8**:129–136.

Frydman, J. 2001. Folding of newly translated proteins *in vivo*: the role of molecular chaperones. *Ann. Rev. Biochem.* **70**:603–649.

Fujita, C., Maeda, M., Fujii, T., *et al.* 2002. Identification of an indispensable amino acid for ppGpp synthesis of *Escherichia coli* SpoT protein. *Biosci. Biotechnol. Biochem.* **66**:2735–2738.

Fujiwara, T., Ito, K., Nakayashiki, T., and Nakamura, Y. 1999. Amber mutaions in ribosome recycling factors of *Escherichia coli* and *Thermus thermophilus*: evidence for C-terminal modulator element. *FEBS Lett.* **447**:297–302.

Gabashvili, I.S., Agrawal, R.K., Grassucci, R., and Frank, J. 1999. Structure and structural variations of the *Escherichia coli* 30S ribosomal subunit as revealed by three-dimensional cryo-electron microscopy. *J. Mol. Biol.* **286**:1285–1291.

Gabashvili, I.S., Agrawal, R.K., Spahn, C.M., *et al.* 2000. Solution structure of the *E. coli* 70S ribosome at 11.5 A resolution. *Cell* **100**:537–549.

Gabashvili, I.S., Gregory, S.T., Valle, M., *et al.* 2001. The polypeptide tunnel system in the ribosome and its gating in erythromycin resistance mutants of L4 and L22. *Mol. Cell* **8**:181–188.

Gale, E.F., Cundliffe, E., Reynolds, P.E., *et al.* 1981. *The Molecular Basis of Antibiotic Action.* John Wiley & Sons, London.

Gallant, J., Bonthuis, P., and Lindsley, D. 2003. Evidence that the bypassing ribosome travels through the coding gap. *Proc. Natl. Acad. Sci. USA* **100**:13430–13435.

Galvani, C., Terry, J., and Ishiguro, E.E. 2001. Purification of the RelB and RelE proteins of *Escherichia coli*: RelE binds to RelB and to ribosomes. *J. Bacteriol.* **183**:2700–2703.

Ganoza, M.C. 1966. Polypeptide chain termination in cell-free extracts of *E. coli*. *Cold Spring Harb. Symp. Quant. Biol.* **31**:273–278.

Gao, H., Sengupta, J., Valle, M., *et al.* 2003. Study of the structural dynamics of the *E. coli* 70S ribosome using real-space refinement. *Cell* **113**:789–801.

Garcia, C., Fortier, P.L., Blanquet, S., *et al.* 1995a. Solution structure of the ribosome-binding domain of *E. coli* translation initiation factor IF3. Homology with the U1A protein of the eukaryotic spliceosome. *J. Mol. Biol.* **254**:247–259.

Garcia, C., Fortier, P.L., Blanquet, S., *et al.* 1995b. ^{1}H and ^{15}N resonance assignments and structure of the N-terminal domain of *Escherichia coli* initiation factor 3. *Eur. J. Biochem.* **228**:395–402.

Garrett, R.A., and Wittmann, H.G. 1973. Structure of bacterial ribosomes. *Adv. Prot. Chem.* **27**:277–347.

Gast, W.H., Kabsch, W., Wittinghofer, A., and Leberman, R. 1976. Crystals of a large tryptic peptide (fragment A) of elongation factor EF-Tu from *Escherichia coli*. *FEBS Lett.* **74**:88–90.

Gavrilova, L.P., and Spirin, A.S. 1971. Stimulation of "non-enzymic" translocation in ribosomes by p-chloromercuribenzoate. *FEBS Lett.* **17**:324–326.

Gavrilova, L.P., and Spirin, A.S. 1972. A modification of the 30S ribosomal subparticle is responsible for stimulation of "non-enzymatic" translocation by p-chloromercuribenzoate. *FEBS Lett.* **22**:91–92.

Gavrilova, L.P., Koteliansky, V.E., and Spirin, A.S. 1974. Ribosomal protein S12 and "non-enzymatic" translocation. *FEBS Lett.* **45**:324–328.

Gavrilova, L.P., Kostiashkina, O.E., Koteliansky, V.E., *et al.* 1976. Factor-free ("non-enzymic") and factor-dependent systems of translation of polyuridylic acid by *Escherichia coli* ribosomes. *J. Mol. Biol.* **101**:537–552.

Geigenmuller, U., and Nierhaus, K.H. 1986. Tetracycline can inhibit tRNA binding to the ribosomal P-site as well as to the E-site. *Eur. J. Biochem.* **161**:723–726.

Gentry, D.R., and Cashel, M. 1996. Mutational analysis of the *E. coli* spoT gene identifies distinct but overlapping regions involved in ppGpp synthesis and degradation. *Mol. Microbiol.* **19**:1373–1384.

Gerbi, S.A. 1996. Expansion segments: regions of variable size that interrupt the universal core secondary structure of ribosomal RNA. In *Ribosomal RNA. Structure, Evolution, Processing and Function in Protein Biosynthesis*, Eds. R.A. Dahlberg and A.E. Dahlberg, pp. 71–87. CRC Press, Boca Raton.

Gesteland, R.F., Cech, T.R., and Atkins, J.F. (Eds.) 1999. *The RNA World*, 2nd edition. Cold Spring Harbor Laboratory Press.

Giege, R., Sissler, M., and Florentz, C. 1998. Universal rules and idiosyncratic features in tRNA identity. *Nucl. Acids Res.* **26**:5017–5035.

Giri, L., Hill, W.E., Wittmann, H.G., and Wittmann-Liebold, B. 1984. Ribosomal proteins: their structure and spatial arrangement in prokaryotic ribosomes. *Adv. Prot. Chem.* **36**:1–78.

Giri, L., and Subramanian, A.R. 1977. Hydrodynamic properties of protein S1 from *Escherichia coli* ribosome. *FEBS Lett.* **81**:199–203.

Girshovich, A.S., Bochkareva, E.S., and Ovchinnikov, Y.A. 1981. Elongation factor G and protein S12 are the nearest neighbors in the *Escherichia coli* ribosomes. *J. Mol. Biol.* **151**:229–243.

Giege, R., Sissler, M., and Florentz, C. 1998. Universal rules and idiosynchratic features in tRNA identity. *Nucl. Acids Res.* **26**:5017–5035.

Gilbert, W. 1963. Polypeptide synthesis in *Escherichia coli*. II. The polypeptide chain and S-RNA. *J. Mol. Biol.* **6**:389–403.

Glick, B.R., Chladek, S., and Ganoza, M.C. 1979. Peptide bond formation stimulated by protein synthesis factor EF-P depends on the aminoacyl moiety of the acceptor. *Eur. J. Biochem.* **97**:23–28.

Glotz, C., and Brimacombe, R. 1980. An experimentally-derived model for the secondary structure of the 16S ribosomal RNA from *Escherichia coli*. *Nucl. Acids Res.* **8**:2377–2395.

Glotz, C., Zwieb, C., Brimacombe, R., *et al.* 1981. Secondary structure of the large subunit ribosomal RNA from *Escherichia coli*, Zea mays chloroplast, and human and mouse mitochondrial ribosomes. *Nucl. Acids Res.* **9**:3287–3306.

Glotz, C., Mussig, J., Gewitz, H.S., *et al.* 1987. Three-dimensional crystals of ribosomes and their subunits from eu- and archae-bacteria. *Biochem. Int.* **15**:953–960.

Gluehmann, M., Zarivach, R., Bashan, A., *et al.* 2001. Ribosomal crystallography: from poorly diffracting microcrystals to high-resolution structures. *Methods* **25**:292–302.

Goldberg, I.H., and Mitsugi, K. 1966. Sparsomycin, an inhibitor of aminoacyl transfer to polypeptide. *Biochem. Biophys. Res. Commun.* **23**:453–459.

Golden, B.L., Hoffman, D.W., Ramakrishnan, V., and White, S.W. 1993a. Ribosomal protein S17: characterization of the three-dimensional structure by 1H and 15N NMR. *Biochemistry* **32**:12812–12820.

Golden, B.L., Ramakrishnan, V., and White, S.W. 1993b. Ribosomal protein-L6 — structural evidence of gene duplication from a primitive RNA binding protein. *EMBO J.* **12**:4901–4908.

Gomez-Lorenzo, M.G., and Garcia-Bustos, J.F. 1998. Ribosomal P-protein stalk function is targeted by sordarin antifungals. *J. Biol. Chem.* **273**:25041–25044.

Gomez-Lorenzo, M.G., Spahn, C.M., Agrawal, R.K., *et al.* 2000. Three-dimensional cryo-electron microscopy localization of EF2 in the *Saccharomyces cerevisiae* 80S ribosome at 17.5 Å resolution. *EMBO J.* **19**:2710–2718.

Gong, F., and Yanofsky, C. 2002. Instruction of translating ribosome by nascent peptide. *Science* **297**:1864–1867.

Gongadze, G.M., Tischenko, S.V., Sedelnikova, S.E., and Garber, M.B. 1993. Ribosomal proteins, TL4 and TL5, from *Thermus thermophilus* form hybrid complexes with 5S ribosomal RNA from different microorganisms. *FEBS Lett.* **386**:46–48.

Gongadze, G.M., Meshcheryakov, V.A., Serganov, A.A., *et al.* 1999. N-terminal domain, residues 1–91, of ribosomal protein TL5 from *Thermus thermophilus* binds specifically and strongly to the region of 5S rRNA containing loop E. *FEBS Lett.* **451**:51–55.

Gonzales, A., Jimenez, A., Vazquez, D., *et al.* 1978. Studies on the mode of action of hygromycin B, an inhibitor of translocation in eucaryotes. *Biochim. Biophys. Acta* **551**:459–469.

Gonzalo, P., Lavergne, J.P., and Reboud, J.-P. 2001. Pivotal role of the P1 N-terminal domain in mammalian ribosome stalk and in the proteosynthetic activity. *J. Biol. Chem.* **276**:19762–19769.

Gonzalo, P., and Reboud, J.-P. 2003. The puzzling lateral flexible stalk of the ribosome. *Biol. Cell* **95**:179–193.

Goody, R.S., Hofmann-Goody, W. 2002. Exchange factors, effectors, GAPs and motor proteins: common thermodynamic and kinetic principles for different functions. *Eur. Biophys. J.* **31**:268–274.

Gordon, J. 1969. Hydrolysis of guanosine 5'-triphosphate associated with binding of aminoacyl transfer ribonucleic acid to ribosomes. *J. Biol. Chem.* **24**:5680–5686.

Gornicki, P., Nurse, K., Hellmann, W., *et al.* 1984. High resolution localization of the tRNA anticodon interaction site on the *Escherichia coli* 30S ribosomal subunit. *J. Biol. Chem.* **259**:10493–10498.

Gotfredsen, M., and Gerdes, K. 1998. The *Escherichia coli* relBE genes belong to a new toxin-antitoxin gene family. *Mol. Microbiol.* **29**:1065–1076.

Gravel, M., Melancon, P., and Brakier-Gringas, L. 1987. Cross-linking of streptomycin to the 16S ribosomal RNA of *Escherichia coli*. *Biochemistry* **26**:6227–6232.

Green, R., Switzer, C., and Noller, H.F. 1998. Ribosome-catalyzed peptide-bond formation with an A-site substrate covalently linked to 23S ribosomal RNA. *Science* **280**:286–289.

Griaznova, O., and Traut, R.R. 2000. Deletion of C-terminal residues of *Escherichia coli* ribosomal protein L10 causes the loss of binding of one L7/L12 dimer: ribosomes with one L7/L12 dimer are active. *Biochemistry* **39**:4075–4081.

Gromadski, K.B., and Rodnina, M.V. 2004. Kinetic determinants of high-fidelity tRNA discrimination on the ribosome. *Mol. Cell.* **13**:191–200.

Gronlund, H., and Gerdes, K. 1999. Toxin-antitoxin systems homologous with relBE of *Escherichia coli* plasmid P307 are ubiquitous in prokaryotes. *J. Mol. Biol.* **285**: 1401–1415.

Gros, F., Gilbert, W., Hiatt, H., *et al.* 1961. Unstable ribonucleic acid revealed by pulse labelling of *Escherichia coli*. *Nature* **190**:581–585.

Grosjean, H.J., de Henau, S., and Crothers, D.M. 1978. On the physical basis for ambiguity in genetic coding interactions. *Proc. Natl. Acad. Sci. USA* **75**:610–614.

Grunberg-Manago, M., Dessen, P., Pantaloni, D., *et al.* 1975. Light-scattering studies showing the effect of initiation factors on the reversible dissociation of *Escherichia coli* ribosomes. *J. Mol. Biol.* **94**:461–478.

Gryaznova, O.I., Davydova, N.L., Gongadze, G.M., *et al.* 1996. A ribosomal protein from *Thermus thermophilus* is homologous to a general shock protein. *Biochimie* **78**:915–919.

Gu, S.Q., Peske, F., Wieden, H.J., *et al.* 2003. The signal recognition particle binds to protein L23 at the peptide exit of *Escherichia coli* ribosome. *RNA* **9**:566–573.

Guarinos, E., Remacha, M., and Ballesta, J.P.G. 2001. Asymmetric interactions between the acidic P1 and P2 proteins in *Saccharomyces cerevisiae* ribosomal stalk. *J. Biol. Chem.* **276**:32474–32479.

Gudkov, A.T. 1997. The L7/L12 domain of the ribosome: structural and functional studies. *FEBS Lett.* **407**:253–256.

Gudkov, A.T., and Behlke, J. 1978. The N-terminal sequence protein of L7/L12 is responsible for its dimerization. *Eur. J. Biochem.* **90**:309–312.

Gudkov, A.T., and Bubunenko, M.G. 1989. Conformational changes in ribosomes upon interaction with elongation factors. *Biochimie* **71**:779–785.

Gudkov, A.T., and Gongadze, G.M. 1984. The L7/L12 proteins change their conformation upon interaction of EF-G with ribosomes. *FEBS Lett.* **176**:32–36.

Gudkov, A.T., Tumanova, L.G., Gongadze, G.M., and Bushuev, V.N. 1980. Role of different regions of ribosomal proteins L7 and L10 in their complex formation and in the interaction with the ribosomal 50S subunit. *FEBS Lett.* **109**:34–38.

Gudkov, A.T., Gongadze, G.M., Bushuev, V.N., and Okon, M.S. 1982. Proton nuclear magnetic resonance study of the ribosomal protein L7/L12 *in situ*. *FEBS Lett.* **138**:229–232.

Gudkov, A.T., Bubunenko, M., and Gryaznova, O. 1991. Overexpression of L7/L12 protein with mutations in its flexible region. *Biochimie* **73**:1387–1389.

Guerrier-Takada, C., Gardiner, K., Marsh, T., Pace, N., and Altman, S. 1983. The RNA moiety of ribonuclease P is the catalytically active subunit of the enzyme. *Cell* **35**:849–857.

Gutmann, S., Haebel, P., Metzinger, L., *et al.* 2003. Crystal structure of the tRNA domain of transfer messenger RNA in complex with SmpB. *Nature* **424**:699–703.

Halic, M., Becker, T., Pool, M.R., *et al.* 2004. Structure of the signal recognition particle interacting with the elongation arrested ribosome, *Nature* **427**:808–814.

Hall, C.E., and Slayter, H.S. 1959. Electron microscopy of nucleoprotein particles from *E. coli*. *J. Mol. Biol.* **1**:329–332.

Hamel, E., Koka, M., and Nakamoto, T. 1972. Requirement of an *E. coli* 50S ribosomal protein component for effective interaction of the ribosome with T and G factors with guanosine triphosphate. *J. Biol. Chem.* **247**:805–814.

Hampl, H., Schulze, H., and Nierhaus, K.H. 1981. Ribosomal components from *Escherichia coli* 50S subunits involved in the reconstitution of peptidyltransferase activity. *J. Biol. Chem.* **256**:2284–2288.

Hang, J.Q., Meier, T.I., and Zhao, G. 2001. Analysis of the interaction of 16S rRNA and cytoplasmic membrane with the C-terminal part of the *Streptococcus pneumoniae* Era GTPase. *Eur. J. Biochem.* **268**:5570–5577.

Hansen, J.L., Ippolito, J.A., Ban, N., *et al.* 2002a. The structures of four macrolide antibiotics bound to the large ribosomal subunit. *Mol. Cell* **10**:117–128.

Hansen, J.L., Schmeing, T.M., Moore, P.B., and Steitz, T.A. 2002b. Structural insights into peptide bond formation. *Proc. Natl. Acad. Sci. USA* **99**:11670–11675.

Hansen, J.L., Moore, P.B., and Steitz, T.A. 2003. Structures of five antibiotics bound at the peptidyl transferase center of the large ribosomal subunit. *J. Mol. Biol.* **330**:1061–1075.

Hanson, C.L., Fucini, P., Ilag, L.L., *et al.* 2003. Dissociation of intact *Escherichia coli* ribosomes is a mass spectrometer. *J. Biol. Chem.* **278**:1259–1267.

Hao, B., Gong, W., Ferguson, T.K., *et al.* 2002. A novel UAG encoded residue in the structure of a methanogen methyltransferase. *Science* **296**:1462–1466.

Harauz, G., and Van Heel, M. 1986. Exact filters for general geometry three-dimensional reconstruction. *Optik* **73**:146–156.

Hardesty, B., and Kramer, G. 2001. Folding of a nascent peptide on the ribosome. *Prog. Nucl. Acid Res. Mol. Biol.* **66**:41–66.

Hardy, S.J.S. 1975. The stoichiometry of the ribosomal proteins of *Escherichia coli*. *Mol. Gen. Genet*. **140**:253–274.

Harms, J., Tocilj, A., Levin, I., *et al.* 1999. Elucidating the medium-resolution structure of ribosomal particles: an interplay between electron cryo-microscopy and X-ray crystallograhy. *Struct. Fold Des*. **7**:931–941.

Harms, J., Schlüenzen, F., Zarivach, R., *et al.* 2001. High resolution structure of the large ribosomal subunit from a mesophilic eubacterium. *Cell* **107**:679–688.

Harris, J.K., Kelley, S.T., Spiegelman, G.B., and Pace, N. 2003. The genetic core of the universal ancestor. *Gen. Res.* **13**:407–412.

Hartl, F.U., and Hayer-Hartl, M. 2002. Molecular chaperones in the cytosol: from nascent chain to folded protein. *Science* **295**:1852–1858.

Haseltine, W.A., and Block, R. 1973. Synthesis of guanosine tetra- and penta-phosphate requires the presence of a codon-specific, uncharged transfer ribonucleic acid in the acceptor site of ribosomes. *Proc. Natl. Acad. Sci. USA* **70**:1564–1568.

Heffron, S.E., and Jurnak, F. 2000. Structure of an EF-Tu complex with a thiazol peptide antibiotic determined at 2.35 Å resolution: atomic basis for GE2270A inhibition of EF-Tu. *Biochemistry* **39**:37–45.

Heimark, R.L., Hersehy, J.W., and Traut, R.R. 1976. Cross-linking of initiation factor IF2 to proteins L7/L12 in 70S ribosomes of *Escherichia coli. J. Biol. Chem.* **251**:779–784.

Heinemeyer, E.A., and Richter, D. 1977. *In vitro* degradation of guanosine tetraphosphate (ppGpp) by an enzyme associated with the ribosomal fraction from *Escherichia coli. FEBS Lett.* **84**:357–361.

Held, W.A., Ballou, B., Mizushima, S., and Nomura, M. 1974. Assembly mapping of 30S ribosomal proteins from *E. coli*. Further studies. *J. Biol. Chem.* **249**: 3103–3111.

Helgstrand, M., Rak, A.V., Allard, P., *et al.* 1999. Solution structure of the ribosomal protein S19 from *Thermus thermophilus. J. Mol. Biol.* **292**:1071–1081.

Helliwell, J.R. 1998. Synchrotron radiation facilities. *Nat. Struct. Biol.* **5**(Suppl):614–617.

Hendrickson, W.A. 1991. Determination of macromolecular structures from anomalous diffraction of synchrotron radiation. *Science* **254**:51–58.

Herendeen, S.L., VanBogelen, R.A., and Neidhardt, F.C. 1979. Levels of major proteins of *Escherichia coli* during growth at different temperatures. *J. Bacteriol.* **139**:185–194.

Herr, A.J., Wills, N.M., Nelson, C.C., *et al.* 2001. Drop-off during ribosome hopping. *J. Mol. Biol.* **311**:445–452.

Herskovits, A.A., and Bibi, E. 2000. Association of *Escherichia coli* ribosomes with the inner membrane requires the signal recognition particle receptor but is independent of the signal recognition particle. *Proc. Natl. Acad. Sci. USA* **97**:4621–4626.

Hesterkamp, T., Hauser, S., Lütcke, H., and Bukau, B. 1996. *Escherichia coli* trigger factor is a prolyl isomerase that associates with nascent polypeptide chains. *Proc. Natl. Acad. Sci. USA* **93**:4437–4441.

Hesterkamp, T., Deuerling, E., and Bukau, B. 1997. The amino-terminal 118 amino acids of *Escherichia coli* trigger factor constitute a domain that is necessary and sufficient for binding to ribosomes. *J. Biol. Chem.* **272**:21865–21871.

Highland, J.H., Howard, G.A., Ochsner, E., *et al.* 1975. Identification of a ribosomal protein necessary for thiostrepton binding to *E. coli* ribosomes. *J. Biol. Chem.* **250**:1141–1145.

Hinck, A.P., Marcus, M.A., Huang, S., *et al.* 1997. The RNA binding domain of ribosomal protein L11: three-dimensional structure of the RNA-bound form of the protein and its interaction with 23S rRNA. *J. Mol. Biol.* **274**:101–113.

Hinnebusch, A.G. 2000. Mechanism and regulation of initiator methionine tRNA binding to ribosomes. In *Translational Control of Gene Expression*, Eds. N. Sonnenberg, J.W. Hershey and M.B. Matthews, pp. 185–243. Cold Spring Harbor Laboratory Press.

Hirashima, A., and Kaji, A. 1972. Factor-dependent release of ribosomes from messenger RNA. Requirement for two heat-stable factors. *J. Mol. Biol.* **65**:43–58.

Hirashima, A., and Kaji, A. 1973. Role of elongation factor G and a protein factor on the release of ribosomes from messenger ribonucleic acid. *J. Biol. Chem.* **248**:7580–7587.

Hoagland, M.B. 2003. Celebrating complementarity. *Ann. Intern. Med.* **138**:583–586.

Hoagland, M.B., Zamecnik, P., and Stephenson, M.L. 1957. Intermediate reactions in protein biosynthesis. *Biochim. Biophys. Acta* **24**:215–216.

Hoffman, D.W., Davies, C., Gerchman, S.E., *et al.* 1994. Crystal structure of prokaryotic ribosomal protein L9: a bi-lobed RNA- binding protein. *EMBO J.* **13**:205–212.

Hoffman, D.W., Cameron, C.S., Davies, C., *et al.* 1996. Ribosomal protein L9: a structure determination by the combined use of X-ray crystallography and NMR spectroscopy. *J. Mol. Biol.* **264**:1058–1071.

Hogg, T., Mechold, U., Malke, H., *et al.* 2004. The structural basis of (p)ppGpp metabolism and the stringent response: two conformations correspond to opposing activity states of a bifunctional RelA/SpoT homolog. Submitted.

Holley, R.W., Apgar, J., Everett, G.A., *et al.* 1965. Structure of a ribonucleic acid. *Science* **147**:1462–1465.

Hope, H., Frolow, F., von Böhlen, K., *et al.* 1989. Cryocrystallography of ribosomal particles. *Acta Cryst.* **B45**:190–199.

Hopfield, J.J. 1974. Kinetic proofreading: a new mechanism for reducing errors in biosynthetic processes requiring high specificity. *Proc. Natl. Acad. Sci. USA.* **71**: 4135–4139.

Horne, J.R., and Erdmann, V.A. 1972. Isolation and characterization of 5S RNA-protein complexes from *Bacillus stearothermophilus* and *Escherichia coli* ribosomes. *Mol. Gen. Genet.* **119**:337–344.

Hosaka, H., Nakagawa, A., Tanaka, I., *et al.* 1997. Ribosomal protein S7: a new RNA-binding motif with structural similarities to a DNA architectural factor. *Structure* **5**:1199–1208.

Hou, Y., Lin, Y.-P., Sharer, D., and March, P. 1994. *In vivo* selection of conditional-lethal mutations in the gene encoding elongation factor G of *Escherichia coli*. *J. Bacteriol.* **176**:123–129.

Huai, Q., Wang, H., Sun, Y., *et al.* 2003. Three-dimensional structures of PDE4D in complex with rolipram and implication on inhibitor selectivity. *Structure* **11**:865–873.

Huang, K.H., Fairclough, R.H., and Cantor, C.R. 1975. Singlet energy transfer studies of the arrangement of proteins in the 30S *Escherichia coli* ribosome. *J. Mol. Biol.* **97**:443–470.

Huang, Y.J., Swapna, G.V.T., Rajan, P.K., *et al.* 2003. Solution NMR structure of ribosome-binding factor A (RbfA), a cold-shock adaptation protein from *Escherichia coli*. *J. Mol.Biol.* **327**:521–536.

Hummel, H., and Böck, A. 1987. 23S ribosomal RNA mutations in halobacteria conferring resistance to the anti-80S ribosomal targeted antibiotic anisomycin. *Nucl. Acids
Res.* **15**:2431–2443.

Hyxley, H.E., and Zubay, G. 1960. Electron microscope observations on the structure of
microsomal particles from *Escherichia coli. J. Mol. Biol.* **2**:10–18.

Ibba, M., and Söll, D. 2001. The renaissance of aminoacyl-tRNA synthesis. *EMBO Rep.*
2:382–387.

Ibba, M., and Söll, D. 2002. Genetic code: introducing pyrrolysine. *Curr. Biol.*
12:R464–R466.

Ibba, M., Morgan, S., Curnow, A.W., *et al.* 1997. A euryarchaeal lysyl-tRNA synthetase:
resemblance to class I synthetases. *Science* **278**:1119–1122.

Ibba, M., Becker, H.D., Stathopoulos, C., *et al.* 2000. The adaptor hypothesis revisited.
Trends Biochem. Sci. **25**:311–316.

Inge-Vechtomov, S., Zhouravleva, G., and Philippe, M. 2003. Eukaryotic release factors
(eRFs) history. *Biol. Cell.* **95**:195–209.

Inoue, K., Alsina, J., Chen, J., and Inouye, M. 2003. Suppression of defective ribosome
assembly in a rbfA deletion mutant by overexpression of Era, an essential GTPase
in *Escherichia coli. Mol. Microbiol.* **48**:1005–1016.

Inoue-Yokosawa, N., Ishikawa, C., and Kaziro, Y. 1974. The role of guanosine triphosphate in translocation reaction catalyzed by elongation factor G. *J. Biol. Chem.*
249:4321–4323.

Ishitani, R., Nureki, O., Nameki, N., *et al.* 2003. Alternative tertiary structure of a tRNA
for recognition by a post-transcriptional modification enzyme. *Cell* **113**:383–394.

Ishitsuka, H., Kuriki, Y., and Kaji, A. 1970. Release of transfer ribonucleic acid from
ribosomes. A G factor and guanosine triphosphate-dependent reaction. *J. Biol.
Chem.* **245**:3346–3351.

Ito, K., Uno, M., and Nakamura, Y. 2000. A tripeptide "anticodon" deciphers stop
codons in messenger RNA. *Nature* **403**:680–684.

Ito, K., Fujiwara, T., Toyoda, T., and Nakamura, Y. 2002. Elongation factor G participates in ribosome disassembly by interacting with ribosome recycling factor at their
tRNA-mimicry domains. *Mol. Cell* **9**:1263–1272.

Izutsu, K., Wada, A., and Wada, C. 2001. Expression of ribosome modulation factor
(RMF) in *Escherichia coli* requires ppGpp. *Genes Cells* **6**:665–676.

Jacob, F., and Monod, J. 1961. Genetic regulatory mechanisms in the synthesis of proteins. *J. Mol. Biol.* **3**:318–356.

Jacquet, E., and Parmeggiani, A. 1988. Structure-function relationships in the GTP
binding domain of EF-Tu: mutation of Val20, the residue homologous to position 12
in p21. *EMBO J.* **7**:2861–2867.

Jakubowski, H., and Goldman, E. 1992. Editing of errors in selection of amino acids for
protein synthesis. *Microbiol. Rev.* **56**:412–429.

James, C.M., Ferguson, T.K., Leykam, J.F., and Krzycki, J.A. 2001. The amber codon in the
gene encoding the monomethylamine methyltransferase isolated from *Methanosarcina
barkeri* is translated as a sense codon. *J. Biol. Chem.* **276**:34252–34258.

Janosi, L., Hara, H., Zhang, S., and Kaji, A. 1996. Ribosome recycling by ribosome
recycling factor (RRF) — an important but overlooked step of protein biosynthesis.
Adv. Biophys. **32**:121–201.

Janosi, L., Mottagui-Tabar, S., Isaksson, L.A., *et al.* 1998. Evidence for *in vivo* ribosome recycling, the fourth step in protein biosynthesis. *EMBO J.* **17**:1141–1151.

Jaishree, T.N., Ramakrishnan, V., and White, S.W. 1996. Solution structure of prokaryotic ribosomal protein S17 by high-resolution NMR spectroscopy. *Biochemistry* **35**: 2845–2853.

Jenni, S., and Ban, N. 2003. The chemistry of protein synthesis and voyage through the ribosome tunnel. *Curr. Opin. Struct. Biol.* **13**:212–219.

Jiang, Y., Nock, S., Nesper, M., *et al.* 1996. Structure and importance of the dimerization domain in elongation factor Ts from *Thermus thermophilus. Biochemistry* **35**: 10269–10278.

Johanson, U., and Hughes, D. 1994. Fusidic acid-resistant mutants define three regions in elongation factor G of *Salmonella typhimurium. Gene* **143**:55–59.

Johanson, U., Ævarsson, A., Liljas, A., and Hughes, D. 1996. The dynamic structure of EF-G studied by fusidic acid resistance and internal revertants. *J. Mol. Biol.* **258**: 420–432.

Johnson, C.H., Kruft, V., and Subramanian, A.R. 1990. Identification of a plastid-specific ribosomal protein in the 30S subunit of chloroplast ribosomes and isolation of the cDNA clone encoding its cytoplasmic precursor. *J. Biol. Chem.* **265**: 12790–12795.

Jones, P.G., and Inouye, M. 1996. RbfA, a 30S ribosomal binding factor, is a cold-shock protein whose absence triggers the cold-shock response. *Mol. Microbiol.* **21**: 1207–1218.

Jones, P.G., Mitta, M., Kim, Y., *et al.* 1996. Cold shock induces a major ribosomal-associated protein that unwinds double-stranded RNA in *Escherichia coli. Proc. Natl. Acad. Sci. USA* **93**:76–80.

Jørgensen, R., Ortiz, P.A., Carr-Schmid, A., *et al.* 2003. Two crystal structures demonstrate large conformational changes in the eukaryotic ribosomal translocase. *Nat. Struct. Biol.* **10**:379–385.

Joseph, S., and Noller, H.F. 2000. Directed hydroxyl radical probing using iron(II) tethered to RNA. *Methods Enzymol.* **318**:175–190.

Joyce, G.F. 2002. The antiquity of RNA-based evolution. *Nature* **418**:214–221.

Justice, M.C., Hsu, M.J., Tse, B., *et al.* 1998. Elongation factor 2 as a novel target for selective inhibition of fungal protein synthesis. *J. Biol. Chem.* **273**:3148–3151.

Justice, M.C., Ku, T., Hsu, M.J., *et al.* 1999. Mutations in ribosomal protein L10e confer resistance to the fungal-specific eukaryotic elongation factor 2 inhibitor sordarin. *J. Biol. Chem.* **274**:4869–4875.

Kaempfer, R. 1972. Initiation factor IF-3: a specific inhibitor of ribosomal subunit association. *J. Mol. Biol.* **71**:583–598.

Kaltschmidt, E., Dzionara, M., Donner, D., and Wittmann, H.G. 1967. Ribosomal proteins I. Isolation, amino acid composition, molecular weights and peptide mapping of proteins from *E. coli* ribosomes. *Mol. Gen. Genet.* **100**:364–373.

Kaltschmidt, E., and Wittmann, H.G. 1970. Ribosomal proteins. XII. Number of proteins in small and large ribosomal subunits of *Escherichia coli* as determined by two-dimensional gel electrophoresis. *Proc. Natl. Acad. Sci. USA* **67**:1276–1282.

Kamtekar, S., Kennedy, W.D., Wang, J., *et al.* 2003. The structural basis of cysteine aminoacylation of tRNAPro by prolyl-tRNA synthetases. *Proc. Natl. Acad. Sci. USA* **100**:1673–1678.

Kappen, L.S., and Goldberg, I.H. 1976. Analysis of the two steps in polypeptide chain initiation inhibited by pactamycin. *Biochemistry* **15**:811–818.

Karimi, R., and Ehrenberg, M. 1994. Dissociation rate of cognate peptidyl-tRNA from the A-site of hyperaccurate and error-prone ribosomes. *Eur. J. Biochem.* **226**:355–360.

Karimi, R., and Ehrenberg, M. 1996. Dissociation rate of peptidyl-tRNA from the P-site of *E. coli* ribosomes. *EMBO J.* **15**:1149–1154.

Karimi, R., Pavlov, M.Y., Buckingham, R.H., and Ehrenberg, M. 1999. Novel roles for classical factors at the interface between translation termination and initiation. *Mol. Cell* **3**:601–609.

Karzai, A.W., Roche, E.D., and Sauer, R.T. 2000. The SsrA-SmpB system for protein tagging, directed degradation and ribosome rescue. *Nat. Struct. Biol.* **7**:449–455.

Kastner, B., Stöffler-Meilicke, M., and Stöffler, G. 1981. Arrangement of the subunits in the ribosome of *Escherichia coli*: demonstration by immuno electron microscopy. *Proc. Natl. Acad. Sci. USA* **78**:6652–6656.

Katunin, V.I., Muth, G.W., Strobel, S., *et al.* 2002. Important contribution to catalysis of peptide bond formation by a single ionizing group within the ribosome. *Mol. Cell* **10**:339–346.

Kawashima, T., Berthet-Colominas, C., Cusack, S., *et al.* 1996. The crystal structure of the *Escherichia coli* EF-Tu.EF-Ts complex at 2.5 Å resolution: mechanism of GDP/GTP exchange. *Nature* **379**:511–518.

Kaziro, Y. 1978. The role of guanosine 5'-triphosphate in polypeptide chain elongation. *Biochim. Biophys. Acta* **505**:95–127.

Keasling, J.D., Bertsch, L., and Kornberg, A. 1993. Guanosine pentaphosphate phosphohydrolase of *Escherichia coli* is a long-chain exopolyphosphatase. *Proc. Natl. Acad. Sci. USA* **90**:7029–7033.

Keeling, P.J., and Doolittle, W.F. 1995. An archaebacterial eIF-1A: new grist for the mill. *Mol. Microbiol.* **17**:399–400.

Keeling, P.J., Fast, N.M., and McFadden, G.I. 1998. Evolutionary relationship between translation initiation factor eIF-2gamma and selenocysteine-specific elongation factor SELB: change of function in translation factors. *J. Mol. Evol.* **47**:649–655.

Keenan, R.J., Freymann, D.M., Walter, P., and Stroud, R.M. 1998. Crystal structure of the signal sequence binding subunit of the signal recognition particle. *Cell* **94**:181–191.

Keenan, R.J., Freymann, D.M., Stroud, R.M., and Walter, P. 2001. The signal recognition particle. *Ann. Rev. Biochem.* **70**:755–775.

Keiler, K.C., Waller, P.R., and Sauer, R.T. 1996. Role of a peptide tagging system in degradation of proteins synthesized from damaged messenger RNA. *Science* **271**:990–993.

Khorana, H.G., Büchi, H., Ghosh, H., *et al.* 1966. Polynucleotide synthesis and the genetic code. *Cold Spring Harb. Symp. Quant. Biol.* **31**:39–49.

Kiel, M.C., Raj, V.S., Kaji, H., and Kaji A. 2003. Release of ribosome-bound ribosome recycling factor by elongation factor G. *J. Biol. Chem.* **278**:48041–48050.

Kim, S.H., Suddath, F.L., Quigley, G.J., *et al.* 1974. Three-dimensional tertiary structure of yeast phenylalanine transfer RNA. *Science* **185**:435–440.

Kim, K.K., Hung, L.W., Yokota, H., *et al.* 1998. Crystal structures of eukaryotic translation initiation factor 5 A from *Methanococcus jannaschii* at 1.8 A resolution. *Proc. Natl. Acad. Sci. USA* **95**:10419–10424.

Kim, K.K., Min, K., and Suh, S.W. 2000. Crystal structure of the ribosome recycling factor from *Escherichia coli*. *EMBO J.* **19**:2362–2370.

Kirsebom, L.A., Amons, R., and Isaksson, L.A. 1986. Primary structures of mutationally altered ribosomal protein L7/L12 and their effects on cellular growth and translational accuracy. *Eur. J. Biochem.* **156**:669–675.

Kischa, K., Möller, W., and Stöffler, G. 1971. Reconstitution of a GTPase activity by a 50S ribosomal protein from *E. coli*. *Nature New Biol.* **233**:62–63.

Kisselev, L.L., and Buckingham, R.H. 2000. Translational termination comes of age. *Trends Biochem. Sci.* **25**:561–566.

Kisselev, L., Ehrenberg, M., and Frolova, L. 2003. Termination of translation: interplay of mRNA, rRNAs and release factors? *EMBO J.* **22**:175–182.

Kjeldgaard, N.O., and Gaussing, K. 1974. Regulation of biosynthesis of ribosomes. In *Ribosomes*, Eds. M. Nomura *et al.*, pp. 369–392. Cold Spring Harbor Laboratory, Cold Spring Harbor, New York.

Kjeldgaard, M., and Nyborg, J. 1992. Refined structure of elongation factor EF-Tu from *Escherichia coli*. *J. Mol. Biol.* **223**:721–742.

Kjeldgaard, M., Nissen, P., Thirup, S., and Nyborg, J. 1993. The crystal structure of elongation factor EF-Tu from *Thermus aquaticus* in the GTP conformation. *Structure* **1**:35–50.

Kjeldgaard, M., Nyborg, J., and Clark, B.F.C. 1996. The GTP binding motif: variations on a theme. *FASEB J.* **10**:1347–1368.

Klaholz, B.P., Pape, T., Zavialov, A.V., *et al.* 2003. Structure of the *Escherichia coli* ribosomal termination complex with release factor 2. *Nature* **421**:90–94.

Klaholz, B.P., Myasnikov, A.G., and van Heel, M. 2004. Visualization of release factor 3 on the ribosome during termination of protein synthesis. *Nature* **427**:862–865.

Klein, D.J., Schmeing, T.M., Moore, P.B., and Steitz, T.A. 2001. The kink-turn: a new RNA secondary structure motif. *EMBO J.* **20**:4214–4221.

Klein, D.J., Moore, P.B., and Steitz, T.A. 2004. The roles of ribosomal proteins in the structure, assembly, and evolution of the large ribosomal subunit. Submitted to *J. Mol. Biol.*

Klug, A., and Schwabe, J.W. 1995. Protein motifs 5. Zinc fingers. *FASEB J.* **9**:597–604.

Knight, R.D., Freeland, S.J., and Landweber, L.F. 2001. Rewiring the keyboard: evolvability of the genetic code. *Nat. Rev. Gen.* **2**:49–58.

Knowles, D.J.C., Foloppe, N., Matassova, N.B., and Murchie, A.I.H. 2002. The bacterial ribosome, a promising focus for structure based drug design. *Curr. Opin. Pharmacol.* **2**:501–506.

Knudsen, C.R., and Clark, B.F. 1995. Site-directed mutagenesis of Arg58 and Asp86 of elongation factor Tu from *Escherichia coli*: effects on the GTPase reaction and aminoacyl-tRNA binding. *Prot. Eng.* **8**:1267–1273.

Kolesnikov, I.V., Protasova, N.Y., and Gudkov, A.T. 1996. Tetracyclines induce changes in accessibility of ribosomal proteins to proteases. *Biochimie* **78**:868–873.

Kolesnikov, A., and Gudkov, A. 2002. Elongation factor G with effector loop from elongation factor Tu is inactive in translocation. *FEBS Lett.* **514**:67–69.

Kolesnikov, A.V., and Gudkov, A.T. 2003. Mutation analysis of the role in ribosomal translocation for loops of domain IV of the elongation factor G. *Mol. Biol.* **37**:611–616.

Kosloff, M., and Selinger, Z. 2003. GTPase catalysis by Ras and other G-proteins: insights from Substrate Directed SuperImposition. *J. Mol. Biol.* **331**: 1157–1170.

Koteliansky, V.E., Domogatsky, S.P., and Gudkov, A.T. 1978. Dimer state of protein L7/L12 and EF-G dependent reactions on ribosomes. *Eur. J. Biochem.* **90**:319–323.

Krab, I.M., and Parmeggiani, A. 1998. EF-Tu, a GTPase odyssey. *Biochim. Biophys. Acta* **1443**:1–22.

Krab, I.M., and Parmeggiani, A. 1999. Mutagenesis of three residues, isoleucine-60, threonine-61, and aspartic acid-80, implicated in the GTPase activity of *Escherichia coli* elongation factor Tu. *Biochemistry* **38**:13035–13041.

Kramer, G., Ramachandiran, V., and Hardesty, B. 2001. Cotranslational folding — omnia mea mecum porto? *Int. J. Biochem. Cell Biol.* **33**:541–553.

Kramer, G., Ramachandiran, V., Horowitz, P.M., and Hardesty, B. 2002a. The molecular chaperone DnaK is not recruited to translating ribosomes that lack trigger factor. *Arch. Biochem. Biophys.* **403**:63–70.

Kramer, G., Rauch, T., Rist, W., *et al.* 2002b. L23 protein functions as a chaperone docking site on the ribosome. *Nature* **419**:171–174.

Kraulis, P. 1991. MOLSCRIPT: a program to produce both detailed and schematic plots of protein structures. *J. Appl. Cryst.* **24**:946–950.

Kristensen, O., and Gajhede, M. 2003. Chaperone binding at the ribosome exit tunnel. *Structure* **11**:1547–1556.

Kristensen, O., Laurberg, M., Liljas, A., and Selmer, M. 2002. Is tRNA binding or tRNA mimicry mandatory for translation factors? *Curr. Prot. Pept. Sci.* **3**:133–141.

Kristensen, O., Laurberg, M., Liljas, A., *et al.* 2004. Structural characterization of the stringent response related exopolyphosphatase/guanosine pentaphosphate phosphohydrolase protein family. *Biochemistry*, Web release date: 19-Jun-2004; (Accelerated Publication) DOI: 10.1021/bi049083c.

Kromayer, M., Wilting, R., Tormay, P., and Böck, A. 1996. Domain structure of the prokaryotic selenocysteine-specific elongation factor SelB. *J. Mol. Biol.* **262**: 413–420.

Kromayer, M., Neuhierl, B., Friebel, A., and Böck, A. 1999. Genetic probing of the interaction between the translation factor SelB and its mRNA binding element in *Escherichia coli*. *Mol. Gen. Genet.* **262**:800–806.

Kukimoto-Niino, M., Murayama, K., Inoue, M., *et al.* 2004. Crystal structure of the protein Obg from *Thermus thermophilus* HB8. *J. Mol. Biol.* **337**:761–770.

Kurland, C.G. 1960. Molecular characterizaion of ribonucleic acid from *E. coli* ribosomes, I. Isolation and molecular weights. *J. Mol. Biol.* **2**:83–91.

Kurland, C.G. 1972. Structure and function of the bacterial ribosome. *Ann. Rev. Biochem.* **41**:377–408.

Kurland, C.G. 1992. Translational accuracy and the fitness of bacteria. *Ann. Rev. Genet.* **26**:29–50.

Kurland, C.G., Hughes, D., and Ehrenberg, M. 1996. Limitations of translational accuracy. In *Escherichia coli and Salmonella typhimurium: Cellular and Molecular Biology*, Vol 1, Eds. F.C. Neidhardt, R. Curtis, III, J.L. Ingraham, pp. 979–1004. American Society for Microbiology Press.

Kyrpides, N.C., and Woese, C.R. 1998a. Universally conserved translation initiation factors. *Proc. Natl. Acad. Sci. USA* **95**:224–228.

Kyrpides, N.C., and Woese, C.R. 1998b. Archaeal translation initiation revisited: the initiation factor 2 and eucaryotic initiation factor 2B α-β-γ subunit families. *Proc. Natl. Acad. Sci. USA* **95**:3726–3730.

Laffler, T., and Gallant, J. 1974. *spoT*, a new genetic locus involved in the stringent response in *Escherichia coli*. *Cell* **1**:27–30.

Lake, J.A. 1976. Ribosome structure determined by electron microscopy of *Escherichia coli* small subunits, large subunits and monomeric ribosomes. *J. Mol. Biol.* **105**:131–139.

Lake, J.A. 1982. Ribosomal subunit orientations determined in the monomeric ribosome by single- and double-labelling immune electron microscopy. *J. Mol. Biol.* **161**:89–106.

Lake, J.A. 1985. Evolving ribosome structure: domains in archaebacteria, eubacteria, eocytes and eucaryotes. *Ann. Rev. Biochem.* **54**:507–530.

Lake, J., Pendergast, M., Kahan, L., and Nomura, M. 1974. Localization of *Escherichia coli* ribosomal proteins S4 and S14 by electron microscopy of antibody-labelled subunits. *Proc. Natl. Acad. Sci. USA* **71**:4688–4692.

Lancaster, L., Kiel, M.C., Kaji, A., and Noller, H.F. 2002. Orientation of ribosome recycling factor in the ribosome from directed hydroxyl radical probing. *Cell* **111**:129–140.

La Teana, A., Gualerzi, C.O., and Dahlberg, A.E. 2001. Initiation factor IF2 binds to the alpha-sarcin loop and helix 89 of *Escherichia coli* 23S ribosomal RNA. *RNA* **7**:1173–1179.

Laughrea, M., and Moore, P.B. 1977. Physical properties of ribosomal protein S1 and its interaction with the 30S ribosomal subunit of *Escherichia coli*. *J. Mol. Biol.* **112**:399–421.

Laurberg, M. 2002. Dynamics in protein synthesis. Structural studies of translation factors. Doctoral thesis from Lund University (ISBN 91-628-5088-1).

Laurberg, M., Kristensen, O., Martemyanov, K., *et al.* 2000. Structure of a mutant EF-G reveals domain III and possibly the fusidic acid binding site. *J. Mol. Biol.* **303**:593–603.

Laurberg, M., Kristensen, O., Martemyanov, K., *et al.* 2004. Crystal structure of a fusidic acid hypersensitive mutant of elongation factor G, G16V. Manuscript in preparation.

Laursen, R.A., L'Italien, J.J., Nagarkatti, S., and Miller, D.L. 1981. The amino acid sequence of elongation factor Tu of *Escherichia coli*. The complete sequence. *J. Biol. Chem.* **256**:8102–8109.

Lazzarini, R., and Dahlberg, A. 1971. The control of ribonucleic acid synthesis during amino acid depravation in *Escherichia coli*. *J. Biol. Chem.* **246**:420–429.

Lecompte, O., Ripp, R., Thierry, J.C., *et al.* 2002. Comparative analysis of ribosomal proteins in complete genomes: an example of reductive evolution at the domain scale. *Nucl. Acids Res.* **30**:5382–5390.

Lee, J.H., Choi, S.K., Roll-Mecak, A., *et al.* 1999. Universal conservation in translational initiation revealed by human and archaeal homologs of bacterial translation initiation factor IF2. *Proc. Natl. Acad. Sci. USA* **96**:4342–4347.

Lee, J.H., Pestova, T.V., Shin, B.-S., *et al.* 2002. Initiation factor eIF5B catalyzes second GTP-dependent step in eukaryotic translation initiation. *Proc. Natl. Acad. Sci. USA* **99**:16689–16694.

Leijonmarck, M., and Liljas, A. 1987. Structure of the C-terminal domain of the ribosomal protein L7/L12 from *Escherichia coli* at 1.7 Å. *J. Mol. Biol.* **195**:555–579.

Leijonmarck, M., Eriksson, S., and Liljas, A. 1980. Crystal structure of a ribosomal component at 2.6 A resolution. *Nature* **286**:824–826.

Leijonmarck, M., Appelt, K., Badger, J., *et al.* 1988. Structural comparison of the procaryotic ribosomal protein L7/L12 and L30. *Proteins* 3:243–248.

Leinfelder, W., Stadtman, T.C., and Böck, A. 1989. Occurrence *in vivo* of selenocysteyl-tRNA(SERUCA) in *Escherichia coli*. Effect of sel mutations. *J. Biol. Chem.* **264**: 9720–9723.

Leipe, D.D., Wolf, Y.I., Koonin, E.V., and Aravind, L. 2002. Classification and evolution of P-loop GTPases and related ATPases. *J. Mol. Biol.* **317**:41–72.

Lentzen, G., Klinck, R., Matassova, N., *et al.* 2003. Structural basis for contrasting activities of ribosome binding thiazole antibiotics. *Chem. Biol.* **10**:769–778.

Levy, S.B., McMurry, L.M., Barbosa, T.M., *et al.* 1999. Nomenclature for new tetracycline resistance determinants. *Antimicrob. Agents Chemother.* **43**:1523–1524.

Li, C., Reches, M., and Engelberg-Kulka, H. 2000. The bulged nucleotide in the *Escherichia coli* minimal selenocysteine insertion sequence participates in interaction with SelB: a genetic approach. *J. Bacteriol.* **182**:6302–6307.

Liljas, A. 1982. Structural studies of ribosomes. *Prog. Biophys. Mol. Biol.* 40:161–228.

Liljas, L. 1986. The structure of spherical viruses. *Prog. Biophys. Mol. Biol.* 48:1–36.

Liljas, A. 1990. Some structural aspects of elongation. In *The Structure, Function and Evolution of Ribosomes*, Eds. W. Hill, P.B. Moore, A. Dahlberg, *et al.*, pp. 309–317. ASM Press, American Society for Mocrobiology.

Liljas, A. 1991. Comparative biochemistry and biophysics of ribosomal proteins. *Int. Rev. Cytol.* **124**:103–136.

Liljas, A. 1996. Imprinting through molecular mimicry. Protein synthesis. *Curr. Biol.* **6**:247–249.

Liljas, A., and Al-Karadaghi, S. 1997. Structural aspects of protein synthesis. *Nat. Struct. Biol.* **4**:767–771.

Liljas, A., and Kurland, C.G. 1976. Crystallization of ribosomal protein L7/L12 from *Escherichia coli*. *FEBS lett.* **71**:130–132.

Liljas, A., Eriksson, S., Donner, D., and Kurland, C.G. 1978. Isolation and crystallization of stable domains of the protein L7/L12 from *Escherichia coli* ribosomes. *FEBS Lett.* **88**:300–304.

Liljas, A., Kirsebom, L.A., and Leijonmarck, M. 1986. Structural studies of the factor binding domain. In *Structure, Function and Genetics of Ribosomes*, Eds. B. Hardesty and G. Kramer, pp. 379–390. Springer-Verlag, New York.

Liljas, A., Kristensen, O., Laurberg, M., *et al.* 2000. The states, conformational dynamics and fusidic acid resistant mutants of EF-G. In *The Ribosome: Structure, Function, Antibiotics and Cellular Interactions*, Eds. R.A. Garrett, S.R. Douthwaite, A. Liljas, *et al.*, pp. 359–365. ASM Press, American Society for Mocrobiology, Washington, D.C.

Lim, V., and Spirin, A.S. 1986. Stereochemical analysis of ribosomal transpeptidation. *J. Mol. Biol.* **188**:565–577.

Lindahl, M., Svensson, L.A., Liljas, A., *et al.* 1994. Crystal structure of the ribosomal protein S6 from *Thermus thermophilus*. *EMBO J.* **13**:1249–1254.

Lipmann, F. 1969. Polypeptide chain elongation in protein biosynthesis. *Science* **164**:1024–1031.

Lipman, R.S., Sowers, K., and Hou, Y.M. 2000. Synthesis of cysteinyl-tRNA[Cys] by a genome that lacks a normal cysteine-tRNA synthase. *Biochemistry* **39**:7792.

Littlechild, J.A., and Malcolm, A.L. 1978. A new method for the purification of 30S ribosomal proteins from *Escherichia coli* using nondenaturing conditions. *Biochemistry* **17**:3363–3369.

Lodmell, J.S., and Dahlberg, A.E. 1997. A conformational switch in *Escherichia coli* 16S ribosomal RNA during decoding of messenger RNA. *Science* **277**:1262–1267.

Loftfield, R.B., and Vanderjagt, D. 1972. The frequency of errors in protein biosynthesis. *Biochem. J.* **128**:1353–1356.

Lu, M., and Steitz, T.A. 2000. Structure of *Escherichia coli* ribosomal protein L25 complexed with a 5S rRNA fragment at 1.8-A resolution. *Proc. Natl. Aad. Sci. USA* **97**:2023–2028.

Lubin, M. 1968. Observations on the shape of the 50S ribosomal subunit. *Proc. Natl. Acad. Sci. USA* **61**:1454–1461.

Luger, K., and Richmond, T.J. 1998. The histone tails of the nucleosome. *Curr. Opin. Genet. Dev.* **8**:140–146.

Luirink, J., ten Hagen-Jongman, C.M., van der Weijden, C.C., *et al.* 1994. An alternative protein targeting pathway in *Escherichia coli*: studies on the role of FtsY. *EMBO J.* **13**:2289–2296.

Maden, B.E.H., Traut, R.R., and Monro, R.E. 1968. Ribosome-catalyzed peptidyl transfer: the polyphenylalanine system. *J. Mol. Biol.* **35**:333–345.

Maki, Y., Yoshida, H., and Wada, A. 2000. Two proteins, YfiA and YhbH, associated with resting ribosomes in stationary phase *Escherichia coli*. *Genes Cells* **5**:965–974.

Malhotra, A., Penczek, P., Agrawal, R.K., *et al.* 1998. *Escherichia coli* 70S ribosome at 15 A resolution by cryo-electron microscopy: localization of fMet-tRNAfMet and fitting of L1 protein. *J. Mol. Biol.* **280**:103–116.

Malkin, L.I., and Rich, A. 1967. Partial resistance to proteolytic digestion due to ribosomal shielding. *J. Mol. Biol.* **26**:329–346.

Mankin, A.S. 1997. Pactamycin resistance mutations in functional sites of 16S RNA. *J. Mol. Biol.* **274**:8–15.

Mankin, A.S., and Garrett, R.A. 1991. Chloramphenicol resistance mutations in the single 23S gene of the archaeon *H. halobium*. *J. Bacteriol.* **173**:3559–3563.

Mao, J.C.H., and Robishaw, E.E. 1971. Effects of macrolides on peptide-bond formation and translocation. *Biochemistry* **10**:2054–2061.

Markus, M.A., Hinck, A.P., Huang, S., *et al.* 1997. High resolution solution structure of ribosomal protein L11-C76, a helical protein with a flexible loop that becomes structured upon binding to RNA. *Nat. Struct. Biol.* **4**:70–77.

Markus, M.A., Gerstner, R.B., Draper, D.E., and Torchia, D.A. 1998. The solution structure of ribosomal protein S4 delta41 reveals two subdomains and a positively charged surface that may interact with RNA. *EMBO J.* **17**:4559–4571.

Martemyanov, K.A., and Gudkov, A.S. 1999. Domain IV of elongation factor G from *T. thermophilus* is strictly required for translation. *FEBS Lett.* **452**:155–159.

Martemyanov, K.A., and Gudkov, A.T. 2000. Domain III of elongation factor G from *Thermus thermophilus* is essential for induction of GTP hydrolysis on the ribosome. *J. Biol. Chem.* **275**:35820–35824.

Martemyanov, K.A., Yarunin, A.S., Liljas, A., and Gudkov, A.T. 1998. An intact conformation at the tip of elongation factor G domain IV is functionally important. *FEBS Lett.* **434**:205–208.

Martemyanov, K.A., Liljas, A., Yarunin, A.S., and Gudkov, A.T. 2001. Mutations in the G-domain of elongation factor G from *Thermus thermophilus* affect both its interaction with GTP and fusidic acid. *J. Biol. Chem.* **276**:28774–28778.

Mason, N., Ciufo, L.F., and Brown, J.D. 2000. Elongation arrest is a physiologically important function of signal recognition particle. *EMBO J.* **19**:4164–4174.

Matadeen, R., Patwardhan, A., Gowen, B., *et al.* 1999. The *Escherichia coli* large ribosomal subunit at 7.5 Å resolution. *Structure* **7**:1575–1583.

Maxwell, I.H. 1967. Partial removal of bound transfer RNA from polysomes engaged in protein synthesis *in vitro* after addition of tetracycline. *Biochim. Biophys. Acta* **138**:337–346.

May, R.P., Nowotny, V., Nowotny, P., *et al.* 1992. Inter-protein distances within the large subunit from *Escherichia coli* ribosomes. *EMBO J.* **11**:373–378.

McCloskey, J.A., and Crain, P.F. 1998. The RNA modification database. *Nucl. Acids Res.* **26**:196–197.

McCutcheon, J.P., Agrawal, R.K., Philips, S.M., *et al.* 1999. Location of translational initiation factor IF3 on the small ribosomal subunit. *Proc. Natl. Acad. Sci. USA* **96**:4301–4306.

McGhee, J., and von Hippel, P. 1977. Formaldehyde as a probe of DNA structure. I. Reaction with exocyclic amino groups of DNA bases. *Biochemistry* **14**:1281–1303.

Meinnel, T., Sacerdot, C., Graffe, M., *et al.* 1999. Discrimination by *Escherichia coli* initiation factor IF3 against initiation on non-canonical codons relies on complementarity rules. *J. Mol. Biol.* **290**:825–837.

Melancon, P., Lemieux, C., and Brakier-Gringas, L. 1988. A mutation in the 530 loop of the *Escherichia coli* 16S ribosomal RNA causes resistance to streptomycin. *Nucl. Acids Res.* **16**:9631–9639.

Menetret, J.F., Neuhof, A., Morgan, D.G., *et al.* 2000. The structure of ribosome-channel complexes engaged in protein translocation. *Mol. Cell* **6**:1219–1232.

Mesters, J.R., Zeef, L.A., Hilgenfeld, R., *et al.* 1994. The structural and functional basis for the kirromycin resistance of mutant EF-Tu species in *Escherichia coli*. *EMBO J.* **13**:4877–4885.

Miller, D.L., and Weissbach, H. 1977. Factors involved in the transfer of aminoacyl-tRNA to the ribosome. In *Molecular Mechanisms of Protein Biosynthesis*, Eds. H. Weissbach and S. Petska, pp. 323–373. Academic Press, New York.

Milligan, R.A., and Unwin, P.N.T. 1986. Location of exit channel for nascent protein in 80S ribosome. *Nature* **319**:693–695.

Mittenhuber, G. 2001. Comparative genomics and evolution of genes encoding bacterial (p)ppGpp synthetases/hydrolases (the Rel, RelA and SpoT proteins). *J. Mol. Microbiol. Biotechnol.* **3**:585–600.

Mizushima, S., and Nomura, M. 1970. Assembly mapping of 30S ribosomal proteins from *E. coli*. *Nature* **226**:1214–1218.

Moazed, D., and Noller, H. 1986. Transfer RNA shields specific nucleotides in 16S ribosomal RNA from attack by chemical probes. *Cell* **47**:985–994.

Moazed, D., and Noller, H. 1987. Interaction of antibiotics with functional sites in 16S ribosomal RNA. *Nature* **327**:389–394.

Moazed, D., and Noller, H. 1989. Interaction of tRNA with 23S RNA in the ribosomal A-, P- and E-sites. *Cell* **57**:586–597.

Moazed, D., and Noller, H.F. 1990. Binding of tRNA to the ribosomal A- and P-sites protects two distinct sets of nucleotides in 16S rRNA. *J. Mol. Biol.* **211**:135–145.

Moazed, D., and Noller, H.F. 1991. Sites of interaction of the CCA-end of peptidyl-tRNA with 23S rRNA. *Proc. Natl. Acad. Sci. USA* **88**:3725–3728.

Moazed, D., Robertson, J.M., and Noller, H.F. 1988. Interaction of elongation factors EF-G and EF-Tu with a conserved loop in 23S RNA. *Nature* **334**:362–364.

Moazed, D., Samaha, R.R., Gualerzi, C., and Noller, H.F. 1995. Specific protection of 16S rRNA by translation initiation factors. *J. Mol. Biol.* **248**:207–210.

Modolell, J., Girbes, T., and Vazquez, D. 1975. Ribosomal translocation promoted by guanylylimido diphosphate and guanylyl-methylene diphosphonate. *FEBS Lett.* **60**:109–113.

Mohr, D., Wintermeyer, W., and Rodnina, M.V. 2000. Arginines 29 and 59 of elongation factor G are important for GTP hydrolysis or translocation on the ribosome. *EMBO J.* **19**:3458–3464.

Mohr, D., Wintermeyer, W., and Rodnina, M.V. 2002. GTPase activation of elongation factors Tu and G on the ribosome. *Biochemistry* **41**:12520–12528.

Monro, R.E. 1967. Catalysis of peptide bond formation by 50S ribosomal subunits from *Escherichia coli. J. Mol. Biol.* **26**:147–151.

Monro, R.E., Cerna, J., and Marcker, K.A. 1968. Ribosome-catalyzed peptidyl transfer: substrate specificity at the P-site. *Proc. Natl. Acad. Sci. USA* **61**:1042–1049.

Montandon, P.E., Wagner, R., and Stutz, E. 1986. *E. coli* ribosomes with a C912 to U base change in the 16S rRNA are streptomycin resistant. *EMBO J.* **5**:3705–3708.

Montesano-Roditis, L., Glitz, D.G., Traut, R.R., and Stewart, P.L. 2001. Cryo-electron microscopic localization of protein L7/L12 within the *Escherichia coli* 70S ribosome by difference mapping and Nanogold labeling. *J. Biol. Chem.* **276**: 14117–14123.

Montoya, G., Svensson, C., Luirink, J., and Sinning, I. 1997. Crystal structure of the NG domain from the signal-recognition particle receptor FtsY. *Nature* **385**:365–368.

Montoya, G., Kaat, K., Moll, R., Schafer, G., and Sinning, I. 2000. The crystal structure of the conserved GTPase of SRP54 from the archaeon *Acidianus ambivalens* and its comparison with related structures suggests a model for the SRP-SRP receptor complex. *Struct. Fold. Des.* **8**:515–525.

Moore, P.B. 1999. Structural motifs in RNA. *Ann. Rev. Biochem.* **67**:287–300.

Moore, P.B., and Steitz, T.A. 2003a. After the ribosome structures: how does peptidyl transferase work? *RNA* **9**:155–159.

Moore, P.B., and Steitz, T.A. 2003b. The structural basis of large ribosomal subunit function. *Ann. Rev. Biochem.* **72**:813–850.

Moras, D., Comarmond, M.B., Fischer, J., *et al.* 1980. Crystal structure of yeast tRNA[Asp]. *Nature* **288**:669–674.

Moreno, J.M., Kildsgaard, J., Siwanowicz, I., *et al.* 1998. Binding of *Escherichia coli* initiation factor IF2 to 30S ribosomal subunits: a functional role for the N-terminus of the factor. *Biochem. Biophys. Res. Commun.* **252**:465–471.

Mori, H., and Ito, K. 2001. The Sec protein-translocation pathway. *Trends Microbiol.* **9**:494–500.

Morrison, C.A., Bradbury, E.M., Littlechild, J., and Dijk, J. 1977. Proton magnetic resonance studies to compare *Escherichia coli* ribosomal proteins prepared by two different methods. *FEBS Lett.* **83**:348–352.

Mueller, F., and Brimacombe, R. 1997. A new model for the three-dimensional folding of *Escherichia coli* 16S ribosomal RNA. II. The RNA-protein interaction data. *J. Mol. Biol.* **271**:545–565.

Mueller, F., Stark, H., van Heel, M., *et al.* 1997. A new model for the three-dimensional folding of *Escherichia coli* 16S ribosomal RNA. III. The topography of the functional centre. *J. Mol. Biol.* **271**:566–587.

Mulder, F.A.A., Bauakaz, L., Lundell, A., *et al.* 2004. Conformation and dynamics of ribosomal stalk protein L12 in solution and on the ribosome. *Biochemistry* **43**:5930–5936.

Munishkin, A., and Wool, I.G. 1997. The ribosome-in-pieces: binding of elongation factor EF-G to oligoribonucleotides that mimic the sarcin/ricin and thiostrepton domains of 23S ribosomal RNA. *Proc. Natl. Acad. Sci. USA* **94**:12280–12284.

Murakami, K.S., Masuda, S., Campbell, E.A., *et al.* 2002. Structural basis of transcription initiation: an RNA polymerase holoenzyme-DNA complex. *Science* **296**: 1285–1290.

Murzin, A.G. 1993. OB (oligonucleotide/oligosaccharide binding)-fold: common structural and functional solution for non-homologous sequences. *EMBO J.* **12**:861–867.

Muth, G.W., Ortoleva-Donnelly, L., and Strobel, S. 2000. A single adenosine with a neutral pKa in the ribosomal peptidyl transferase center. *Science* **289**:947–950.

Muth, G.W., Chen, L., Kosek, A., and Strobel, S. 2001. pH-dependent conformational flexibility within the ribosomal peptidyl transferase center. *RNA* **7**:1403–1415.

Nagaev, I., Bjorkman, J., Andersson, D.I., and Hughes, D. 2001. Biological cost and compensatory evolution in fusidic acid-resistant *Staphylococcus aureus*. *Mol. Microbiol.* **40**:433–439.

Nakatogawa, H., and Ito, K. 2001. Secretion monitor, SecM, undergoes selftranslation arrest in the cytosol. *Mol. Cell* **7**:185–192.

Nakatogawa, H., and Ito, K. 2002. The ribosomal exit tunnel functions as a discriminating gate. *Cell* **108**:629–636.

Nakagawa, A., Nakashima, T., Taniguchi, M., *et al.* 1999. The three-dimensional structure of the RNA-binding domain of ribosomal protein L2; a protein at the peptidyl transferase center of the ribosome. *EMBO J.* **18**:1459–1467.

Nakamura, Y., and Ito, K. 2003. Making sense of mimic in translation termination. *TIBS*, **28**:99–105.

Nevskaya, N., Tishchenko, S., Nikulin, A., *et al.* 1998. Crystal structure of ribosomal protein S8 from *Thermus thermophilus* reveals a high degree of structural conservation of a specific RNA binding site. *J. Mol. Biol.* **279**:233–244.

Nevskaya, N., Tishchenko, S., Fedorov, R., *et al.* 2000. Archaeal ribosomal protein L1: the structure provides new insights into RNA binding of the L1 protein family. *Structure* **8**:363–371.

Nevskaya, N., Tishchenko, S., Paveliev, M., *et al.* 2002. Structure of ribosomal protein L1 from *Methanococcus thermolithotrophicus*. Functionally important structural invariants on the L1 surface. *Acta Cryst. D Biol. Cryst.* **58**:1023–1029.

Nielsen, R.C., Kristensen, O., Kjeldgaard, M., *et al.* 2004. Kirromycin induces conformational changes in the ternary complex that mimic ribosome binding. Manuscript submitted.

Nierhaus, K. 1990. Reconstitution of ribosomes. In *Ribosomes and Protein Synthesis*. pp. 161–189, IRL Press, Oxford, New York, Tokyo.

Nikonov, S., Nevskaya, N., Eliseikina, I. 1996. Crystal structure of the RNA binding ribosomal protein L1 from *Thermus thermophilus*. *EMBO J.* **15**:1350–1359.

Nikulin, A., Eliseikina, I., and Tishchenko, S. 2003. Structure of the L1 protuberance in the ribosome. *Nat. Struct. Biol.* **10**:104–108.

Ninio, J. 1974. A semi-quantitative treatment of missense and nonsense suppression in the strA and ram ribosomal mutants of *Escherichia coli*. Evaluation of some molecular parameters of translation *in vivo*. *J. Mol. Biol.* **84**:297–313.

Nirenberg, M.W., and Matthaei, J.H. 1961. The dependence of cell-free protein synthesis in *E. coli* upon naturally occuring or synthetic polynucleotides. *Proc. Natl. Acad. Sci. USA* **47**:1588–1602.

Nirenberg, M.W., and Leder, P. 1964. RNA code words and protein synthesis: the effect of trinucleotides upon the binding of sRNA to ribosomes. *Science* **145**:1399–1407.

Nishizuka, Y., and Lipmann, F. 1966. Comparison of guanosine triphosphate split and polypeptide synthesis with a purified *E. coli* system. *Proc. Natl. Acad. Sci. USA* **55**:212–219.

Nissen, P., Kjeldgaard, M., and Thirup, S. 1995. Crystal structure of the ternary complex of Phe-tRNA^{Phe}, EF-Tu, and a GTP analog. *Science* **270**:1464–1472.

Nissen, P., Thirup, S., Kjeldgaard, M., and Nyborg, J. 1999. The crystal structure of Cys-tRNACys-EF-Tu-GDPNP reveals general and specific features in the ternary complex and in tRNA. *Struct. Fold Des.* **7**:143–156.

Nissen, P., Kjeldgaard, M., and Nyborg, J. 2000. Macromolecular mimicry. *EMBO J.* **19**:489–495.

Nissen, P., Hansen, J., Ban, N., *et al*. 2000. The structural basis of ribosome activity in peptide bond synthesis. *Science* **289**:920–930.

Nissen, P., Ippolito, J.A., Ban, N., *et al*. 2001. RNA tertiary interactions in the large ribosomal subunit: the A-minor motif. *Proc. Natl. Acad. Sci. USA* **98**:4899–4903.

Noller, H.F. 1991. Ribosomal RNA and translation. *Ann. Rev. Biochem.* **60**:191–227.

Noller, H.F. 1993. Peptidyl transferase — protein, ribonucleoprotein, or RNA? *J. Bacteriol.* **175**:5297–5300.

Noller, H.F., and Chaires, J.B. 1972. Functional modification of 16S ribosomal RNA by kethoxal. *Proc. Natl. Acad. Sci. USA* **69**:3115–3118.

Noller, H.F., and Woese, C.R. 1981. Secondary structure of 16S ribosomal RNA. *Science* **212**:403–411.

Noller, H.F., Kop, J., Wheaton, V., *et al*. 1981. Secondary structure model for 23S ribosomal RNA. *Nucl. Acids Res.* **9**:6167–6189.

Noller, H.F., Hoffarth, V., and Zimniak, L. 1992. Unusual resistance of peptidyl transferase to protein extraction procedures. *Science* **256**:1416–1419.

Nomura, M. 1973. Assembly of bacterial ribosomes. *Science* **179**:864–873.

Nomura, M. 1990. History of ribosome research: a personal account. In *The Ribosome. Structure, Function and Evolution*. Eds. W.E. Hill, A. Dahlberg, R.A. Garrett, *et al.*, pp. 3–55. ASM Press, Washington.

Nomura, M., Yates, J.L., Dean, D., and Post, L.E. 1980. Feedback regulation of ribosomal protein gene expression in *Escherichia coli*: structural homology of ribosomal RNA and ribosomal protein mRNA. *Proc. Natl. Acad. Sci. USA* **77**:7084–7088.

Nomura, M., Gourse, R., and Baughman, G. 1984. Regulation of the synthesis of ribosomes and ribosomal components. *Ann. Rev. Biochem.* **53**:75–117.

Nureki, O., Vassylev, D.G., Tateno, M., *et al.* 1998. Enzyme structure with two catalytic sites for double-sieve selection of substrate. *Science* **280**:578–582.

O'Connor, M., Gregory, S.T., Rajbhandary, U.L., and Dahlberg, A.E. 2001. Altered discrimination of start codons and initiator tRNAs by mutant initiation factor 3. *RNA* **7**:969–978.

Odom, O.W., Kramer, G., Henderson, A.B., *et al.* 1978. GTP hydrolysis during methionyl-tRNAf binding to 40S ribosomal subunits and the site of edeine inhibition. *J. Biol. Chem.* **253**:1807–1816.

Odom, O.W., Picking, W.D., Tsalkova, T., and Hardesty, B. 1991. The synthesis of polyphenylalanine on ribosomes to which erythromycin is bound. *Eur. J. Biochem.* **198**:713–722.

Öfverstedt, L.G., Zhang, K., Tapio, S., *et al.* 1994. Starvation *in vivo* for aminoacyl-tRNA increases the spatial separation between the two ribosomal subunits. *Cell* **79**:629–638.

Ogle, J.M., Brodersen, D.E., Clemons, W.M. Jr., *et al.* 2001. Recognition of cognate transfer RNA by the 30S ribosomal subunit. *Science* **292**:897–902.

Ogle, J.M., Murphy, F.V., Tarry, M.J., and Ramakrishnan, V. 2002. Selection of tRNA by the ribosome requires a transition from an open to a closed form. *Cell* **111**:721–732.

Ogle, J.M., Carter, A.P., and Ramakrishnan, V. 2003. Insights into the decoding mechanism from recent ribosome structures. *Trends Biochem. Sci.* **28**:259–266.

Ohman, A., Rak, A., Dontsova, M., *et al.* 2003. NMR structure of the ribosomal protein L23 from *Thermus thermophilus*. *J. Biomol. NMR* **26**:131–137.

Oleinikov, A.V., Jokhadze, G.G., and Traut, R.R. 1998. A single-headed dimer of *Escherichia coli* ribosomal protein L7/L12 supports protein synthesis. *Proc. Natl. Acad. Sci. USA* **95**:4215–4218.

Okamoto, T., and Takanami, M. 1963. Interaction of ribosomes and natural polyribonucleotides. *Biochim. Biophys. Acta* **76**:266–274.

Orengo, C.A., and Thornton, J.M. 1993. Alpha plus beta folds revisited: some favoured motifs. *Structure* **1**:105–120.

Österberg, R., Sjöberg, B., Liljas, A., and Petterson, I. 1976. Small-angle X-ray scattering and crosslinking study of the protein L7/L12 from *E. coli* ribosomes. *FEBS Lett.* **66**:48–51.

Österberg, R., Sjöberg, B., Pettersson, I., *et al.* 1997. Small-angle scattering study of the protein complex of L7/L12 and L10 from *Escherichia coli* ribosomes. *FEBS Lett.* **73**:22–24.

Ovchinnikov, Yu. A., Alakhov, Yu. B., Bundulis, Yu. P., *et al.* 1982. The primary structure of elongation factor G from *Escherichia coli*: a complete amino acid sequence. *FEBS Lett.* **139**:130–135.

Palade, G.E. 1955. A small particulate component of the cytoplasm. *J. Biophys. Biochem. Cytol.* **1**:59–68.

Pandit, S.B., and Srinivasan, N. 2003. Survey for g-proteins in the prokaryotic genomes: prediction of functional roles based on classification. *Proteins* **52**:585–597.

Pape, T., Wintermeyer, W., and Rodnina, M.V. 1998. Complete kinetic mechanism of elongation factor Tu-dependent binding of aminoacyl-tRNA to the A-site of the *E. coli* ribosome. *EMBO J.* **17**:7490–7497.

Pape, T., Wintermeyer, W., and Rodnina, M. 1999. Induced fit in initial selection and proof-reading of aminoacyl-tRNA on the ribosome. *EMBO J.* **18**:3800–3807.

Pape, T., Wintermeyer, W., and Rodnina, M. 2000. Conformational switch in the decoding region of 16S rRNA during aminoacyl-tRNA selection on the ribosome. *Nat. Struct. Biol.* **7**:104–107.

Parker, J., Watson, R.J., and Friesen, J.D. 1976. A relaxed mutant with an altered ribosomal protein L11. *Mol. Gen. Genet.* **144**:111–114.

Parmeggiani, A., and Swart, G.W. 1985. Mechanism of action of kirromycin-like antibiotics. *Ann. Rev. Microbiol.* **39**:557–577.

Parsons, L., Eisenstein, E., and Orban, J. 2001. Solution structure of HI0257, a bacterial ribosome binding protein. *Biochemistry* **40**:10979–10986.

Patzelt, H., Rüdiger, S., Brehmer, D., *et al.* 2001. Binding specificity of *Escherichia coli* trigger factor. *Proc. Natl. Acad. Sci. USA* **98**:14244–14249.

Perutz, M.F. 1962. *Proteins and Nucleic Acids. Structure and Function*, Eighth Weizmann Memorial Lecture Series. Elsevier Publishing Company, Amsterdam.

Peske, F., Matassova, N.B., Savelsbergh, A., *et al.* 2000. Conformationally restricted elongation factor G retains GTPase activity but is inactive in translocation on the ribosome. *Mol. Cell* **6**:501–505.

Pestka, S. 1969. Studies on the formation of transfer ribonucleic acid-ribosome complexes. VI. Oligopeptide synthesis and translocation on ribosomes in the presence and absence of soluble transfer factors. *J. Biol. Chem.* **244**:1533–1539.

Pestka, S. 1972. Studies on transfer ribonucleic acid-ribosome complexes. Effects on antibiotics on peptidyl puromycin synthesis on polyribosomes from *Escherichia coli*. *J. Biol. Chem.* **247**:4669–4678.

Pestova, T.V., and Hellen, C.U. 2000. The structure and function of initiation factors in eucaryotic protein synthesis. *Cell. Mol. Life Sci.* **57**:651–674.

Pestova, T.V., Lomakin, I.B., Lee, J.H., *et al.* 2000. The joining of ribosomal subunits in eukaryotes requires eIF5B. *Nature* **403**:332–335.

Petrelli, D., Garofalo, C., Lammi, M., *et al.* 2003. Mapping the active sites of bacterial translation initiation factor IF3. *J. Mol. Biol.* **331**:541–556.

Pettersson, I., and Kurland, C.G. 1980. Ribosomal protein L7/L12 is required for optimal translation. *Proc. Natl. Acad. Sci. USA* **77**:4007–4010.

Pettersson, I., Hardy, S.J., and Liljas, A. 1976. The ribosomal protein L8 is a complex of L7/L12 and L10. *FEBS Lett.* **64**:135–138.

Pfennig, P.L., and Flower, A.M. 2001. BipA is required for growth of *Escherichia coli* K12 at low temperature. *Mol. Genet. Gen.* **266**:313–317.

Piepenburg, O., Pape, T., Pleiss, J.A., *et al.* 2000. Intact aminoacyl-tRNA is required to trigger GTP hydrolysis by elongation factor Tu on the ribosome. *Biochemistry* **39**:1734–1738.

Pinard, R., Peyant, C., Melancon, P., and Brakier-Gringas, L. 1993. The 5′ proximal helix of 16S rRNA is involved in the binding of streptomycin to the ribosome. *FASEB J.* **7**:173–176.

Pioletti, M., Schlünzen, F., Harms, J., *et al.* 2001. Crystal structures of complexes of the small ribosomal subunit with tetracycline, edeine and IF3. *EMBO J.* **20**:1829–1839.

Planta, R.J., and Mager, W.H. 1998. The list of cytoplasmic ribosomal proteins of *Saccharomyces cerevisiae*. *Yeast* **14**:471–477.

Plumbridge, J.A., Deville, F., Sacerdot, C., *et al.* 1985. Two translational initiation sites in the infB gene are used to express initiation factor IF2 alpha and IF2 beta in *Escherichia coli*. *EMBO J.* **4**:223–229.

Polacek, N., Gaynor, M., Yassin, A., and Mankin, A.S. 2001. Ribosomal peptidyl transferase can withstand mutations at the putative catalytic nucleotide. *Nature* **411**:498–501.

Polacek, N., Gomez, M.J., Ito, K., *et al.* 2003. The critical role of the universally conserved A2602 of 23S RNA in the release of the nascent peptide during translation termination. *Mol. Cell* **11**:103–112.

Polekhina, G., Thirup, S., Kjeldgaard, M., *et al.* 1996. Helix unwinding in the effector region of elongation factor EF-Tu-GDP. *Structure* **4**:1141–1151.

Pool, M.R., Stumm, J., Fulga, T.A., *et al.* 2002. Distinct modes of signal recognition particle interaction with the ribosome. *Science* **297**:1345–1348.

Poritz, M.A., Strub, K., and Walter, P. 1988. Human SRP RNA and *E. coli* 4.5S RNA contain a highly homologous structural domain. *Cell* **55**:4–6.

Porse, B.T., Kirillov, S.V., Awayez, M.J., *et al.* 1999. Direct crosslinking of the antitumor antibiotic sparsomycin, and its derivatives, to A2602 in the peptidyltransferase center of 23S like rRENA within ribosome-tRNA complexes. *Proc. Natl. Acad. Sci. USA* **96**:9003–9008.

Poulsen, S.M., Kofoed, C., and Vester, B. 2000. Inhibition of the ribosomal peptidyltransferase reaction by the mycarose moiety of the antibiotics carbomycin, spiramycin and tylosin. *J. Mol. Biol.* **304**:471–481.

Powers, T., and Noller, H.F. 1990. Dominant lethal mutations in a conserved loop in 16S rRNA. *Proc. Natl. Acad. Sci. USA* **87**:1042–1046.

Powers, T., and Noller, H.F. 1995. Hydroxyl radical footprinting of ribosomal proteins on 16S rRNA. *RNA* **1**:194–209.

Prince, J.B., Taylor, B.H., Thurlow, D.L., *et al.* 1982. Covalent crosslinking of tRNA1Val to 16S RNA at the ribosomal P-site: identification of crosslinked residues. *Proc. Natl. Acad. Sci. USA* **79**:5450–5454.

Quiggle, K., Kumar, G., Ott, T.W., *et al.* 1981. Donor site of ribosomal peptidyltransferase: investigation of substrate specificity using 2'(3')-O-(N-acylaminoacyl)dinucleoside phosphates as models of the 3' terminus of N-acylaminoacyl transfer ribonucleic acid. *Biochemistry* **20**:3480–3485.

Ramagopal, S. 1976. Accumulation of free ribosomal proteins S1, L7 and L12 in *E. coli*. *Eur. J. Biochem.* **69**:289–297.

Ramakrishnan, V., and Moore, P.B. 2001. Atomic structure at last: the ribosome in 2000. *Curr. Opin. Struct. Biol.* **11**:144–154.

Ramakrishnan, V., and White, S.W. 1992. The structure of ribosomal protein S5 reveals sites of interaction with 16S rRNA. *Nature* **358**:768–771.

Ramakrishnan, V., and White, S.W. 1998. Ribosomal protein structures: insights into the architecture, machinery and evolution of the ribosome. *TIBS* **23**:208–212.

Rapoport, T.A., Jungnickel, B., and Kutay, U. 1996. Protein transport across eucaryoticcendoplasmic reticulum and bacterial inner membrane. *Ann. Rev. Biochem.* **65**: 271–303.

Rao, A.R., and Varshney, U. 2001. Specific interactions between the ribosome recycling factor and the elongation factor G from *Mycobacterium tuberculosis* mediates peptidyl-tRNA release and ribosome recycling in *Escherichia coli*. *EMBO J.* 20:2977–2986.

Rawat, U.B.S., Zavialov, A.V., Sengupta, J., Valle, M., *et al.* 2003. A cryo-electron micrtoscopic study of ribosome-bound termination factor RF2. *Nature* 421:87–90.

Remacha, M., Jimenez-Diaz, A., Bermejo, B., *et al.* 1995. Ribosomal acidic phosphoproteins P1 and P2 are not required for cell viability but regulate the pattern of protein expression in *Saccharomyces cerevisiae*. *Mol. Cell. Biol.* 15:4754–4762.

Retsema, J., and Fu, W. 2001. Macrolides: structures and microbial targets. *Int. J. Antimicrob. Agents* 18(suppl. 1):S3–S10.

Rheinberger, H.J., Sternbach, H., and Nierhaus, K.H. 1981. Three tRNA binding sites on *Escherichia coli* ribosomes. *Proc. Natl. Acad. Sci. USA* 78:5310–5314.

Ribas de Pouplana, L., and Schimmel, P. 2001. Aminoacyl-tRNA synthetases: potential markers of genetic code development. *TIBS* 26:591–596.

Rich, A. 2001. RNA structure and the roots of protein synthesis. *Cold Spring Harb. Symp. Quant. Biol.* 66:1–16.

Robertus, J.D., Ladner, J.E., Finch, J.T., *et al.* 1974. Structure of yeast phenylalanine tRNA at 3 Å resolution. *Nature* 250:546–551.

Rodnina, M., and Wintermeyer, W. 2003. peptide bond formation on the ribosome: structure and mechanism. *Curr. Opin. Struct. Biol.* 13:334–340.

Rodnina, M.V., Fricke, R., and Wintermeyer, W. 1993. Kinetic fluorescence study on EF-Tu-dependent binding of Phe-tRNA[Phe] to the ribosomal A-site. In *The Translational Apparatus: Structure, Function, Regulation and Evolution*, Eds. K.H. Nierhaus, F. Franceschi, A.R. Subramanian, *et al.*, pp. 317–326. Plenum Publishing Corp. New York.

Rodnina, M.V., Fricke, R., and Wintermeyer, W. 1994. Transient conformational states of aminoacyl-tRNA during ribosome binding catalyzed by elongation factor Tu. *Biochemistry* 33:12267–12275.

Rodnina, M.V., Fricke, R., Kuhn, L., and Wintermeyer, W. 1995a. Codon-dependent conformational change of elongation factor Tu preceding GTP hydrolysis on the ribosome. *EMBO J.* 14:2613–2619.

Rodnina, M.V., Pape, T., Fricke, R., and Wintermeyer, W. 1995b. Elongation factor Tu, a GTPase triggered by codon recognition on the ribosome: mechanism and GTP consumption. *Biochem. Cell Biol.* 73:1221–1227.

Rodnina, M.V., Pape, T., Fricke, R., *et al.* 1996. Initial binding of the elongation factor Tu.GTP. aminoacyl-tRNA complex preceeding codon recognition on the ribosome. *J. Biol. Chem.* 271:646–652.

Rodnina, M.V., Savelsbergh, A., Katunin, V.I., and Wintermeyer, W. 1997. Hydrolysis of GTP by elongation factor G drives tRNA movement on the ribosome. *Nature* 385:37–41.

Rodnina, M.V., Savelsbergh, A., Matassova, N.B., *et al.* 1999. Thiostrepton inhibits the turnover but not the GTPase of elongation factor G on the ribosome. *Proc. Natl. Acad. Sci. USA* 96:9586–9590.

Rodnina, M.V., Stark, H., Savelsbergh, A., *et al.* 2000. GTPases mechanisms and functions of translation factors on the ribosome. *Biol. Chem.* 381:377–387.

Röhl, R., and Nierhaus, K.H. 1982. Assembly map of the large subunit (50S) of *Escherichia coli* ribosomes. *Proc. Natl. Acad. Sci. USA* 79:729–733.

Roll-Mecak, A., Cao, C., Dever, T.E., and Burley, S.K. 2000. X-ray structures of the universal translation initiation factor IF2/eIF5B: conformational changes on GDP and GTP binding. *Cell* **103**:781–792.

Roll-Mecak, A., Shin, B.S., Dever, T.E., and Burley, S.K. 2001. Engaging the ribosome: universal Ifs of translation. *Trends Biochem. Sci.* **26**:705–709.

Romby, P., and Springer, M. 2003. Bacterial translational control at atomic resolution. *Trends Genet.* **19**:155–161.

Ross, J.I., Eady, E.A., Cove, J.H., and Cunliffe, W.J. 1998. 16S rRNA mutation associated with tetracycline resistance in a gram-positive bacterium. *Antimicrob. Agents Chemother.* **42**:1702–1705.

Rossmann, M.G., and Blow, D.M. 1962. The detection of sub-units within the crystallographic asymmetric unit. *Acta Cryst.* **15**:24–31.

Rossmann, M.G., and Johnson, J.E. 1989. Icosahedral RNA virus structure. *Ann. Rev. Biochem.* **58**:533–573.

Rostom, A.A., Fucini, P., Benjamin, D.R., *et al.* 2000. *Proc. Natl. Acad. Sci. USA* **97**:5185–5190.

Rudinger, J., Hillenbrandt, R., Sprinzl, M., and Giege, R. 1996. Antideterminants present in minihelix[Sec] hinder its recognition by prokaryotic elongation factor Tu. *EMBO J.* **15**:650–657.

Ruusala, T., Ehrenberg, M., and Kurland, C.G. 1982. Is there proofreading during polypeptide synthesis. *EMBO J.* **1**:741–745.

Saarma, U., Remme, J., Ehrenberg, M., and Bilgin, N. 1997. An A to U transversion at position 1067 of 23S rRNA from *Escherichia coli* impairs EF-Tu and EF-G function. *J. Mol. Biol.* **272**:327–335.

Sabatini, D.D., Tashiro, Y., and Palade, G.E. 1966. On the attachment of ribosomes to microsomal membranes. *J. Mol. Biol.* **19**:503–524.

Sacerdot, C., Vachon, G., Laalami, S., *et al.* 1992. Both forms of translational factor 2 (α and β) are required for maximal growth of *Escherichia coli*. Evidence for two translational initiation codons for IF2. *J. Mol. Biol.* **225**:67–80.

Salyers, A.A., Speer, B.S., and Shoemacher, N.B. 1990. New perspectives in tetracycline resistance. *Mol. Microbiol.* **4**:151–156.

Samaha, R.R., Green, R., and Noller, H.F. 1995. A base pair between tRNA and 23S rRNA in the peptidyl transferase centre of the ribosome. *Nature* **377**:309–314. Erratum in: *Nature* **378**:419.

Sanchez-Pescador, R., Brown, J.T., Roberts, M., and Urdea, M.S. 1988. Homology of the TetM with translational elongation factors: implications for potential modes of tetM-conferred tetracycline resistance. *Nucl. Acids Res.* **16**:1218.

Sankaranarayanan, R., and Moras, D. 2001. The fidelity of the translation of the genetic code. *Acta Biochim. Pol.* **48**:323–335.

Santos, C., and Ballesta, J.P.G. 1994. Ribosomal phosphoprotein P0, contrary to phosphoproteins P1 and P2, is required for *S. cerevisiae* viability and ribosome assembly. *J. Biol. Chem.* **269**:15689–15696.

Santos, C., and Ballesta, J.P.G. 1995. The highly conserved protein P0 carboxyl end is essential for ribosome activity only in the absence of proteins P1 and P2. *J. Biol. Chem.* **270**:20608–20614.

Sanyal, C.S., and Liljas, A. 2000. The end of the beginning: structural studies of ribosomal proteins. *Curr. Opin. Struct. Biol.* **10**:633–636.

Saraste, M., Sibbald, P.R., and Wittinghofer, A. 1990. The P-loop, a common motif in ATP- and GTP-binding proteins. *Trends Biochem. Sci.* **15**:430-434.

Savelsbergh, A., Matassova, N.B., Rodnina, M.V., and Wintermeyer, W. 2000a. Role of domains 4 and 5 in elongation factor G functions on the ribosome. *J. Mol. Biol.* **300**:951–961.

Savelsbergh, A., Mohr, D., Wilden, B., *et al.* 2000b. Stimulation of the GTPase activity of translation elongation factor G by ribosomal protein L7/12. *J. Biol. Chem.* **275**:890–894.

Savelsbergh, A., Katunin, V.I., Mohr, D., *et al.* 2002. An elongation factor G-induced ribosome rearrangement precedes tRNA-mRNA translocation. *Mol. Cell.* **11**: 1517–1523.

Savelsbergh, A., Katunin, V.I., Mohr, D., *et al.* 2003. An elongation factor G-induced ribosome rearrangement precedes tRNA-mRNA translocation. *Mol. Cell.* **11**: 1517–1523.

Sawaya, M.R., Prasad, R., Wilson, S.H., *et al.* 1997. Crystal structures of human DNA polymerase β complexed with gapped and nicked DNA: evidence for an induced fit mechanism. *Biochemistry* **36**:11205–11215.

Schaffitzel, E., Rudiger, S., Bukau, B., and Deuerling, E. 2001. Functional dissection of trigger factor and DnaK: interactions with nascent polypeptides and thermally denatured proteins. *Biol. Chem.* **382**:1235–1243.

Schilling-Bartetzko, S., Franceschi, F., Sternbach, H., and Nierhaus, K.H. 1992. Apparent association constants of tRNAs for the ribosomal A-, P-, and E-sites. *J. Biol. Chem.* **267**:4693–4702.

Schimmel, P., and Ribas de Pouplana, L. 2001. Formation of two classes of tRNA synthetases in relation to editing functions and genetic code. *Cold Spring Harb. Symp. Quant. Biol.* **66**:161–166.

Schimmel, P., Giege, R., Moras, D., and Yokoyama, S. 1993. An operational RNA code for amino acids and possible relationship to the genetic code. *Proc. Natl. Acad. Sci. USA* **90**:8763–8768.

Schlünzen, F., Tocilj, A., Zariwach, R., *et al.* 2000. Structure of functionally activated small ribosomal subunit at 3.3 Å resolution. *Cell* **102**:615–623.

Schlünzen, F., Zariwach, R., Harms, J., *et al.* 2001. Structural basis for the interaction of antibiotics with the peptidyltransferase centre in eubacteria. *Nature* **413**:814–821.

Schlünzen, F., Harms, J.M., Franceschi, F., *et al.* 2003. Structural basis for the antibiotic activity of ketolides and azalides. *Structure* **11**:329–338.

Schmeing, T.M., Seila, A.C., Hansen, J.L., *et al.* 2002. A pre-translocational intermediate in protein synthesis observed in crystals of enzymatically active 50S subunits. *Nat. Struct. Biol.* **9**:225–230.

Schmidt, J., Srinivasa, B., Weglohner, W., and Subramanian, A.R. 1993. A small novel chloroplast ribosomal protein (S31) that has no apparent counterpart in the *E. coli* ribosome. *Biochem. Mol. Biol. Int.* **29**:25–31.

Schmitt, E., Blanquet, S., and Mechulam, Y. 2002. The large subunit of initiation factor aIF2 is a close structural homologue of elongation factors. *EMBO J.* **21**:1821–1832.

Schmitz, U., James, T.L., Lukavsky, P., and Walter, P. 1999. Structure of the most conserved internal loop in SRP RNA. *Nat. Struct. Biol.* **6**:634–638.

Schop, R.N., and Massen, J.A. 1982. Characterization of the region on protein L7/L12 involved in binding to ribosomal particles. *Eur. J. Biochem.* **128**:371–375.

Schürer, H., Schiffer, S., Marchfelder, A., and Mörl, M. 2001. This is the end: processing, editing and repair at the tRNA 3′-terminus. *Biol. Chem.* **382**:1147–1156.

Schweins, T., Geyer, M., Scheffzek, K., *et al.* 1995. Substrate assisted catalysis as a mechanism for GTP hydrolysis of ras-p21 and other GTP-binding proteins. *Nat. Struct. Biol.* **2**:36–44.

Scolnick, E., Tompkins, R., Caskey, T., and Nirenberg, M. 1968. Release factors differing in specificity for terminator codons. *Proc. Natl. Acad. Sci. USA* **61**:768–774.

Scott, J.M., Ju, J., Mitchell, T., and Haldenwang, W.G. 2000. The *Bacillus subtilis* GTP binding protein Obg and regulators of the σ^B stress response transcription factor cofractionate with ribosomes. *J. Bacteriol.* **182**:2771–2777.

Seit-Nebi, A., Frolova, L., Justesen, J., and Kisselev, L. 2001. Class-1 translation termination factors: invariant GGQ minidomain is essential for release activity and ribosome binding but not for stop codon recognition. *Nucl. Acids Res.* **29**:3982–3987.

Selivanova, O.M., Shiryaev, V.M., Tiktopulo, E.I., *et al.* 2003. Compact globular structure of *Thermus thermophilus* ribosomal protein S1 in solution. *J. Biol. Chem.* **278**:36311–36314.

Selmer, M. 2002. Protein-RNA interplay in translation. Structural studies of RRF, SelB and L1. Doctoral thesis from Lund University. (ISBN 91-7874-176-9).

Selmer, M., Al-Karadaghi, S., Hirokawa, G., *et al.* 1999. Crystal structure of *Thermatoga maritima* ribosome recycling factor: a tRNA mimic. *Science* **286**:2349–2352.

Selmer, M., and Su, X.-D. 2002. Crystal structure of an mRNA binding fragment of *Moorella thermoacetica* elongation factor SelB. *EMBO J.* **21**:4145–4153.

Sengupta, J., Agrawal, R.K., and Frank, J. 2001. Visualisation of protein S1 whithin the 30S ribosomal subunit and its interaction with mRNA. *Proc. Natl. Sci. Acad. USA* **98**:11991–11996.

Sergiev, P., Leonov. A., Dokudovskaya, S., *et al.* 2001. Correlating the X-ray structures for halo- and thermophilic ribosomal subunits with biochemical data for the *Escherichia coli* ribosome. *Cold Spring Harb. Symp. Quant. Biol.* **66**:87–100.

Sette, M., van Tilborg, P., Spurio, R., *et al.* 1997. The structure of the translational initiation factor IF1 from *E. coli* contains an oligomer-binding motif. *EMBO J.* **16**: 1436–1443.

Sharma, D., Southworth, D.R., and Green, R. 2004. EF-G-independent reactivity of a pre-translocation-state ribosome complex with the aminoacyl tRNA substrate puromycin supports an intermediate (hybrid) state of tRNA binding. *RNA* **10**:102–113.

Shastry, M., Nielsen, J., Ku, T., *et al.* 2001. Species-specific inhibition of fungal protein synthesis by sordarin: identification of a sordarin-specificity region in eukaryotic elongation factor 2. *Microbiology* **147**:383–390.

Shatsky, I.N., Bakin, A.V., Bogdanov, A.A., and Vasiliev, V.D. 1991. How does the mRNA pass through the ribosome? *Biochimie* **73**:937–945.

Shevack, A., Gewitz, H.S., Hennemann, B., *et al.* 1985. Characterization and crystallization of ribosomal particles from *Haloarcula marismortui*. *FEBS Lett.* **184**:68–71.

Shimmin, L.C., Ramirez, G., Matheson, A.T., and Dennis, P.P. 1989. Sequence alignment and evolutionary comparison of the L10 equivalent and L12 equivalent ribosomal proteins from archaebacteria, eubacteria, and eucaryotes. *J. Mol. Evol.* **29**:448–462.

Shin, B.-S., Maag, D., Roll-Mecak, A., *et al.* 2002. Uncoupling of initiation factor eIF5B/IF2 GTPase and translational activities by mutations that lower ribosome affinity. *Cell* **111**:1015–1025.

Shine, J., and Dalgarno, L. 1974. The 3'-terminal sequence of *Escherichia coli* 16S ribosomal RNA: complementarity to nonsense triplets and ribosome binding sites. *Proc. Natl. Acad. Sci. USA* **71**:1342–1346.

Shiryaev, V.M., Selivanova, O.M., Hartsch, T., *et al.* 2002. Ribosomal protein S1 from *Thermus thermophilus*: its detection, identification and overproduction. *FEBS Lett.* **525**:88–92.

Shultzaberger, R.K., Bucheimer, R.E., Rudd, K.E., and Schneider, T.D. 2001. Anatomy of *Escherichia coli* ribosome binding sites. *J. Mol. Biol.* **313**:215–228.

Sievers, A., Beringer, M., Rodnina, M.V., and Wolfenden, R. 2004. The ribosome as an entropy trap. *Proc. Natl. Acad. Sci. USA* **101**:7897–7901.

Sillers, I.Y., and Moore, P.B. 1981. Position of protein S1 in the 30S ribosomal subunit of *Escherichia coli*. *J. Mol. Biol.* **153**:761–780.

Silvian, L.F., Wang, J., and Steitz, T.A. 1999. Insights into editing from an Ile-tRNA synthetase structure with tRNAIle and mupirocin. *Science* **285**:1074–1077.

Sköld, S.E. 1983. Chemical crosslinking of elongation factor G to the 23S RNA in 70S ribosomes from *Escherichia coli*. *Nucl. Acids Res.* **11**:4923–4932.

Skouloubris, S., Ribas de Pouplana, L., de Reuse, H., and Hendrickson, T.L. 2003. A noncognate aminoacyl-tRNA synthetase that may resolve a missing link in protein evolution. *Proc. Natl. Acad. Sci. USA* **100**:11297–11302.

Smith, D.R., Doucette-Stamm, L.A., Deloughery, C., *et al.* 1997. Complete genome sequence of *Methanobacterium thermoautotrophicum* deltaH: functional analysis and comparative genomics. *J. Bacteriol.* **179**:7135–7155.

Smrt, J., Kemper, W., Caskey, T., and Nirenberg, M. 1970. Template activity of modified terminator codons. *J. Biol. Chem.* **245**:2753–2757.

Sonenberg, N., Wilchek, M., and Zamir, A. 1975. Identification of a region in 23S rRNA located at the peptidyl transferase center. *Proc. Natl. Acad. Sci. USA* **72**:4332–4336.

Song, H., Mugnier, P., Das, A.K., *et al.* 2000. The crystal structure of human eukaryotic release factor eRF1 — mechanism of stop codon recognition and peptidyl-tRNA hydrolysis. *Cell* **100**:311–321.

Southworth, D.R., Brunelle, J.L., and Green, R. 2002. EFG-independent translocation of the mRNA:tRNA complex is promoted by modification of the ribosome with thiol-specific reagents. *J. Mol. Biol.* **324**:611–623.

Spahn, C.M., Blaha, G., Agrawal, R.K., *et al.* 2001. Localization of the ribosomal protection protein Tet(O) on the ribosome and the mechanism of tetracycline resistance. *Mol. Cell.* **7**:1037–1045.

Spahn, C.M., Gomez-Lorenzo, M.G., Grassucci, R.A., *et al.* 2004. Domain movements of elongation factor eEf2 and the eukaryotic 80S ribosome facilitate tRNA translocation. *EMBO J.* **23**:1008–1019.

Spirin, A.S. 1968. On the mechanism of ribosome function. The hypothesis of locking-unlocking of subparticles. *Dokl. Akad. Nauk SSSR* **179**:1467–1470.

Spirin, A.S. 1969. A model of the functioning ribosome: locking and unlocking of the ribosome particles. *Cold Spring Harb. Symp. Quant. Biol.* **34**:197–207.

Spirin, A.S. 1978. Energetics of the ribosome. *Prog. Nucl. Acid Res. Mol. Biol.* **21**:39–62.

Spirin, A.S. 1985. Ribosomal translocation: facts and models. *Prog. Nucl. Acid. Res. Mol. Biol.* **32**:75–114.

Spirin, A.S. 1999. In *Ribosomes*, Ed. P. Siekevitz. Kluwer Academic/Plenum Publishers, New York, Boston, Dordrecht, London, Moscow.

Spirin, A.S. 2002. Ribosome as a molecular machine. *FEBS Lett.* **514**:2–10.

Spirin, A.S., and Vasiliev, V.D. 1989. Localization of functional centers on the prokaryotic ribosome: immuno-electron microscopy approach. *Biol. Cell* **66**:215–223.

Song, H., Mugnier, P., Das, A.K., *et al.* 2000. The crystal structure of human eukaryotic release factor eRF1 — mechanism of stop codon recognition and peptidyl-tRNA hydrolysis. *Cell* **100**:311–321.

Southworth, D.R., and Green, R. 2003. Ribosomal translocation: spasomycin pushes the button. *Curr. Biol.* **13**:R652–R654.

Southworth, D.R., Brunelle, J.L., and Green, R. 2002. EFG-independent translocation of the mRNA:tRNA complex is promoted by modifications of the ribosome with thiol-specific reagents. *J. Mol. Biol.* **324**:611–623.

Srinivasan, G., James, C.M., and Krzycki, J.A. 2002. Pyrrolysine encoded by UAG in archaea: charging of a UAG-decoding specialized tRNA. *Science* **296**:1459–1462.

Stahl, G., McCarty, G.P., and Farabaugh, P.J. 2002. Ribosome structure: revisiting the connection between translational accuracy and unconventional decoding. *Trends Biochem. Sci.* **27**:178–183.

Stanzel, M., Schon, A., and Sprinzl, M. 1994. Discrimination against misacylated tRNA by chloroplast elongation factor Tu. *Eur. J. Biochem.* **219**:435–439.

Stark, H., Rodnina, M.V., Rinke-Appel, J., *et al.* 1997. Visualization of elongation factor Tu on the *Escherichia coli* ribosome. *Nature* **389**:403–406.

Stark, H., Rodnina, M.V., Wieden, H.J., *et al.* 2000. Large scale movement of elongation factor G and extensive conformational change of the ribosome during translocation. *Cell* **100**:301–309.

Stark, H., Rodnina, M.V., Wieden, H.J., *et al.* 2002. Ribosome interactions of aminoacyl-tRNA and elongation factor Tu in the codon-recognition complex. *Nat. Struct. Biol.* **9**:849–854.

Stathopouplos, C., Li, T., Longman, R., *et al.* 2000. One polypeptide with two aminoacyl-tRNA synthetase activities, *Science* **287**:479–482.

Stathopouplos, C., Ahel, I., Ali, K., *et al.* 2001. Aminoacyl-tRNA synthesis: a postgenomic perspective. *Cold Spring Harb. Symp. Quant. Biol.* **66**:175–183.

Steitz, J.A. 1969. Polypeptide chain initiation: nucleotide sequences of the three ribosomal binding sites in bacteriophage R17 RNA. *Nature* **224**:957–964.

Stent, G.S., and Brenner, S. 1961. A genetic locus for the regulation of ribonucleic acid synthesis. *Proc. Natl. Acad. Sci. USA* **47**:2005–2014.

Stern, S., Moazed, D., and Noller, H.F. 1988. Structural analysis of RNA using chemical and enzymatic probing monitored by primer extension. *Methods Enzymol.* **164**:481–489.

Stern, S., Powers, T., Changchien, L.M., and Noller, H.F. 1989. RNA-protein interactions in 30S ribosomal subunits: folding and function of 16S rRNA. *Science* **244**:783–790.

Stringer, E.A., Sarkar, P., and Maitra, U. 1977. Function of initiation factor 1 in the binding and release of initiation factor 2 from ribosomal initiation complexes in *Escherichia coli. J. Biol. Chem.* **252**:1739–1744.

Stöffler, G., and Stöffler-Meilicke, M. 1984. Immunoelectron microscopy of ribosomes. *Ann. Rev. Biophys. Bioeng.* **13**:303–330.

Stoldt, M., Wöhnert, J., Gorlach, M., and Brown, L.R. 1998. The NMR structure of *Escherichia coli* ribosomal protein L25 shows homology to general stress proteins and glutaminyl-tRNA synthetases. *EMBO J.* **17**:6377–6384.

Stoldt, M. Wöhnert, J., Ohlenschläger, O., et al. 1999. The NMR structure of the 5S rRNA E-domain-protein L25 complex shows preformed and induced recognition. *EMBO J.* **18**:6508–6521.

Stoller, G., Ruecknagel, K.P., Nierhaus, K.H., et al. 1995. A ribosome-associated peptidyl-prolyl cis/trans isomerase identified as the trigger factor. *EMBO J.* **14**:4939–4948.

Stout, C.D., Mizuno, H., Rubin, J., et al. 1976. Atomic coordinates and molecular conformation of yeast phenylalanyl tRNA. An independent investigation. *Nucl. Acids Res.* **3**:1111–1123.

Strychartz, W.A., Nomura, M., and Lake, J.A. 1978. Ribosomal protein L7/L12 localized at a single region of the large subunit by immune microscopy. *J. Mol. Biol.* **126**:123–140.

Suarez, D., and Merz, K.M. Jr. 2001. Quantum chemical study of ester aminolysis catalyzed by a single adenine: a reference reaction for the ribosomal peptide synthesis. *J. Am. Chem. Soc.* **123**:7687–7690.

Subramanian, A.R. 1983. Structure and functions of ribosomal protein S1. *Prog. Nucl. Acid Res. Mol. Biol.* **28**:101–142.

Subramanian, A.R., and Davis, B.D. 1970. Activity of initiation factor F3 in dissociating *Escherichia coli* ribosomes. *Nature* **228**:1273–1275.

Subramanian, A.R., and van Duin, J. 1977. Exchange of individual ribosomal proteins between ribosomes as studied by heavy isotope-transfer experiments. *Mol. Gen. Genet.* **158**:1–9.

Suh, W.C., Lu, C.Z., and Gross, C.A. 1999. Structural features required for the interaction of the Hsp70 molecular chaperone DnaK with its cochaperone DnaJ. *J. Biol. Chem.* **274**:30534–30539.

Suppmann, S., Persson, B.C., and Böck, A. 1999. Dynamics and efficiency *in vivo* of UGA-directed selenocysteine insertion at the ribosome. *EMBO J.* **18**:2284–2293.

Sundari, R., Stringer, E.A., Schulman, L.H., and Maitra, U., 1976. Interaction of bacterial initiation factor 2 with initiator tRNA. *J. Biol. Chem.* **251**:3338–3345.

Sy, J. 1977. *In vitro* degradation of guanosine 5'-diphosphate, 3'-diphosphate. *Proc. Natl. Acad. Sci. USA* **74**:5529–5533.

Sy, J., and Lipmann, F. 1973. Identification of the synthesis of guanosine tetraphosphate (MS I) as an insertion of a pyrophosphoryl group into the 3'-position in guanosine 5'-diphosphate. *Proc. Natl. Acad. Sci. USA* **70**:306–309.

Takanami, M., and Zubay, G. 1964. An estimate of the size of the ribosomal site for messenger RNA binding. *Proc. Natl. Acad. Sci. USA* **51**:834–839.

Tarn, W.Y., Steitz, J.A. 1997. Pre-mRNA splicing: the discovery of a new spliceosome doubles the challenge. *Trends Biochem. Sci.* **22**:132–137.

Tan, J., Jakob, U., and Bardwell, J.C. 2002. Overexpression of two different GTPases rescues a null mutation in a heat-induced rRNA methyltransferase. *J. Bacteriol.* **184**:2692–2698.

Taylor, D.E., and Chau, A. 1996. Tetracycline resistance mediated by ribosomal protection. *Antimicrob. Agents Chemother.* **40**:1–5.

Taylor, D.E., Trieber, C.A., Trescher, G., and Bekkering, M. 1998. Host mutations (miaA and rpsL) reduce tetracycline resistance mediated by Tet(O) and Tet(M). *Antimicrob. Agents Chemother.* **42**:59–64.

Tchorzewski, M. 2002. The acidic ribosomal P-proteins. *Int. J. Biochem. Cell Biol.* **1265**:1–5.

Tchorzewski, M., Boldyreff, B., Issinger, O.-G., and Grankowski, N. 2000a. Analysis of the protein-protein interactions between the human acidic ribosomal P-proteins: evaluation by the two hybrid system. *Int J. Biochem. Cell Biol.* **32**:737–746.

Tchorzewski, M., Boguszewska, A., Dukowski, P., and Grankowski, N. 2000b. Oligomeric properties of the acidic ribosomal P-proteins from *Saccharomyces cerevisiae*: effect of P1A protein phosphorylation on the formation of the P1A-P2B hetero-complex. *Biochim. Biophys. Acta* **1499**:63–73.

Tchorzewski, M., Krokowski, D., Boguszewska, A., *et al.* 2003. Structural characterization of yeast acidic ribosomal P-proteins forming the hetero-complex P1A/P2B. *Biochemistry* **42**:3399–3408.

Tenson, T., and Ehrenberg, M. 2002. Regulatory nascent peptides in the ribosome tunnel. *Cell* **108**:591–594.

Terhorst, C., Wittmann-Liebold, B., and Möller, W. 1972. 50S ribosomal proteins. Peptide studies on two acidic proteins, A1 and A2, isolated from 50S ribosomes of *Escherichia coli. Eur. J. Biochem.* **25**:13–19.

Teter, S.A., Houry, W.A., Ang, D., *et al.* 1999. Polypeptide flux through bacterial Hsp70: DnaK cooperates with trigger factor in chaperoning nascent chains. *Cell* **97**:755–765.

Thanbichler, M., Böck, A., and Goody, R.S. 2000. Kinetics of the interaction of translation factor SelB from *Escherichia coli* with guanosine nucleotides and selenocysteine insertion sequence RNA. *J. Biol. Chem.* **275**:20458–20466.

Thompson, J., Schmidt, F., and Cundliffe, E. 1982. Site of action of a ribosomal RNA methylase conferring resistance to thiostrepton. *J. Biol. Chem.* **257**:7915–7917.

Thompson, J., Kim, D.F., O'Connor, M., *et al.* 2001. Analysis of mutations at residues A2451 and G2447 of 23S rRNA in the peptidyltransferase active site of the 50S ribosomal subunit. *Proc. Natl. Acad. Sci. USA* **98**:9002–9007.

Tiennault-Desbordes, E., Cenatiempo, Y., and Laalami, S. 2001. Initiation factor 2 of *Myxococcus xanthus*, a large version of prokaryotic translation initiation factor 2. *J. Bacteriol.* **183**:207–213.

Tischendorf, G.W., Zeichhardt, H., and Stöffler, G. 1974. Location of proteins S5, S13, S14 on the surface of the 30S ribosomal subunit from *Escherichia coli* as determined by immune electron microscopy. *Mol. Gen. Genet.* **134**:209–223.

Tissieres, A., Schlessinger, D., and Gros, F. 1960. Amino acid incorporation into proteins by *Escherichia coli* ribosomes. *Proc. Natl. Acad. Sci. USA* **46**:1450–1456.

Tittawella, I., Yasmin, L., and Baranov, V. 2003. Mitochondrial ribosomes in a trypanosome. *Biochem. Biophys. Res. Commun.* 307:578–583.

Tocilj, A., Schlünzen, F., Janell, D., *et al.* 1999. The small ribosomal subunit from *Thermus thermophilus* at 4.5 A resolution: pattern fittings and the identification of a functional site. *Proc. Natl. Acad. Sci. USA* 96:14252–14257.

Tomsic, J., Vitali, L.A., Daviter, T., *et al.* 2000. Late events of translation initiation in bacteria: a kinetic analysis. *EMBO J.* 19:2127–2136.

Toyoda, T., Tin, O.F., Ito, K., *et al.* 2000. Crystal structure combined with genetic analysis of the *Thermus thermophilus* ribosome recycling factor shows that a flexible hinge may act as a functional switch. *RNA* 6:1432–1444.

Trakhanov, S.D., Yusupov, M.M., Agalarov, S.C., *et al.* 1987. Crystallization of 70S ribosomes and 30S ribosomal subunits from *Thermus thermophilus*. *FEBS Lett.* 220: 319–322.

Traut, R.R., and Monro, R.E. 1964. The puromycin reaction and its relation to protein synthesis. *J. Mol. Biol.* 10:63–72.

Traut, R.R., Tewari, D., Sommer, A., *et al.* 1986. Protein topography of ribosomal functional domains: effect of monoclonal antibodies to different epitopes in *E. coli* protein L7/L12 on ribosome function and structure. In *Structure, Function and Genetics of Ribosomes*, Eds. B. Hardesty, G. Kramer, pp. 286–308. Springer-Verlag Press, New York.

Traut, R.R., Dey, D., Bochkariov, D., *et al.* 1995. Location and domain structure of *Escherichia coli* ribosomal protein L7/L12: site specific cysteine crosslinking and attachment of fluorescent probes. *Biochem. Cell Biol.* 73:949–958.

Trieber, C.A., Burkhardt, N., Nierhaus, K.H., and Taylor, D.E. 1998. Ribosomal protection from tetracycline mediated by Tet(O): Tet(O) interaction with ribosomes is GTP-dependent. *Biol. Chem.* 379:847–855.

Tritton, T.R. 1980. Proton NMR observation of the *Escherichia coli* ribosome. *FEBS Lett.* 120:141–144.

Tsalkova, T., Odom, O.W., Kramer, G., and Hardesty, B. 1998. Different conformations of nascent peptides on ribosomes. *J. Mol. Biol.* 278:713–723.

Tumanova, L.G., Gongadze, G.M., Venyaminov, S. Yu., *et al.* 1983. Physical properties of ribosomal proteins isolated under different conditions from the *Escherichia coli* 50S subunit. *FEBS Lett.* 157:85–90.

Tumbula, D.L., Becker, H.D., Chang, W.Z., and Soll, D. 2000. Domain-specific recruitment of amide amino acids for protein synthesis. *Nature* 407:106–110.

Turner, C.F., and Moore, P.B. 2004. The solution structure of ribosomal protein L18 from *Bacillus stearothermophilus*. *J. Mol. Biol.* 335:679–684.

Uchiumi, T., Wahba, A.J., and Traut, R.R. 1987. Topography and stoichiometry of acidic proteins in large ribosomal subunits from *Artemia salina* as determined by crosslinking. *Proc. Natl. Acad. Sci. USA* 84:5580–5584.

Ullers, R.S., Houben, E.N., Raine, A., *et al.* 2003. Interplay of signal recognition particle and trigger factor at L23 near the nascent chain exit site on the *Escherichia coli* ribosome. *J. Cell Biol.* 161:679–684.

Unge, J., Al-Karadaghi, S., Liljas, A., *et al.* 1997. A mutant form of the ribosomal protein L1 reveals conformational flexibility. *FEBS Lett.* 411:53–59.

Unge, J., Åberg, A., Al Karadaghi, S., *et al.* 1998. The crystal structure of ribosomal protein L22 from *Thermus thermophilus*: insights into the mechanism of erythromycin resistance. *Structure* 6:1577–1586.

Valentini, S.R., Casolari, J.M., Oliveira, C.C., *et al.* 2002. Genetic interactions of yeast eukaryotic translation initiation factor 5A (eIF5A) reveal connections to poly(A)-binding protein and protein kinase C signaling. *Genetics* **160**:393–405.

Valle, M., Sengupta, J., Swami, N.K., *et al.* 2002. Cryo-EM reveals an active role for aminoacyl-tRNA in the accommodation process. *EMBO J.* **21**:3557–3567.

Valle, M., Zavialov, A., Li, W., *et al.* 2003a. Incorporation of aminoacyl-tRNA into the ribosome as seen by cryo-electron microscopy. *Nat. Struct. Biol.* **10**:899–906.

Valle, M., Zavialov, A., Sengupta, J., *et al.* 2003b. Locking and unlocking of ribosomal motions. *Cell* **114**:123–134.

Valle, M., Gillet, R., Kaur, S., *et al.* 2003c. Visualizing tmRNA entry into a stalled ribosome. *Science* **300**:127–130.

van Agthoven, A., Maassen, J.A., Schrier, P.I., and Möller, W. 1975. Inhibition of EF-G dependent GTPase by an amino-terminal fragment of L7/L12. *Biochim. Biophys. Res. Commun.* **64**:1184–1191.

van Agthoven, A., Kriek, J., Amons, R., and Möller, W. 1976. Isolation and characterization of the acidic phosphoproteins of 60S ribosomes from *Artemia salina* and rat liver. *Eur. J. Biochem.* **91**:553–556.

VanBogelen, R.A., and Neidhardt, F.C. 1990. Ribosomes are sensors of heat and cold shock in *Escherichia coli*. *Proc. Natl. Acad. Sci. USA* **87**:5589–5593.

van den Berg, B., Clemons, W.M., Collinson, I., *et al.* 2004. X-ray structure of a protein conducting channel. *Nature* **427**:36–44.

van Doorn, L.J., Giesendorf, B.A., Bax, R., *et al.* 1997. Molecular discrimination between *Campylobacter jejuni*, *Campylobacter coli*, *Campylobacter lari* and *Campylobacter upsaliensis* by polymerase chain reaction based on a novel putative GTPase gene. *Mol. Cell Probes* **11**:177–185.

van Heel, M. 1987. Angular reconstruction: *a posteriori* assignment of projection direction for 3D reconstruction. *Ultramicroscopy* **21**:111–114.

VanLoock, M.S., Agrawal, R.K., Gabashvili, L.Q., *et al.* 2000. Movement of the decoding region of the 16S ribosomal RNA accompanies tRNA translocation. *J. Mol. Biol.* **304**:507–515.

Vasiliev, V.D. 1974. Morphology of the ribosomal 30S subparticle according to electron microscopic data. *Acta Biol. Med. Ger.* **33**:779–793.

Vazquez, D. 1974. Inhibitors of protein synthesis. *FEBS Lett.* **40**(Suppl.):S63–S84.

Vazquez, D. 1979. Inhibitors of protein synthesis. *Mol. Biochem. Biochem. Biophys.* **30**:1–312.

Vester, B., and Douthwaite, S. 2001. Macrolide resistance conferred by base substitutions in 23S RNA. *Antimicrob. Agents Chemother.* **45**:1–12.

Vester, B., and Garrett, R.A. 1988. The importance of highly conserved nucleotides in the binding region of chloramphenicol at the peptidyl transfer center of *E. coli* 23S ribosomal RNA. *EMBO J.* **7**:3577–3588.

Vestergaard, B., Van, L.B., Andersen, G.R., *et al.* 2001. Bacterial polypeptide release factor RF2 is structurally distinct from eucayotic eRF1. *Mol. Cell* **8**:1375–1382.

Vetter, I.R., and Wittinghofer, A. 2001. The guanine nucleotide-binding switch in three dimensions. *Science* **294**:1299–1304.

Vila-Sanjurjo, A., Ridgeway, W.K., Seymaner, V., *et al.* 2003. X-ray crystal structures of the WT and a hyper-accurate ribosome from *Escherichia coli*. *Proc. Natl. Acad. Sci. USA* **100**:8682–8687.

Vogeley, L., Palm, G.J., Mesters, J.R., and Hilgenfeld, R. 2001. Conformational change of elongation factor Tu (EF-Tu) induced by antibiotic binding. Crystal structure of the complex between ET-Tu.GDP and aurodox. *J. Biol. Chem.* **276**:17149–17155.

von Bohlen, K., Makowski, I., Hansen, H.A., *et al.* 1991. Characterization and preliminary attempts for derivatization of crystals of large ribosomal subunits from *Haloarcula marismortui* diffracting to 3 A resolution. *J. Mol. Biol.* **222**:11–15.

von Heijne, G. 1985. Signal sequences. The limits of variation. *J. Mol. Biol.* **184**:99–105.

Völker, U., Engelmann, S., Maul, B., *et al.* 1994. Analysis of the induction of general stress proteins of *Bacillus subtilis*. *Microbiology* **140**:741–752.

Wabl, M.R. 1974. Electron microscopic location of two proteins on the surface of the 50S ribosomal subunit of *Escherichia coli* using specific antibody markers *J. Mol. Biol.* **84**:241–247.

Wada, A. 1998. Growth phase coupled modulation of *Escherichia coli* ribosomes. *Genes Cells* **3**:203–208.

Wada, A., Yamazaki, Y., Fujita, N., and Ishihama, A. 1990. Structure and probable genetic location of a "ribosome modulation factor" associated with 100S ribosomes in stationary phase *Escherichia coli* cells. *Proc. Natl. Acad. Sci. USA* **87**:2657–2661.

Wahl, M.C., Bourenkov, G.P., Bartunik, H.D., and Huber, R. 2000. Flexibility, conformational diversity and two dimerization modes in complexes of ribosomal protein L12. *EMBO J.* **19**:174–186.

Wahl, M.C., and Möller, W. 2002. Structure and function of the acid ribosomal stalk proteins. *Curr. Prot. Pept. Sci.* **3**:93–106.

Walker, J.E., Saraste, M., Runswick, M.J., and Gay, N.J. 1982. Distantly related sequences in the alpha- and beta-subunits of ATP synthase, myosin, kinases and other ATP-requiring enzymes and a common nucleotide binding fold. *EMBO J.* **1**:945–951.

Wang, Y., Jiang, Y., Meyering-Voss, M., *et al.* 1997. Crystal structure of the EF-Tu.EF-Ts complex from *Thermus thermophilus*. *Nat. Struct. Biol.* **4**:650–656.

Walter, P., and Blobel, G. 1980. Purification of a membrane-associated protein complex required for protein translocation across the endoplasmic reticulum. *Proc. Natl. Acad. Sci. USA* **77**:112–116.

Walter, P., and Blobel, G. 1982. Signal recognition particle contains a 7S RNA essential for protein translocation across the endoplasmic reticulum. *Nature* **299**:691–698.

Warner, J.R., and Rich, A. 1964. The number of soluble RNA molecules on reticulocyte polyribosomes. *Proc. Natl. Acad. Sci. USA* **51**:1134–1141.

Watson, J.D. 1964. The synthesis of proteins upon ribosomes. *Bull. Soc. Chim. Biol.* **46**:1399–1425.

Watson, J.D., and Crick, F. 1953. Molecular structure of nucleic acids. *Nature* **171**:737–738.

Wedekind, J.E., Dance, G.S., Sowden, M.P., and Smith, H.C. 2003. Messenger RNA editing in mammals: new members of the APOBEC family seeking roles in the family business. *Trends Genet.* **19**:207–216.

Welch, M., Chastang, J., and Yarus, M. 1995. An inhibitor of ribosomal peptidyltransferase using transition-state analogy. *Biochemistry* **34**:385–390.

Wendrich, T.M., and Marahiel, M.A. 1997. Cloning and characterization of a relA/spoT homologue from *Bacillus subtilis*. *Mol. Microbiol.* **26**:65–79.

Wendrich, T.M., Blaha, G., Wilson, D.N., *et al.* 2002. Dissection of the mechanism for the stringent factor RelA. *Mol. Cell* **10**:779–788.

Westhof, E., and Fritsch, V. 2000. RNA folding: beyond Watson-Crick pairs. *Struct. Fold. Des.* **8**:R55–R65.

Wieden, H.J., Gromadski, K., Rodnin, D., and Rodnina, M.V. 2002. Mechanism of elongation factor (EF)-Ts-catalyzed nucleotide exchange in EF-Tu. Contribution of contacts at the guanine base. *J. Biol. Chem.* **277**:6032–6036.

Wilcox, M. 1969. Gamma-glutamyl phosphate attached to glutamine-specific tRNA. A precursor of glutaminyl-tRNA in *Bacillus subtilis*. *Eur. J. Biochem.* **11**:405–412.

Willie, G.R., Richman, N., Godtfredsen, W.O., and Bodley, J.W. 1975. Some characteristics and structural requirements for the interaction of 24, 25-dihydrofusidic acid with ribosome elongation factor G complexes. *Biochemistry* **14**:1713–1718.

Willumeit, R., Diedrich, G., Forthmann, S., *et al.* 2001. Mapping proteins of the 50S subunit from *Escherichia coli* ribosomes. *Biochim. Biophys. Acta.* **1520**:7–20.

Wilson, K.S., and Noller, H.F. 1998. Mapping the position of translational elongation factor EF-G in the ribosome by directed hydroxyl radical probing. *Cell* **92**:131–139.

Wilson, K.S., Appelt, K., Badger, J., *et al.* 1986. Crystal structure of a prokaryotic ribosomal protein. *Proc. Natl. Acad. Sci. USA* **83**:7251–7255.

Wimberly, B.T., White, S.W., and Ramakrishnan, V. 1997. The structure of ribosomal protein S7 at 1.9 Å resolution reveals a beta-hairpin motif that binds double-stranded nucleic acids. *Structure* **5**:1187–1198.

Wimberly, B.T., Guymon, R., McCutcheon, J.P., *et al.* 1999. A detailed view of a ribosomal active site: the structure of the L11-RNA complex. *Cell* **97**:491–502.

Wimberly, B.T., Brodersen, D.F., Clemons, W.M. Jr., *et al.* 2000. Structure of the 30S ribosomal subunit. *Nature* **407**:327–339.

Wintermeyer, W., Savelsbergh, A., Semenkov, Y.P., *et al.* 2001. Mechanism of elongation factor G function in tRNA translocation on the ribosome. *Cold Spring Harb. Symp. Quant. Biol.* **66**:449–458.

Wittinghofer, A., and Pai, E.F. 1991. The structure of Ras protein: a model for a universal molecular switch. *Trends Biochem. Sci.* **16**:382–387.

Wittmann, H.G. 1982. Components of bacterial ribosomes. *Ann. Rev. Biochem.* **51**:155–183.

Wittmann-Liebold, B., and Greuer, B. 1978. The primary structure of protein S5 from the small subunit of the *Escherichia coli* ribosome. *FEBS Lett.* **95**:91–98.

Woese, C. 1970. Molecular mechanics of translation: a reciprocating ratchet mechanism. *Nature* **226**:817–820.

Woese, C. 1998. The universal ancestor. *Proc. Natl. Acad. Sci. USA* **95**:6854–6859.

Woese, C. 2001. Translation: in retrospect and prospect. *RNA* **7**:1055–1067.

Woese, C.R. 2002. On the evolution of cells. *Proc. Natl. Acad. Sci. USA* **99**:8742–8747.

Woese, C.R., and Fox, G.E. 1977. Phylogenetic structure of the procaryotic domain. The primary kingdoms. *Proc. Natl. Acad. Sci. USA* **74**:5088–5090.

Woese, C.R., Kandler, O., and Wheelis, M.L. 1990. Towards a natural system of organisms: proposal for the domains archaea, bacteria and eucarya. *Proc. Natl. Acad. Sci. USA* **84**:4576–4579.

Woestenenk, E.A., Gongadze, G.M., Shcherbakov, D.V., *et al.* 2002. The solution structure of ribosomal protein L18 from *Thermus thermophilus* reveals a conserved RNA-binding fold. *Biochem. J.* **363**:553–561.

Wolf, H., Chinali, G., and Parmeggiani, A. 1974. Kirromycin, an inhibitor of protein biosynthesis that acts on elongation factor Tu. *Proc. Natl. Acad. Sci. USA* **71**: 4910–4914.

Wolf, H., Chinali, G., and Parmeggiani, A. 1977. Mechanism of the inhibition of protein synthesis by kirromycin. Role of elongation factor Tu and ribosomes. *Eur. J. Biochem.* **75**:67–75.

Woodcock, J., Moazed, D., Cannon, M., et al. 1991. Interaction of antibiotics with A-site specific and P-site specific bases in 16S ribosomal RNA. *EMBO J.* **10**:3099–3103.

Wool, I.G., Chan, Y.L., Gluck, A., and Suzuki, K. 1991. The primary structure of rat ribosomal proteins P0, P1 and P2 and a proposal for a uniform nomenclature for mammalian and yeast ribosomal proteins. *Biochemie* **73**:861–870.

Worbs, M., Huber, R., and Wahl, M.C. 2000. Crystal structure of ribosomal protein L4 shows RNA binding sites for ribosome incorporation and feedback control of the S10 operon. *EMBO J.* **19**:807–818.

Wower, I.K., Wower, J., and Zimmermann, R.A. 1998. Ribosomal protein L27 participates in both 50S subunit assembly and the peptidyltransferase reaction. *J. Biol. Chem.* **273**:19847–19852.

Wower, J., Kirillov, S.V., Wower, I.K., et al. 2000. Transit of tRNA through the *Escherichia coli* ribosome. Cross-linking of the 3'-end of tRNA to specific nucleotides of the 23S ribosomal RNA at the A-, P- and E-sites. *J. Biol. Chem.* **275**:37887–37894.

Wriggers, W., Agrawal, R.K., Drew, D.L., et al. 2000. Domain motions of EF-G bound to the 70S ribosome: insights from a hand-shaking between multi-resolution structures. *Biophys. J.* **79**:1670–1678.

Xing, Y., Guha Takurta, D., and Draper, D.E. 1997. The RNA binding domain of ribosomal protein L11 is structurally similar to homeodomains. *Nat. Struct. Biol.* **4**:24–27.

Xiong, L., Polacek, N., Sander, P., et al. 2001. pKa of adenine 2451 in the ribosomal peptidyltransferase center remains elusive. *RNA* **7**:1365–1369.

Xu, R.X., Hassel, A.M., Vanderwall, D., et al. 2000. Atomic structure of PDE4: insights into phosphodiesterase mechanism and specificity. *Science* **288**:1822–1825.

Yamagishi, M., Matsushima, H., Wada, A., et al. 1993. Regulation of the *Escherichia coli* rmf gene encoding the ribosome modulation factor: growth phase- and growth rate-dependent control. *EMBO J.* **12**:625–630.

Yaremchuk, A., Cusack, S., and Tukalo, 2000. Crystal structure of an eucaryote/archaea-like prolyl-tRNA synthetase at 2.4 Å resolution and its complex with tRNA^{Pro} (CGG). *EMBO J.* **19**:4745–4758.

Yaremchuk, A., Tukalo, M., Grotli, M., and Cusack, S. 2001. A succession of substrate induced conformational changes ensures the amino acid specificity of *Thermus thermophilus* prolyl-tRNA synthase: comparison with histidyl-tRNA synthase. *J. Mol. Biol.* **309**:989–1002.

Yarmolinsky, M.B., and de la Haba, G.L. 1959. Inhibition by puromycin of amino acid incorporation into protein. *Proc. Natl. Acad. Sci. USA* **45**:1721–1729.

Ye, K., Serganov, A., Hu, W., et al. 2002. Ribosome-associated factor Y adopts a fold resembling a double-stranded RNA binding domain scaffold. *Eur. J. Biochem.* **269**:5182–5191.

Yegian, C.D., Stent, G.S., and Martin, E.M. 1966. Intracellular condition of *Escherichia coli* transfer RNA. *Proc. Natl. Acad. Sci. USA* **55**:839–846.

Yokosawa, H., Kawakita, M., Arai, K., *et al.* 1975. Binding of aminoacyl-tRNA to ribosomes promoted by elongation factor Tu. Studies on the role of GTP hydrolysis. *J. Biochem.* (Tokyo). **77**:719–728.

Yonath, A. 2003. Structural insight into functional aspects of ribosomal RNA targeting. *Chem. Bio. Chem.* **4**:1008–1017.

Yonath, A., and Franceschi, F. 1998. Functional universality and evolutionary diversity: insights from the structure of the ribosome. *Structure* **6**:679–684.

Yonath, A., Mussig, J., Teshe, B., *et al.* 1980. Crystallization of the large ribosomal subunit from *B. stearothermophilus. Biochem. Int.* **1**:428–435.

Yonath, A., Mussig, J., and Wittmann, H.G. 1982. Parameters for crystal growth of ribosomal subunits. *J. Cell Biochem.* **19**:145–155.

Yonath, A., Leonard, K.R., and Wittmann, H.G.A. 1987. Tunnel in the large ribosomal subunit revealed by three-dimensional image reconstruction. *Science* **236**:813–816.

Yonath, A., Harms, J., Hansen, H.A.S., *et al.* 1998. Crystallographic studies on the ribosome, a large macromolecular assembly exhibiting severe nonisomorphism, extreme beam sensitivity and no internal symmetry. *Acta Cryst.* **A54**:945–955.

Yoshida, T., Uchiyama, S., Nakano, H., *et al.* 2001. Solution structure of the ribosome recycling factor from *Aquifex aeolicus. Biochemistry* **40**:2387–2396.

Yoshizawa, S., Fourmy, D., and Puglisi, J.D. 1998. Structural origins of of gentamin antibiotic interaction. *EMBO J.* **17**:6437–6448.

Yoshizawa, S., Fourmy, D., and Puglisi, J.D. 1999. Recognition of the codon-anticodon helix by ribosomal RNA. *Science* **285**:1722–1725.

Yusupov, M.M., and Spirin, A.S. 1986. Are there proteins between the ribosomal subunits? Hot tritium bombardment experiments. *FEBS Lett.* **197**:229–233.

Yusupov, M.M., Yusupova, G.Z., Baucom, A., *et al.* 2001. Crystal structure of the ribosome at 5.5 Å resolution. *Science* **292**:883–896.

Yusupova, G.Z., Yusupov, M.M., Cate, J.H., and Noller, H.F. 2001. The path of messenger RNA through the ribosome. *Cell* **106**:233–241.

Zantema, A., Maassen, J.A., Kriek, J., and Moller, W. 1982a. Fluorescence studies on the location of L7/L12 relative to L10 in the 50S ribosome of *Escherichia coli. Biochemistry* **21**:3077–3082.

Zantema, A., Maassen, J.A., Kriek, J., and Moller, W. 1982b. Preparation and characterization of fluorescent 50S ribosomes. Specific labeling of ribosomal proteins L7/L12 and L10 of *Escherichia coli. Biochemistry* **21**:3069–3076.

Zavialov, A.V., Buckingham, R.H., and Ehrenberg, M. 2001. A post termination ribosomal complex is the guanine nucleotide exchange factor for peptide release factor RF3. *Cell* **107**:115–124.

Zavialov, A.V., Mora, L., Buckingham, R.H., and Ehrenberg, M. 2002. Release of peptide promoted by the GGQ motif of class 1 release factors regulates the GTPase activity of RF3. *Mol. Cell* **10**:789–798.

Zavialov, A.V., and Ehrenberg, M. 2003. Peptidyl-tRNA regulates the GTPase activity of translation factors. *Cell* **114**:113–122.

Zeidler, W., Egle, C., Ribeiro, S., *et al.* 1995. Site-directed mutagenesis of *Thermus thermophilus* elongation factor Tu. Replacement of His85, Asp81 and Arg300. *Eur. J. Biochem.* **229**:596–604.

Zeidler, W., Schirmer, N.K., Egle, C., *et al.* 1996. Limited proteolysis and amino acid replacements in the effector region of *Thermus thermophilus* elongation factor Tu. *Eur. J. Biochem.* **239**:265–271.

Zhou, D.X., and Mache, R. 1989. Presence in the stroma of chloroplasts of a large pool of a ribosomal protein not structurally related to any *Escherichia coli* ribosomal protein. *Mol. Gen. Genet.* **219**:204–208.

Zhou, Y.N., and Lin, D.J. 1998. The rpoB mutants destabilizing initiation complexes at stringently controlled promoters behave like "stringent" RNA polymerases in *Escherichia coli*. *Proc. Natl. Acad. Sci. USA* **95**:2908–2913.

Zhouravleva, G., Frolova, L., Le Goff, X., *et al.* 1995. Termination of translation in eukaryotes is governed by two interacting polypeptide chain release factors, eRF1 and eRF3. *EMBO J.* **14**:4065–4072.

Zimmermann, R.A. 1980. Interactions among protein and RNA components of the ribosome. In *Ribosomes. Structure, Function and Genetics*, Eds. G. Chambliss, *et al.*, pp. 135–169. University Park Press, Baltimore.

Zinker, S. 1980. P5/P5′, the acidic ribosomal phosphoprotein from *S. cerevisiae. Biochim. Biophys. Acta* **606**:76–82.

Zinker, S., and Warner, J.R. 1976. The ribosomal proteins of *Saccharomyces cerevisiae*. Phosphorylated and exchangeable proteins. *J. Biol. Chem.* **251**:1799–1807.

Zucker, F.H., and Hershey, J.W. 1986. Binding of *Escherichia coli* protein synthesis initiation factor IF1 to 30S ribosomal subunits measured by fluorescence polarization. *Biochemistry* **25**:3682–3690.

Zurdo, J., Parada, P., van den Berg, A., *et al.* 2000. Assembly of *Saccharomyces cerevisiae* ribosome stalk: binding of P1 proteins is required for the interaction nof P2 proteins *Biochemistry* **39**:8929–8934.

Index

A

affinity label 14
aIF5A 119
ALS. *See* tRNA: anticodon stem loop
amino acid starvation 151, 153–155
aminoglycoside 163, 164, 171
A-minor motif 57, 83
antibiotics 8, 53, 72, 87, 89, 90, 91, 93, 97, 101, 124, 134, 139, 151, 156, 159–160, 162–164, 167,169–171, 182, 186, 196, 217
anticodon mimicry 205
assembly 8, 13, 41, 43–46, 62, 63, 72, 158
ATPase 104, 198, 210–212
AUG. *See* Initiation codon 22, 24, 77, 113, 162, 178, 179
aurodox. *See* kirromycin
azithromycin 161, 170

B

BABE. See Fe-EDTA

C

carbomycin A 161, 169, 196
chaperone 100, 208–211
chloramphenicol 161, 167, 169, 196
clarithromycin 170
codon 2, 7, 9, 21–25, 27, 32, 35, 76–79, 80–83, 85–86, 94, 95, 97, 108, 113, 116, 120, 126, 130–133, 144, 162, 171, 176, 179–186, 188, 195, 196, 203, 217, 218
codon-anticodon interaction 80, 82, 183–186, 203
crosslinking 9, 12–14, 41, 67, 69, 87, 163

281

S

T